Stable Homotopy and Generalised Homology

Chicago Lectures in Mathematics Series
Robert J. Zimmer, Series Editor
J. Peter May, Spencer J. Bloch, Norman R. Lebovitz, William Fulton, and Carlos Kenig, editors

Other *Chicago Lectures in Mathematics* titles available from the University of Chicago Press:

Simplicial Objects in Algebraic Topology, by J. Peter May (1967)

Fields and Rings, Second Edition, by Irving Kaplansky (1969, 1972)

Lie Algebras and Locally Compact Groups, by Irving Kaplansky (1971)

Several Complex Variables, by Raghavan Narasimhan (1971)

Torsion-Free Modules, by Eben Matlis (1973)

Rings with Involution, by I. N. Herstein (1976)

Theory of Unitary Group Representation, by George V. Mackey (1976)

Infinite-Dimensional Optimization and Convexity, by Ivar Ekeland and Thomas Turnbull (1983)

Commutative Semigroup Rings, by Robert Gilmer (1984)

Navier-Stokes Equations, by Peter Constantin and Ciprian Foias (1988)

Essential Results of Functional Analysis, by Robert J. Zimmer (1990)

Fuchsian Groups, by Svetlana Katok (1992)

Unstable Modules over the Steenrod Algebra and Sullivan's Fixed Point Set Conjecture, by Lionel Schwartz (1994)

Topological Classification of Stratified Spaces, by Shmuel Weinberger (1994)

The University of Chicago Press, Chicago 60637
The University of Chicago Press, Ltd., London

Printed in the United States of America
International Standard Book Number: 0-226-00524-0 (paperbound)
Library of Congress Catalog Card Number: 72-5735
99 98 97 96 95   6 5 4 3

This book is printed on acid-free paper.

Chicago Lectures in Mathematics

stable homotopy and generalised homology

J. F. Adams

The University of Chicago Press
Chicago and London

CONTENTS

PREFACE

The three sections of this book represent courses of lectures which I delivered at the University of Chicago in 1967, 1970 and 1971 respectively; and the three sections are of slightly different characters. The 1967 lectures dealt with part of Novikov's work on complex cobordism while that work was still new--they were prepared before I had access to a translation of Novikov's full-length paper, Izvestija Akademii Nauk SSSR, Serija Matematičeskaja 31 (1967) 855-951. They were delivered as seminars to an audience assumed to be familiar with algebraic topology. The 1970 lectures also assumed familiarity, but were a longer series attempting a more complete exposition; I aimed to cover Quillen's work on formal groups and complex cobordism. Finally, the 1971 lectures were a full-length ten-week course, aiming to begin at the beginning and cover many of the things a graduate student needs to know in the area of stable homotopy and generalised homology theories. They form two-thirds of the present book.

No attempt has been made to rewrite the three sections to impose uniformity, whether of notation or of anything else. Each section has its own introduction, where the reader may find more details of the topics considered. Each section has its own system of references; in Part I the references are given where they are needed; in Part II the references are collected at the end, with Part I as reference [2]; in Part III the references are again at the end, with Part II as reference [2]. However, the page numbers given in references to [2] refer--I hope-- to pages in the present book.

Although I have not tried to impose uniformity by rewriting, a certain unity of theme is present. Among the notions with which familiarity is assumed near the beginning of Part I, I note the following: spectra, products, and the derived functor of the inverse limit. All these matters are treated in Part III--in sections 2-3, 9 and 8. Similarly, near the beginning of Part II I assume it known that a spectrum determines a generalised homology theory and a generalised cohomology theory; this is set out in Part III, section 6. Again, at the end of Part I, section 2 (page 7) the reader is referred to the literature for information on $\pi_*(MU)$; he could equally well go to Part II, section 8 (page 75). Perhaps one should infer that in my choice of material, methods and results for my later courses, I was influenced by the applications I had already lectured on, as well as others I knew.

I am conscious of other places where the three parts of this book overlap, but perhaps the reader can profit by analysing these overlaps for himself; and certainly he should feel free to read the parts in an order reflecting his own taste. I need hardly direct the expert; a newcomer to the subject would probably do best to begin by taking what he needs from the first ten sections of Part III.

I would like to express my thanks to my hosts in the University of Chicago, and to R. Ming for taking the original notes of Part III.

PART I

S. P. NOVIKOV'S WORK ON OPERATIONS ON COMPLEX COBORDISM

1. INTRODUCTION

The work of S.P. Novikov which is in question was presented at the International Congress of Mathematicians, Moscow, 1966, in a half-hour lecture, in a seminar and in private conversations. It has also been announced in the Doklady of the Academy of Sciences of the USSR, vol. 172 (1967) pp. 33-36. Some of Novikov's results have been obtained independently by P.S. Landweber (to appear in the Transactions of the AMS).

The object of these seminar notes is to give an exposition of that part of Novikov's work which deals with operations on complex cobordism. I hope that this will be useful, because I believe that the cohomology functor provided by complex cobordism is now ripe for exploitation. I therefore aim to present the material in sufficient detail, so that a reader who has a concrete application in mind can make his own calculations. In particular, I will give certain formulae which are not made explicit in the sources cited above.

These notes will not deal with any of the other topics which are mentioned in the sources cited above. These include the following.

(i) Generalizations of the Adams spectral sequence in which ordinary cohomology is replaced by generalized (extraordinary) cohomology.

(ii) Connections between these studies for complex cobordism $\Omega_U^*(X, Y)$ and the corresponding studies for complex K-theory $K^*(X, Y)$.

(iii) The cohomology functor $\Omega_U^*(X, Y) \otimes Q_p$ (where Q_p is the ring of rational numbers a/b with b prime to p); and the splitting of this functor into direct summands.

2. COBORDISM GROUPS

Let ξ be a $U(n)$-bundle over the CW-complex X. Let E and E_o be the total spaces of the associated bundles whose fibers are respectively the unit disc $E^{2n} \subset C^n$ and the unit sphere $S^{2n-1} \subset C^n$. Then the Thom complex is by definition the quotient space E/E_o; it is a CW-complex with base point. In particular, if we take ξ to be the universal $U(n)$-bundle over $BU(n)$, then the resulting Thom complex $M(\xi)$ is written $MU(n)$.

<u>Example 2.1.</u> There is a homotopy equivalence $MU(1) \sim BU(1)$.

<u>Proof.</u> Since E is a bundle with contractible fibers, the projection $p: E \to BU(1)$ and the zero cross-section $s_o: BU(1) \to E$ are mutually inverse equivalences. Since $S^1 = U(1)$ and E_o is the total space of the universal $U(1)$-bundle over $BU(1)$, E_o is contractible, and the quotient map $E \to E/E_o$ is a homotopy equivalence.

We have an obvious map $S^2 MU(n) \xrightarrow{i_n} MU(n+1)$. In this way the sequence of spaces

$$(MU(0), MU(1), MU(2), \ldots, MU(n), \ldots)$$

and maps i_n becomes a spectrum. Associated with this spectrum we have a cohomology functor, as in G. W. Whitehead, "Generalized homology theories," Trans. Amer. Math. Soc. 102 (1962), pp 227-283.

The groups of this cohomology functor are written $\Omega_U^q(X, Y)$, and called complex cobordism groups. For other accounts, see M.F. Atiyah, "Bordism and Cobordism," Proc. Camb. Phil. Soc. 57 (1961) pp 200-208, and P.E. Conner and E.E. Floyd, "The Relation of Cobordism to K-Theories," Springer, Lecture Notes in Mathematics No. 28, 1966, pp 25-28.

We will generally suppose that this cohomology functor is defined on some category of spectra or stable objects. This assumption can easily be removed, if the reader wishes, at the cost of making some of the proofs more complicated; one would have to replace the appropriate spectra by sequences of complexes approximating to them.

Next we wish to discuss the cup-products in this cohomology theory. We therefore wish to introduce the product map

$$\mu: MU \wedge MU \rightarrow MU.$$

Here "$\wedge$" means the smash product, and we assume that $MU \wedge MU$ can be formed in our stable category. We further assume that $MU \wedge MU$ has skeletons $(MU \wedge MU)^q$, in a suitable sense, so that we have a short exact sequence

$$0 \rightarrow \operatorname*{Lim}_q{}^1 [S(MU \wedge MU)^q, MU] \rightarrow [MU \wedge MU, MU] \rightarrow \operatorname*{Lim}_q{}^0 [(MU \wedge MU)^q, MU] \rightarrow 0.$$

(Here Lim^0 means the inverse limit, Lim^1 means the first derived functor of the inverse limit, and $[X, Y]$ means the group of stable homotopy classes of maps from X to Y in our stable category.) In this exact sequence, the group $\operatorname*{Lim}_q{}^1 [S(MU \wedge MU)^q, MU]$ is zero. (This follows from the facts that $H_r(MU \wedge MU) = 0$ for r odd and $\pi_r(MU) = 0$ for r odd--see below. Thus the spectral sequence

$$H^*(MU \wedge MU, \quad \pi_*(MU)) \Longrightarrow [MU \wedge MU, MU]$$

has all its differentials zero.) It will therefore be sufficient to give an element of $\lim_q{}^0 [(MU \wedge MU)^q, MU]$.

Now, we have a map

$$BU(n) \times BU(m) \to BU(n+m),$$

namely the classifying map for the Whitney sum of the universal bundles over $BU(n)$ and $BU(m)$. Over this map we have a map

$$\mu_{n,m}: MU(n) \wedge MU(m) \to MU(n+m).$$

The maps $\mu_{n,m}$ yield an element of $\lim_q{}^0 [(MU \wedge MU)^q, MU]$, and therefore they yield a unique homotopy class of maps

$$\mu: MU \wedge MU \to MU.$$

The map μ is commutative and associative (up to homotopy).

Using the map μ, one introduces products in cobordism. More precisely, one has a product

$$\Omega_U^q(X) \otimes \Omega_U^r(Y) \to \Omega_U^{q+r}(X \wedge Y)$$

where X and Y are spectra, and therefore a similar product for the reduced groups $\tilde{\Omega}_U^*$ where X and Y are spaces. For spaces we have also an external product

$$\Omega_U^q(X, A) \otimes \Omega_U^r(Y, B) \to \Omega_U^{q+r}(X \times Y, A \times Y \cup X \times B)$$

and an internal product

$$\Omega_U^q(X, A) \otimes \Omega_U^r(X, B) \to \Omega_U^{q+r}(X, A \cup B).$$

The products satisfy the axioms which products should satisfy, that is, naturality, associativity, anticommutativity, existence of a unit, and behavior with respect to suspension or coboundary.

Next we must mention the Thom isomorphism. For each $U(n)$-bundle ξ over X the classifying map for ξ induces a map

$$\gamma : M(\xi) \rightarrow MU(n).$$

The map γ represents a canonical element g in $\Omega_U^{2n}(E, E^0)$. We define the Thom isomorphism

$$\varphi : \Omega_U^q(X) \rightarrow \Omega_U^{q+2n}(E, E^0)$$

by $\varphi(x) = (p^* x)g$, as usual. (See A. Dold, "Relations between Ordinary and Extraordinary Cohomology," Colloquium on Algebraic Topology, Aarhus 1962.)

Only one thing remains before we have a fair grasp on the cohomology functor Ω_U; we need to know the coefficient groups $\Omega_U^q(P)$, where P is a point. In fact $\Omega_U^*(P)$ is a polynomial ring

$$Z[x_1, x_2, \ldots, x_i, \ldots],$$

where $x_i \in \Omega_U^{-2i}(P)$. A good grasp on $\Omega_U^*(P)$ is provided by the following authors: J. Milnor, "On the Cobordism Ring Ω^* and a Complex Analogue," Amer. Jour. Math. 82 (1960) pp 505-521; R. Stong, "Relations among Characteristic Numbers. I," Topology 4 (1965) pp 267-281; A. Hattori, "Integral characteristic numbers for weakly almost complex manifolds," Topology 5 (1966) pp 259-280.

3. HOMOLOGY

The Novikov operations are closely related to certain polynomials in the Conner-Floyd Chern classes. (These classes may be found in P.E. Conner and E.E. Floyd, loc. cit. pp 48-52.) It is convenient to begin by introducing the corresponding polynomials in the ordinary Chern classes.

The Whitney sum map $BU(n) \times BU(m) \rightarrow BU(n+m)$ defines products in $H_*(BU)$. We have $BU(1) = CP^\infty$, so $H^*(BU(1))$ has a

Z-base consisting of elements $1, x, x^2, x^3, \ldots$, where $x \in H^2(BU(1))$ is the generator. Take the dual base in $H_*(BU(1))$ and call it $b_o, b_1, b_2, b_3, \ldots$. The injection $BU(1) \to BU$ maps these elements into $H_*(BU)$, where they can be multiplied. $H_*(BU)$ has a Z-base consisting of the monomials

$$b_1^{\nu_1} b_2^{\nu_2} b_3^{\nu_3} \ldots \qquad (b_o = 1).$$

Take the dual base in $H^*(BU)$ and call its elements c_ν; here the index ν runs through sequences of integers

$$\nu = (\nu_1, \nu_2, \nu_3, \ldots)$$

in which all but a finite number of terms are zero. We have $c_\nu \in H^{2|\nu|}(BU)$, where

$$|\nu| = \nu_1 + 2\nu_2 + 3\nu_3 + \cdots.$$

If we take $\nu = (i, 0, 0, \ldots)$, we obtain the classical i^{th} Chern class c_i.

We have thus given a base of $H^*(BU)$ which is well related to the Whitney sum map. This is obviously profitable in considering MU, because in $H^*(MU)$ we have a Whitney sum map but not a cup-product map.

For later use, we describe $H_*(MU)$, which is defined by

$$H_{2i}(MU) = \lim_{n \to \infty} H_{2n+2i}(MU(n)).$$

The Whitney sum map $MU(n) \wedge MU(m) \to MU(n+m)$ defines products in $H_*(MU)$. The Thom isomorphism

$$\varphi: H^q(BU(n)) \to H^{q+2n}(MU(n))$$

passes to the limit and gives an isomorphism

$$\varphi: H^q(BU) \to H^q(MU),$$

and similarly for homology. In particular, we have a "Thom isomorphism"

$$\varphi: H_*(BU) \to H_*(MU),$$

which commutes with the products. Thus the ring $H_*(MU)$ is a polynomial ring on generators $b'_1, b'_2, b'_3, \ldots$, corresponding to $b_1, b_2, b_3, \ldots$ under the Thom isomorphism. It is equivalent, of course, to describe these generators as follows: take the generators $b_i \in H_{2i}(BU(1))$, take their images $b'_1 \in H_{2i+2}(MU(1))$ under the Thom isomorphism, and apply the injection

$$H_{2i+2}(MU(1)) \longrightarrow H_{2i}(MU).$$

Under the equivalence $MU(1) \sim BU(1)$, the class $b'_i \in H_{2i+2}(MU(1))$ corresponds to $b_{i+1} \in H_{2i+2}(BU(1))$.

4. THE CONNER-FLOYD CHERN CLASSES

Conner and Floyd take a $U(n)$-bundle ξ over a CW-complex X, and undertake to assign to it characteristic classes which lie, not in ordinary cohomology $H^*(X)$, but in $\Omega_U^*(X)$.

THEOREM 4.1. To each ξ over X and each $\alpha = (\alpha_1, \alpha_2, \alpha_3, \ldots)$ we can assign classes $cf_\alpha(\xi) \in \Omega_U^{2|\alpha|}(X)$, called the Conner-Floyd Chern classes, with the following properties:

(i) $cf_0(\xi) = 1$.

(ii) Naturality: $cf_\alpha(g^*\xi) = g^* cf_\alpha(\xi)$.

(iii) Whitney sum formula: $cf_\alpha(\xi \oplus \eta) = \sum_{\beta+\gamma=\alpha} (cf_\beta \xi)(cf_\gamma \eta)$.

(iv) Let ξ be a $U(1)$-bundle over X, classified by a map $X \xrightarrow{f} BU(1)$, and let the composite $X \xrightarrow{f} BU(1) \longrightarrow MU(1)$ represent the element $\omega \in \Omega^2(X)$. Then

$$cf_\alpha(\xi) = \sum_{i \geq 0} (c_\alpha, b_i)\omega^i.$$

<u>Explanations.</u> In (iii), the addition of the sequences β and γ

is done term-by-term; that is, if

$$\beta = (\beta_1, \beta_2, \beta_3, \ldots),$$

$$\gamma = (\gamma_1, \gamma_2, \gamma_3, \ldots),$$

then

$$\beta + \gamma = (\beta_1 + \gamma_1, \beta_2 + \gamma_2, \beta_3 + \gamma_3, \ldots).$$

The multiplication of $(cf_\beta \xi)$ and $(cf_\gamma \eta)$ is done in the ring $\Omega_U^*(X)$.

In (iv), the map $BU(1) \longrightarrow MU(1)$ is the equivalence provided by Example 2.1. The integer (c_α, b_i) is defined by the Kronecker pairing of $H^*(BU)$ and $H_*(BU)$ to Z. The sum over i is illusory; a non-zero contribution can arise only for $i = |\alpha|$. The formula merely means that $cf_\alpha(\xi)$ is $\omega^{|\alpha|}$ if α has the form $(0, 0, 0, \ldots)$ or $(0, 0, \ldots, 0, 1, 0, \ldots)$, and otherwise zero. The use of coefficients like (c_α, b_i) is however convenient for doing algebra, and saves dividing cases.

Sketch proof of Theorem 4.1. The Grothendieck method for defining the ordinary Chern classes works just as well in generalized cohomology, and defines $cf_1, cf_2, cf_3, \ldots$. (See Conner and Floyd, loc. cit.). Of course, Conner and Floyd restrict their spaces to be finite CW-complexes (although their arguments apply unchanged to finite-dimensional CW-complexes.) It is therefore necessary to argue that

$$\lim_q{}^1 \, \Omega_U^*((BU(n))^q) = 0,$$

so that cf_i defines an element of $\Omega_U^*(BU(n))$ (or of $\Omega_U^*(BU)$, if required). Therefore cf_i is defined on all $U(n)$-bundles, by naturality. The same means is employed to extend the scope of conclusions (iii) and (iv) beyond the case considered by Conner and Floyd.

It works because the appropriate Lim^1 groups for $BU(n) \times BU(m)$ and $BU(1)$ are zero.

So far we have only considered the classes $cf_1, cf_2, cf_3, \ldots$. Now, each element in $H^*(BU)$ can be written as a unique polynomial in the ordinary Chern classes $c_1, c_2, c_3, \ldots$; say

$$c_\alpha = P_\alpha(c_1, c_2, c_3, \ldots).$$

Define cf_α to be the same polynomial in $cf_1, cf_2, cf_3, \ldots$; that is,

$$cf_\alpha = P_\alpha(cf_1, cf_2, cf_3, \ldots).$$

Of course, one of the advantages claimed for the treatment above is that it avoids mentioning the algebra of symmetric polynomials. At the insistence of my friends, I explain the connection of the P_α with symmetric polynomials. Let $\sigma_1, \sigma_2, \sigma_3, \ldots$ be the elementary symmetric functions in a sufficiency of variables $x_1, x_2, \ldots, x_n$; then

$$P_\alpha(\sigma_1, \sigma_2, \sigma_3, \ldots) = \sum x_1^{m_1} x_2^{m_2} \ldots x_n^{m_n},$$

where the sum runs over n-tuples $(m_1, m_2, \ldots, m_n)$ such that α_1 of the m's are 1, α_2 of the m's are 2, and so on, while the rest of the m's are 0.

Both for practical calculation and conceptual work I recommend the study of the dual rings $H_*(BU)$ and $H^*(BU)$ above the study of symmetric polynomials.

Now that we have defined the classes cf_α, the Whitney sum formula (iii) is deduced from the special case

$$cf_k(\xi \oplus \eta) = \sum_{i+j=k} cf_i(\xi)\, cf_j(\eta)$$

by pure algebra, and similarly the behavior on line bundles (iv) is deduced by pure algebra from the special case

$$cf_i(\xi) = \begin{cases} 1 & (i = 0) \\ \omega & (i = 1) \\ 0 & (i > 1). \end{cases}$$

5. THE NOVIKOV OPERATIONS

The basic analogy which Novikov follows is now: as the Steenrod squares are to the Stiefel-Whitney classes, so the Novikov operations are to the Conner-Floyd characteristic classes. This will be made precise in Theorem 5.1 (vii) below.

THEOREM 5.1. (S.P. Novikov): For each $\alpha = (\alpha_1, \alpha_2, \alpha_3, \ldots)$ there exists an operation

$$s_\alpha : \Omega_U^q(X, Y) \to \Omega_U^{q+2|\alpha|}(X, Y)$$

with the following properties:

(i) $s_o = 1$, the identity operation.

(ii) s_α is natural: $s_\alpha f^* = f^* s_\alpha$.

(iii) s_α is stable: $s_\alpha \delta = \delta s_\alpha$.

(iv) s_α is additive: $s_\alpha(x+y) = (s_\alpha x) + (s_\alpha y)$.

(v) Cartan formula:

$$s_\alpha(xy) = \sum_{\beta+\gamma=\alpha} (s_\beta x)(s_\gamma y).$$

(vi) Suppose that an element $\omega \in \Omega^2(x)$ is represented by a map $X \xrightarrow{g} MU(1)$. Then

$$s_\alpha(\omega) = \sum_i (c_\alpha, b_i)\omega^{i+1} .$$

(vii) Suppose that ξ is an $U(n)$-bundle over X, and consider the following diagram.

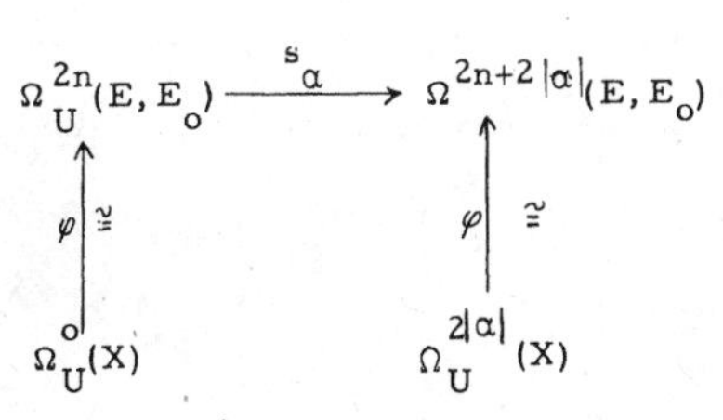

(Here the pair E, E_o is as in §2, and φ is the Thom isomorphism for Ω_U^*.) Then we have

$$cf_\alpha(\xi) = \varphi^{-1} s_\alpha \varphi 1.$$

Explanations: In (v), the addition of the sequences β and γ is done term-by-term. The cup-product xy may be taken in any one of the three senses explained above, and then the cup-product $(s_\beta x)(s_\gamma y)$ is to be taken in the same sense.

For the coefficient (c_α, b_i) in (vi), see the note on Theorem 4.1 (iv).

Sketch proof. We take (vii) as our guide. We have a Thom isomorphism

$$\varphi: \Omega_U^*(BU(n)) \to \tilde{\Omega}_U^*(MU(n)).$$

Consider the elements $\varphi cf_\alpha \in \tilde{\Omega}_U^{2n+2|\alpha|}(MU(n))$. They yield a unique element $s_\alpha \in \Omega^{2|\alpha|}(MU)$ (the Lim^1 argument again). This element defines an operation on the cohomology theory Ω_U^*.

Property (vii) results immediately from the definition, and properties (ii), (iii) and (iv) are trivial. For example, if $x, y: X \to MU$ are maps, and if we represent s_α by a map $s: MU \to S^{2a}MU$, then the maps $s(x+y)$ and $(sx)+(sy): X \to S^{2a}MU$ are homotopic, since we are working in a stable category.

Properties (i), (v) and (vi) are deduced from the corresponding properties (i), (iii) and (iv) of the Conner-Floyd classes (Theorem 4.1) by using appropriate properties of the Thom isomorphism φ. For example: in proving (v), it is sufficient to consider the case in which x and y are both the identity map $i: MU \to MU$, so that xy is the product map $\mu: MU \wedge MU \to MU$. Using the Lim^1 argument again, it is sufficient to consider the case in which x and y are the generators for $\tilde{\Omega}_U^{2n}(MU(n))$, $\tilde{\Omega}_U^{2m}(MU(m))$. Now we use the fact that if ξ is a $U(n)$-bundle over X and η is a $U(m)$-bundle over Y the following diagram is commutative.

$$\begin{array}{ccc}
\tilde{\Omega}_U^{p+2n}(M(\xi)) \otimes \tilde{\Omega}_U^{q+2m}(M(\eta)) & \xrightarrow{\text{product}} & \tilde{\Omega}^{p+q+2n+2m}(M(\xi) \wedge M(\eta)) \\
 & & \| \\
\uparrow \varphi_\xi \otimes \varphi_\eta & & \tilde{\Omega}^{p+q+2n+2m}(M(\xi \times \eta)) \\
 & & \uparrow \varphi_{\xi \times \eta} \\
\Omega_U^p(X) \otimes \Omega_U^q(Y) & \xrightarrow{\text{product}} & \Omega_U^{p+q}(X \times Y)
\end{array}$$

The application, of course, is with ξ the universal bundle over $BU(\eta)$ and η the universal bundle over $BU(m)$.

For (vi) we need to know that for the universal $U(1)$-bundle over $BU(1)$, the homomorphism

$$\Omega_U^{2i}(BU(1)) \to \tilde{\Omega}_U^{2i+2}(MU(1)) = \Omega_U^{2i+2}(MU(1)) \qquad (i \geq 0)$$

carries $\bar{\omega}^i$ to $\bar{\omega}^{i+1}$. (Here $\bar{\omega}$ is the universal element in $\Omega_U^2(BU(1))$ or $\Omega_U^2(MU(1))$.)

Since s_α is a homotopy class of maps

$$MU \to S^{2|\alpha|}MU,$$

it induces a homomorphism

$$s_\alpha: H_q(MU) \to H_{q-2|\alpha|}(MU).$$

It is reasonable to ask for this homomorphism to be made explicit. Since we have seen in §3 that $H_*(MU)$ is a polynomial ring, it is reasonable to ask (i) how s_α acts on products, and (ii) how s_α acts on the generators b_i'. Set $b' = \sum_{i=0}^{\infty} b_i'$; then it is sufficient to know $s_\alpha(b')$, since one can separate the components again.

THEOREM 5.2. (i) If $x, y \in H_*(MU)$, then

$$s_\alpha(xy) = \sum_{\beta+\gamma=\alpha} (s_\beta x)(s_\gamma y).$$

(ii) $s_\alpha(b') = \sum_{i\geq 0} (c_\alpha, b_i)(b')^{i+1}$.

Sketch proof. Part (i). By Theorem 5.1 (v), we have the following commutative diagram.

$$\begin{array}{ccc} MU \wedge MU & \xrightarrow{\mu} & MU \\ \Big\downarrow {\scriptstyle \sum_{\beta+\gamma=\alpha} s_\beta \wedge s_\gamma} & & \Big\downarrow {\scriptstyle s_\alpha} \\ \bigvee_{\beta+\gamma=\alpha} S^{2|\beta|}MU \wedge S^{2|\gamma|}MU & \xrightarrow{\mu} & S^{2|\alpha|}MU \end{array}$$

Pass to induced maps of homology.

Part (ii). Since the generators b'_t come from $MU(1)$, we can make use of Theorem 5.1 (vi). If ω is the canonical element of $\Omega^2(MU(1))$, we wish to compute the effect on homology of the element $\omega^{i+1} \in \Omega^{2i+2}(MU(1))$, that is, the effect of the following composite map.

$$\begin{array}{ccc} MU(1) \xrightarrow{\Delta} & MU(1) \wedge MU(1) \wedge \ldots \wedge MU(1) & \quad (i+1 \text{ factors}) \\ & \Big\downarrow {\scriptstyle \mu} & \\ & MU(i+1) & \end{array}$$

Now, the diagonal map

$$BU(1) \xrightarrow{\Delta} BU(1) \times BU(1) \times \ldots \times BU(1)$$

induces a map of cohomology given by

$$\Delta^*(x^{u_1} \otimes x^{u_2} \otimes \ldots \otimes x^{u_{i+1}}) = x^{u_1+u_2+\ldots+u_{i+1}} \quad ;$$

therefore it induces a map of homology given by

$$\Delta_* b_t = \sum_{u_1+u_2+\ldots+u_{i+1}=t} b_{u_1} \otimes b_{u_2} \otimes \ldots \otimes b_{u_{i+1}} \quad .$$

The map of $\tilde{H}_*$ induced by

$$BU(1) \xrightarrow{\Delta} BU(1) \wedge BU(1) \wedge \dots \wedge BU(1)$$

is given by the same formula, provided we now interpret b_o as 0. Next recall that b'_t in $MU(1)$ corresponds to b_{t+1} in $BU(1)$. We deduce that

$$\Delta_* b'_t = \sum_{u_1+u_2+\dots+u_{i+1}=t-i} b'_{u_1} \otimes b'_{u_2} \otimes \dots \otimes b'_{u_{i+1}}$$

and

$$\mu_* \Delta_* b'_t = \sum_{u_1+u_2+\dots+u_{i+1}=t-i} b'_{u_1} b'_{u_2} \dots b'_{u_{i+1}} .$$

Adding, we see that

$$\mu_* \Delta_* b' = (b')^{i+1} .$$

By Theorem 5.1 (vi), we have the following commutative diagram.

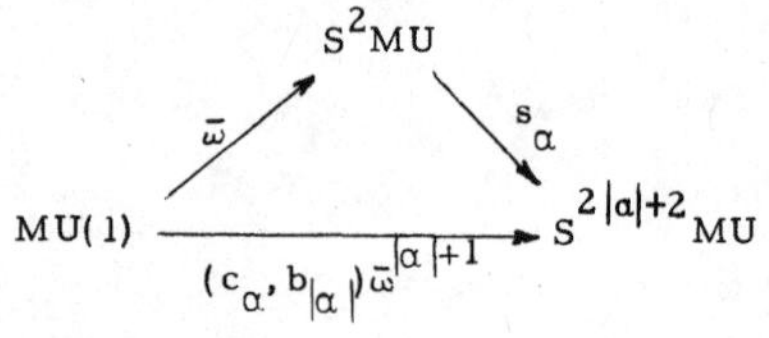

$$\begin{array}{ccc} & S^2MU & \\ {}^{\bar{\omega}}\nearrow & & \searrow^{s_\alpha} \\ MU(1) & \xrightarrow{(c_\alpha, b_{|\alpha|})\bar{\omega}^{|\alpha|+1}} & S^{2|\alpha|+2}MU \end{array}$$

Pass to induced maps of homology.

COROLLARY 5.3. $s_\alpha : H^o(MU) \to H^{2|\alpha|}(MU)$ is given by

$$s_\alpha \varphi 1 = \varphi c_\alpha .$$

<u>Proof.</u> By Theorem 5.2 (ii),

$$s_\alpha(b'_i) = \begin{cases} 0 & \text{if } i < |\alpha| \text{ (trivially)} \\ (c_\alpha, b_i) 1 & \text{if } i = |\alpha| . \end{cases}$$

Using Theorem 5.2 (i) we have

$$s_\alpha(b'_{i_1} b'_{i_2} \dots b'_{i_r}) = \sum_{\beta_1+\beta_2+\dots+\beta_r=\alpha} (s_{\beta_1} b'_{i_1})(s_{\beta_2} b'_{i_2}) \dots (s_{\beta_r} b'_{i_r}).$$

If we assume that $i_1+i_2+\dots+i_r = |\alpha|$, then the only terms which can contribute to this sum are those with

$$|\beta_1| = i_1, \quad |\beta_2| = i_2, \dots, |\beta_r| = i_r ,$$

and we obtain

$$\sum (c_{\beta_1}, b_{i_1})(c_{\beta_2}, b_{i_2}) \ldots (c_{\beta_r}, b_{i_r}) 1$$

where the sum runs over such $\beta_1, \beta_2, \ldots, \beta_r$. This of course yields

$$(c_\alpha, b_{i_1} b_{i_2} \ldots b_{i_r}) 1 .$$

We have shown that

$$s_\alpha(\varphi x) = (c_\alpha, x) 1$$

for $x \in H_{2|\alpha|}(BU)$. Transposing to cohomology, we obtain

$$s_\alpha \, \varphi 1 = \varphi c_\alpha .$$

6. THE ALGEBRA OF ALL OPERATIONS

Next we need to consider a much more trivial sort of operation. Let x be a fixed element in $\Omega_U^p(P)$. Let X, Y be a pair, and let $c: X \to P$ be the constant map; thus $c^*(x) \in \Omega_U^p(X)$. For each $y \in \Omega_U^q(X, Y)$, we define

$$t(y) = (c^* x) y \in \Omega_U^{p+q}(X, Y).$$

This defines a cohomology operation

$$t: \Omega_U^q(X, Y) \to \Omega_U^{p+q}(X, Y).$$

In fact, we can say that $\Omega_U^*(P)$ acts on all our groups $\Omega_U^*(X, Y)$, acting on the left.

Now suppose that we fix a dimension d (positive, negative or zero), and for each index $\alpha = (\alpha_1, \alpha_2, \alpha_3, \ldots)$ we choose an element

$$x_\alpha \in \Omega_U^{d-2|\alpha|}(P).$$

(We do not require that all but a finite number of the x_α are zero; they may all be non-zero if they wish.) For each x_α we have a corresponding operation

$$t_\alpha : \Omega_U^{q+2|\alpha|}(X, Y) \to \Omega_U^{q+d}(X, Y).$$

We now consider the infinite sum

$$\sum_\alpha t_\alpha s_\alpha : \Omega_U^q(X, Y) \to \Omega_U^{q+d}(X, Y).$$

(Here we are assuming, as usual that X, Y is a CW-pair of finite homological dimension.)

THEOREM 6.1 (Novikov).

(i) This sum converges, in the sense that all but a finite number of the terms $t_\alpha s_\alpha$ yield zero.

(ii) This sum defines a cohomology operation on Ω_U^* which is natural and stable.

(iii) Every cohomology operation on Ω_U^* which is natural and stable can be written in this form.

(iv) This way of writing a cohomology operation on Ω_U^* is unique; if

$$t_\alpha s_\alpha = 0 : \Omega_U^q(X, Y) \to \Omega_U^{q+d}(X, Y)$$

for all X, Y and q, then $x_\alpha = 0$ for all α.

<u>Sketch proof.</u> Part (i) is trivial: the group $\Omega_U^{q+2\alpha}(X, Y)$ is zero if $|\alpha|$ is large compared with the homological dimension of the pair X, Y. Part (ii) is also trivial.

For parts (iii) and (iv), consider the spectral sequence

$$H^*(MU, \Omega_U^*(P)) \Longrightarrow \Omega_U^*(MU).$$

It follows from Corollary 5.3 that the elements $s_\alpha \in \Omega_U^*(MU)$ constitute an $\Omega_U^*(P)$-base for the E_2 term of this spectral sequence.

There is an alternative method of proving part (iv), as follows.

<u>Remark 6.2</u> (Novikov). The operations $\sum_\alpha t_\alpha s_\alpha$ are distinguished by their values on the classes

$$\omega_1\omega_2\ldots\omega_m \in \Omega_U^{2m}(CP^n \times CP^n \times \ldots \times CP^n)$$

(where m and n run over all positive integers).

Sketch proof. It is easily seen that $\Omega_U^*(CP^n \times CP^n \times \ldots \times CP^n)$ is free over $\Omega_U^*(P)$, with an $\Omega_U^*(P)$-base consisting of the monomials

$$\omega_1^{i_1}\omega_2^{i_2}\ldots\omega_m^{i_m}$$

with $0 \leq i_r \leq n$ for all r; the remaining monomials are zero. We have

$$s_\alpha(\omega_1\omega_2\ldots\omega_m) = \sum_{i_1, i_2, \ldots, i_m} (c_\alpha, b_{i_1} b_{i_2} \ldots b_{i_m})\omega_1^{i_1+1}\omega_2^{i_2+1}\ldots\omega_m^{i_m+1}.$$

This will of course be zero if $\alpha_1 + \alpha_2 + \alpha_3 + \ldots > m$ or if $\alpha_i > 0$ for any i with $i+1 > n$; but the remaining elements $s_\alpha(\omega_1\omega_2\ldots\omega_m)$ are linearly independent over $\Omega_U^*(P)$.

Note. With the foundations indicated above, the use of CP^∞ instead of CP^n gives no trouble.

Next we need to know how to compute the composite of two operations $t_\alpha s_\alpha$, $t'_\beta s_\beta$. This breaks up into three problems.

(i) We need to write $s_\alpha t'_\beta$ in the form $\sum_\gamma t''_\gamma s_\gamma$. This reduces to computing the action of s_α on $\Omega_U^*(P)$, for

$$s_\alpha((c^* x)y) = \sum_{\beta+\gamma=\alpha} (s_\beta c^* x)(s_\gamma y) = \sum_{\beta+\gamma=\alpha} (c^* s_\beta x)(s_\gamma y).$$

This writes the operation in the required form.

Now we have $\Omega_U^*(P) = \pi_*(MU)$, and by Milnor (loc. cit.) the Hurewicz homomorphism

$$\pi_*(MU \rightarrow H_*(MU)$$

is monomorphic. Therefore, in principle it is sufficient to know the action of s_α on $H_*(MU)$, which has been given in Theorem 5.2.

We will return later to the action of s_α on $\Omega_U^*(P)$.

(ii) We need to compute the composite $t_\alpha t''_\gamma$. This is trivial; just multiply the corresponding elements of $\Omega^*_U(P)$.

(iii) We need to compute the composite $s_\gamma s_\beta$. This is done by the following theorem.

THEOREM 6.3. The set S of Z-linear combinations of the s_α is closed under composition. The ring S is a Hopf algebra over Z, whose dual S_* is the polynomial algebra on generators $b''_1, b''_2, b''_3, \ldots$, where $(s_\alpha, b''_i) = (c_\alpha, b_i)$. Set $b'' = \sum_{i=0}^{\infty} b''_i$, where $b''_0 = 1$; then the diagonal in S_* is given by

$$\Delta b'' = \sum_{i \geq 0} (b'')^{i+1} \otimes b''_i .$$

<u>Explanation.</u> By separating this formula into components we obtain the value of $\Delta b''_k$; this determines the diagonal on the whole of S_*, and hence determines the product in S. The situation is similar to that arising in Milnor's work on the dual of the Steenrod algebra.

Theorem 6.3 is due to Novikov, except that he does not give the explicit formula for the diagonal in S_*.

<u>Sketch proof.</u> In $\Omega^*_U(CP^n \times CP^n \times \ldots \times CP^n)$, $s_\beta(\omega_1 \omega_2 \ldots \omega_m)$ is a Z-linear combination of monomials $\omega_1^{i_1} \omega_2^{i_2} \ldots \omega_m^{i_m}$, and hence $s_\alpha s_\beta(\omega_1 \omega_2 \ldots \omega_m)$ is a Z-linear combination of monomials $\omega_1^{j_1} \omega_2^{j_2} \ldots \omega_m^{j_m}$. By the proof following Remark 6.2, $s_\alpha s_\beta$ is a Z-linear combination of operations s_γ.

We next wish to calculate $\Delta b''_k$, that is, to find $s_\alpha s_\beta(\omega)$ for each α, β, where ω is the generator in $\Omega^2(CP^\infty)$. We have

$$s_\beta \omega = \sum_i (s_\beta, b''_i) \omega^{i+1}$$

and therefore

$$s_\alpha s_\beta \omega = \sum_{i, j_1, j_2, \ldots, j_{i+1}} (s_\alpha, b''_{j_1} b''_{j_2} \ldots b''_{j_{i+1}})(s_\beta, b'_i)\omega^{i+j_1+j_2+\ldots+j_{i+1}+1}.$$

We conclude that

$$\Delta b''_k = \sum_{i+j_1+j_2+\ldots+j_{i+1}=k} b''_{j_1} b''_{j_2} \ldots b''_{j_{i+1}} \otimes b''_i.$$

Summing over b, we obtain the formula given.

Note. Now that we have introduced the dual Hopf algebra S_*, we can reformulate Theorem 5.2. Recall that S acts on $H_*(MU)$, acting on the left; therefore it acts on the right on $H^*(MU)$; that is, we have a product map

$$\mu: H^*(MU) \otimes S \to H^*(MU).$$

Transposing again, we have a coproduct map

$$\Delta: H_*(MU) \to H_*(MU) \otimes S_*.$$

This is related to the original action of S on $H_*(MU)$ as follows: if

$$\Delta h = \sum_i h_i \otimes s_i^*$$

then

$$sh = \sum_i h_i(s_i^* s)$$

for all $s \in S$. The map

$$\Delta: H_*(MU) \to H_*(MU) \otimes S_*$$

may be described as follows.

PROPOSITION 6.4. Δ preserves products, and

$$\Delta b' = \sum_{i \geq 0} (b')^{i+1} \otimes b''_i .$$

This is a trivial reformulation of Theorem 5.2.

The analogy between this formula and that in Theorem 6.3 should be noted.

At this point we possess a firm grasp of the algebra of operations on Ω_U^*.

7. SCHOLIUM ON NOVIKOV'S EXPOSITION

In Moscow, Novikov made a careful distinction, which is maintained in his Doklady note, between $s_\omega : \Omega_U^*(P) \to \Omega_U^*(P)$ and a certain homomorphism $\sigma_\omega^* : \Omega_U^*(P) \to \Omega_U^*(P)$. It is necessary to observe that they coincide, and for this purpose it is necessary to analyse Theorem 3 of Novikov's Doklady note.

First observe that in Novikov's Doklady note, M_U and Ω_U are different names for the same thing, since both are defined to be $\Omega_U^*(P)$ (p. 33 line 4 of Section II; p. 35 line 8). Next recall that Novikov writes A^U for the algebra of operations, and observe that the isomorphism

$$\mathrm{Hom}_{A^U}(A^U, M_U) \xrightarrow{\cong} \Omega_U$$

which he has in mind is precisely the standard isomorphism θ given by

$$\theta(h) = h(1).$$

Next consider Novikov's map $d : A^U \to A^U$. Since it is asserted to induce a map

$$d^* : \mathrm{Hom}_{A^U}(A^U, M_U) \to \mathrm{Hom}_{A^U}(A^U, M_U),$$

it is implicit that d must be a map of left A-modules. Since it is asserted to satisfy $d(1) = s_\omega$, it must be given by

$$d(a) = as_\omega .$$

Now consider the following diagram.

$$\begin{array}{ccc}
\mathrm{Hom}_{A^U}(A^U, M_U) & \xrightarrow{d^*} & \mathrm{Hom}_{A^U}(A^U, M_U) \\
\downarrow \theta & & \downarrow \theta \\
\Omega_U & \xrightarrow{x} & \Omega_U
\end{array}$$

It is trivial to check that it is commutative if we define x by $x(y) = s_\omega y$. But Novikov asserts that it is commutative if we define x to be σ_ω^*. Therefore $\sigma_\omega^*(y) = s_\omega y$.

8. COMPLEX MANIFOLDS

Next it is necessary to recall that a stably almost-complex manifold M^n defines an element $[M^n]$ of $\Omega_U^{-n}(P)$. If we are given such a stably almost-complex manifold M^n, it is natural to ask for the value of $s_\alpha[M^n]$. It is especially reasonable to ask this for the manifolds CP^n, since these manifolds are familiar and are known to provide a set of generators for the polynomial ring $\Omega_U^*(P) \otimes Q$ (where Q is the ring of rational numbers).

THEOREM 8.1. $s_\alpha[CP^n] = (c_\alpha, b^{-n-1})[CP^{n-|\alpha|}]$ where $b = \sum_{i=0}^{\infty} b_i$.

Explanation. Since the element b is a formal series with first term 1, it is invertible. The integer (c_α, b^{-n-1}) is the Kronecker product of an element in $H^{2|\alpha|}(BU)$ and an element in $\prod_q H_q(BU)$. This time we have used the algebra in §3 to write down a coefficient which isn't necessarily 0 or 1.

Theorem 8.1 is due to Novikov, except that he does not give the explicit formula for the coefficient of $[CP^{n-|\alpha|}]$.

Sketch Proof. To preserve the character of the arguments, we will show how to deduce this from Theorem 5.2 by pure algebra.

The letter χ will always mean the canonical anti-automorphism of the relevant Hopf algebra. In CP^n, the tangent bundle τ satisfies $\tau \oplus 1 = (n+1)\xi$, and so for the normal bundle ν we have

$$\begin{aligned} c_\alpha(\nu) &= (\chi c_\alpha)\tau \\ &= (\chi c_\alpha)((n+1)\xi) \\ &= \sum_{i_1, i_2, \ldots i_{n+1}} (\chi c_\alpha, b_{i_1} b_{i_2} \ldots b_{i_{n+1}}) x^{i_1 + i_2 + \ldots + i_{n+1}} \\ &= \sum_{i_1, i_2, \ldots i_{n+1}} (c_\alpha, \chi(b_{i_1} b_{i_2} \ldots b_{i_{n+1}})) x^{i_1 + i_2 + \ldots + i_{n+1}} . \end{aligned}$$

The terms with $i_1 + i_2 + \ldots + i_{n+1} = n$ give the normal characteristic numbers of CP^n. Therefore the class of $[CP^n]$ in $H_{2n}(MU)$ is

$$\begin{aligned} \varphi \sum_{i_1 + i_2 + \ldots + i_{n+1} = n} & \chi(b_{i_1} b_{i_2} , , , b_{i_{n+1}}) \\ &= \varphi\chi(b^{n+1})_n , \end{aligned}$$

where the subscript n means the $2n$-dimensional component. But since $\Delta b = b \otimes b$, we have $\chi b = b^{-1}$ and $\chi(b^{n+1}) = b^{-n-1}$. We conclude that the class of $[CP^n]$ in $H_{2n}(MU)$ is

$$((b')^{-n-1})_n .$$

Now by 5.2 (ii) we have the formula

$$s_\alpha(b') = \sum_{i \geq 0} (c_\alpha, b_i)(b')^{i+1} .$$

From this we will deduce

(Formula 8.2) $$s_\alpha(b')^{-1} = \sum_{j \geq 0} (c_\alpha, \chi b_j)(b')^{j-1} .$$

It is easy to see that this checks; for it yields

$$s_\alpha(b'\cdot(b')^{-1}) = \sum_{\substack{\beta+\gamma=\alpha \\ i\geq 0, j\geq 0}} (c_\beta, b_i)(c_\gamma, \chi b_j)(b')^{i+j}$$

$$= \sum_{i\geq 0, j\geq 0} (c_\alpha, b_i\cdot\chi b_j)(b')^{i+j}$$

$$= (c_\alpha, b_o)\ 1,$$

as it should. But this manipulation allows one to prove the formula for $s_\alpha((b')^{-1})_d$ by double induction over $|\alpha|$ and d, starting from the trivial cases $|\alpha| = 0$ and $d = 0$.

From (8.2) we deduce that

$$s_\alpha((b')^{-n-1})_n = \sum_{i_1+i_2+\ldots+i_{n+1}=|\alpha|} (c_\alpha, \chi(b_{i_1} b_{i_2} \ldots b_{i_{n+1}}))((b')^{|\alpha|-n-1})_{n-|\alpha|}.$$

This is the class of $[CP^{n-|\alpha|}]$ in $H^{2n-2|\alpha|}(MU)$, up to a factor (c_α, b^{-n-1}). Now the result follows from the fact that the Hurewicz homomorphism

$$\pi_*(MU) \rightarrow H_*(MU)$$

is monomorphic.

From a geometrical point of view the proof just given is uncouth and perverse; Theorem 8.1 should be deduced from an elegant formula of Novikov. Before starting this, we will recall some material from ordinary cohomology.

Let M, N be oriented manifolds of dimension m, n, and let $f: M \rightarrow N$ be a continuous map. The "Umkehrunghomomorphismus" or "forward homomorphism"

$$f_!: H^q(M) \rightarrow H^{n-m+q}(N)$$

is defined to be the following composite.

$$\begin{array}{ccc} H^q(M) & & H^{n-m+q}(N) \\ d \downarrow \cong & & d \downarrow \cong \\ H_{m-q}(M) & \xrightarrow{f_*} & H_{m-q}(N) \end{array}$$

Here d is the Poincare duality isomorphism.

A similar construction may be given in which H^* is replaced by Ω_U^*, provided we assume that M and N are stably almost-complex manifolds and replace d by the Atiyah duality isomorphism

$$D: \Omega_U^q(M) \to \Omega_{m-q}^U(M).$$

Here Ω_{m-q}^U means complex bordism; see Atiyah, loc. cit., for real bordism and the corresponding duality theorem.

We shall in fact only have to apply the homomorphism $f_!$ in the case when N is a point P and f is the constant map $c: M \to P$. It will make both the proof and the exposition easier if we give an alternative definition of $c_!$, which does not require the introduction of bordism.

Suppose that we embed the manifold M in a high-dimensional sphere S^{m+2p}, with unitary normal bundle ν. Define $c_!$ to be the following composite.

$$\begin{array}{ccc} \Omega_U^q(M) & & \Omega_U^{q-m}(P) \\ \varphi \downarrow \cong & & \\ \Omega_U^{q+2p}(E, E_o) & & \downarrow \cong \\ \uparrow \cong & & \\ \Omega_U^{q+qp}(S^{m+2p}, C\,\mathrm{Int}\,E) & \xrightarrow{j^*} & \Omega_U^{q+2p}(S^{m+2p}, D^{m+2p}) \end{array}$$

Here φ is the Thom isomorphism; E and E_o refer to the normal bundle ν of M; and $C\,\mathrm{Int}\,E$ is the complement of the interior of E. (If one wished one could replace $\Omega_U^{q+2p}(S^{m+2p}, C\,\mathrm{Int}\,E)$ by

$\Omega_U^{q+2p}(S^{m+2p}, C M)$; this would make it clearer that this group is standing in for a bordism group of M, via Alexander duality or S-duality.) Further, D^{m+2p} is a small disc contained in C Int E, and the right-hand vertical arrow is the usual iterated suspension; this may be viewed as the analogue of the left-hand column, with M replaced by P.

We will accept this composite as our definition of $c_!$. If any reader who is familiar with bordism prefers a different definition, we may leave it to him to reconcile his definition with this one.

Now we come to Novikov's formula. Take a stably almost complex manifold M^m, representing an element $[M^m] \in \Omega_U^{-m}(P)$. Let ν be its stable normal bundle; thus $cf_\alpha(\nu) \in \Omega_U^{2|\alpha|}(M^m)$ and $c_! cf_\alpha(\nu) \in \Omega^{2|\alpha|-m}(P)$.

THEOREM 8.3 (Novikov). $s_\alpha[M^m] = c_! cf_\alpha(\nu)$.

This result follows easily from the definition of s_α in §5, by naturality.

PART II

QUILLEN'S WORK ON FORMAL GROUPS AND COMPLEX COBORDISM

0. INTRODUCTION

These notes derive from a series of lectures which I gave in Chicago in April 1970. It is a pleasure to thank my hosts for an enjoyable and stimulating visit.

In §§1-8, I have tried to give a connected account, beginning from first principles and working up to Milnor's calculation of $\pi_*(MU)$ (8.1) and Quillen's theorem that $\pi_*(MU)$ is isomorphic to Lazard's universal ring L (8.2). The structure of L is obtained from first principles (7.1). This is done by relating the notion of a formal group to the notion of a Hopf algebra. The material has been so arranged that algebraists who are interested in the subject can obtain a fairly self-contained account by reading §§1, 3, 5, 7.

The remaining sections deal with related matters. In [3, Lecture 3], I have shown that for suitable spectra E, $E_*(E)$ can be given the structure of a Hopf algebra analogous to the dual of the Steenrod algebra. The structure of this Hopf algebra is described for the spectrum MU in §11, for the BU-spectrum in §13, and for the Brown-Peterson spectrum in §16. Sections 15 and 16 are devoted to Quillen's work on the Brown-Peterson spectrum [14]. §14 is devoted to the Hattori-Stong theorem.

1. FORMAL GROUPS

We may understand formal groups by an analogy. Let G be a real Lie group of dimension 1. By choosing a chart, we may identify a neighborhood of the unit in G with a neighborhood of zero in R^1, so that the unit in G corresponds to zero. The product in G is then given by a power-series:

(1.1) $$\mu(x, y) = \sum_{i,j\geq 0} a_{ij}x^i y^j .$$

This power-series is convergent for small x and y, and satisfies the following conditions.

(1.2) $$\mu(x, 0) = x, \quad \mu(0, y) = y.$$

(1.3) $$\mu(x, \mu(y, z)) = \mu(\mu(x, y), z).$$

Now let R be any commutative ring with unit. Then a "formal product" (over R) is a formal power-series of the form (1.1), but with coefficients a_{ij} in R, satisfying (1.2) and (1.3).

We have two trivial examples.

(1.4) $$\mu(x, y) = x + y.$$

(1.5) $$\mu(x, y) = x + y + xy.$$

For example, suppose that we consider the Lie group G of positive real numbers under multiplication, and use the chart under which $x \in R^1$ corresponds to $(1 + x) \in G$; we obtain formula (1.5).

Let us return to the general case; there are a few obvious comments. Condition (1.2) is equivalent to:

$$(1.6) \qquad a_{i0} = \begin{cases} 1 & (i = 1) \\ 0 & (i \neq 1) \end{cases} \qquad a_{0j} = \begin{cases} 1 & (j = 1) \\ 0 & (j \neq 1) \end{cases}$$

So we may write our formal power-series in the following form.

$$(1.7) \qquad \mu(x, y) = x + y + \sum_{i,j \geq 1} a_{ij} x^i y^j .$$

Condition (1.3) involves substituting one formal power-series in another, but this involves no difficulty since our formal power-series have their constant terms zero.

We observe that so far we are only discussing the case of dimension 1. That is, in the general case one would start from a Lie group of dimension n, and proceed by analogy.

Given a formal product μ, a formal inverse ι is a formal power-series

$$(1.8) \qquad \iota x = \sum_{j \geq 1} a'_j x^j$$

(with coefficients a'_j in our ring R) such that

$$(1.9) \qquad \mu(x, \iota x) = 0, \qquad \mu(\iota x, x) = 0.$$

LEMMA 1.10. Given any formal product μ, there is a formal inverse ι, and it is unique.

The proof is trivial.

We have two examples; with the "additive product" of (1.4) we have

$$\iota(x) = -x ,$$

and with the "multiplicative product" of (1.5) we have

$$\iota(x) = -x + x^2 - x^3 + x^4 \ldots .$$

So far, a "formal product" is like a grin without a Cheshire cat behind it. A "formal group" must, of course, be a group object in a suitable category; I take this notion as known. If X is to be a group object in the category C, then Cartesian products such as X^n must exist in C for $n = 0, 1, 2, 3$; and X must be provided with structure maps in the category C, namely a product map $m:X^2 \to X$, a unit map $e:X^0 \to X$ and an inverse map $i:X \to X$. These maps must satisfy the obvious conditions. For example, consider the category of smooth manifolds and smooth maps; a group in this category is a Lie group. Again, consider the category of commutative rings and homomorphisms of rings, and let C be the opposite category; with a little goodwill C may be regarded as the category of affine algebraic varieties. A group in this category is an "algebraic group".

Now consider the category in which the objects are filtered commutative algebras over R, which are complete and Hausdorff for the filtration topology; the morphisms are filtration-preserving homomorphisms. Let C be the opposite category. The ring of formal power-series

$$R[[x_1, x_2, \ldots, x_n]],$$

with the obvious filtration, is an object in C. The objects $R[[x_1, x_2, \ldots, x_n]]$ and $R[[y_1, y_2, \ldots, y_m]]$ have a Cartesian product in C, namely

$$R[[x_1, x_2, \ldots, x_n, y_1, y_2, \ldots, y_m]].$$

Let X be the object $R[[x]]$ in C; then a map $m:X^2 \to X$ in C is a filtration-preserving homomorphism

$$m:R[[x]] \to R[[x_1, x_2]];$$

such a map m is determined by giving $m(x)$, which is a formal power-series $\mu(x_1, x_2)$ with zero constant term. It is now easy to check that each "formal product" μ determines a structure map m which makes $R[[x]]$ into a group object, and conversely. (The unit map $e:R[[x]] \to R$ is defined by $e(\sum_{i \geq 0} c_i x^i) = c_o$; inverse maps come free of charge by Lemma 1.10). It is now clear how to proceed in dimension n; we have to consider the object $R[[x_1, x_2, \ldots, x_n]]$, and study the ways of making it into a group-object in C. A "formal group", then, is a group-object in the category C, whose underlying object is $R[[x_1, x_2, \ldots, x_n]]$.

We now revert to the case of dimension 1. Let $\theta: R \to S$ be a homomorphism of rings with unit. Then θ induces a map

$$\theta_*: R[[x_1, x_2]] \to S[[x_1, x_2]]$$

which carries any formal product μ over R into a formal product $\theta_*\mu$ over S. However, this is not the definition of a homomorphism between formal groups. Such a homomorphism is, of course, a map in our category, with the obvious property. That is, if G is a formal group $(R[[x]], \mu)$, and H is a formal group $(R[[y]], \nu)$, then a homomorphism $\theta: G \to H$ is a formal power-series

$$y = f(x) = \sum_{i \geq 1} c_i x^i$$

(with coefficients c_i in R) such that

$$\nu(f(x_1), f(x_2)) = f\mu(x_1, x_2) .$$

The analogy with the case of a Lie group is obvious. If the coeffieient c_1 is invertible in R, then f^{-1} exists, and f is a isomorphism.

In our applications we are interested only in the case of dimension 1, and moreover only in commutative formal groups. That is, our formal products will satisfy

$$\mu(x, y) = \mu(y, x), \tag{1.11}$$

or equivalently

(1.12) $$a_{ij} = a_{ji}.$$

Our applications arise in algebraic topology.

2. EXAMPLES FROM ALGEBRAIC TOPOLOGY

In this section we will explain how examples of formal products arise in studying generalized cohomology theories. According to [17], generalized cohomology theories are closely connected with stable homotopy theory and the study of spectra. For convenience we will suppose that we are working in a suitable category of spectra, such as that constructed by Boardman [5, 6], so that we can form smash-products of spectra. A ring-spectrum is a spectrum E provided with a product map $\mu: E \wedge E \to E$. All our ring-spectra will be associative and commutative up to homotopy, and will be provided with a map $i: S^0 \to E$ which acts as a unit up to homotopy. We shall suppose known the work of G. W. Whitehead [17], according to which a ring-spectrum determines a generalized homology theory E_* and a generalized cohomology theory E^*. These theories admit all the usual products. The coefficient groups for these two theories are given by

$$E^{-n}(pt) = E_n(pt) = \pi_n(E) = [S^n, E].$$

Initially we are interested in three examples. First, the Eilenberg-Mac Lane spectrum for the group of integers. Since the corresponding homology and cohomology theories are universally written H_* and H^*, we will write H for this spectrum. Secondly, the BU-spectrum; since the corresponding homology and cohomology theories are called K-theory, and written K_*, K^* (and since we have just dispensed with the use of K for the Eilenberg-Mac Lane spectrum) we will write K for the BU-spectrum. (Note that we would anyway have to find different notation for the BU-space and the BU-spectrum, since we have to distinguish between them.) Thirdly, the Milnor spectrum [12]; this is always written MU; the corresponding homology and cohomology theories are complex bordism and complex cobordism.

We do not need homology and cohomology with coefficients until §15, but it seems best to deal with the matter now. Let G be an abelian group; then we can construct a Moore spectrum $M = M(G)$ so that

$$\pi_r(M) = 0 \qquad \text{for } r < 0$$

$$\pi_0(M) \cong G$$

$$H_r(M) = 0 \qquad \text{for } r > 0.$$

We define a "spectrum with coefficients" by

$$EG = E \wedge M.$$

For example, HG is an Eilenberg-Mac Lane spectrum for the group G. The homology and cohomology theories associated with EG are written EG_*, EG^*.

We will study spectra E which are provided with "orientations", in the following sense (which owes much to a seminar by A. Dold).

(2.1) There is given an element $x \in \tilde{E}^*(CP^\infty)$ such that $\tilde{E}^*(CP^1)$ is a free module over $\pi_*(E)$ on the generator i^*x, where $i: CP^1 \longrightarrow CP^\infty$ is the inclusion map.

We know, of course that CP^1 may be identified with S^2, and that $\tilde{E}^*(S^2)$ is free over $\pi_*(E)$ on one generator γ, which lies in $\tilde{E}^2(S^2)$, and is represented by the unit map $S^0 \longrightarrow E$; but we do not insist that i^*x is this generator, or even that it lies in $\tilde{E}^2(S^2)$. Our assumption says only that $i^* x = u\gamma$, where u is an invertible element in $\pi_*(E)$.

If we have more than one spectrum in sight, we write x^E for the generator in $\tilde{E}^*(CP^\infty)$, and u^E for u.

We make a blanket assumption that the objects to be studied are pairs (E, x^E); any E which appears in what follows is supposed to be provided with a class x^E.

Examples. (2.2). $E = H$. We take $x^H \in H^2(CP^\infty)$ to be the usual generator.

(2.3). $E = K$. We identify CP^∞ with $BU(1)$, we take ξ to be the universal line bundle over $BU(1)$, and we take

$$x^K = \xi - 1 \in \tilde{K}^0(CP^\infty).$$

Notes. It is justifiable to take x^K in $\tilde{K}^0(CP^\infty)$ instead of $\tilde{K}^2(CP^\infty)$, because it makes the "n-th Chern class in K-cohomology" lie in dimension 0 instead of dimension 2n, so that it is more conveniently related to bundles and representation-theory. Also we get a better formula at (2.9) below. The unit u^K is the usual generator in $\pi_2(K)$; this provides some justification for writing $i^* x = u\gamma$ rather that $\gamma = ui^* x$.

(2.4). $E = MU$. We have a canonical homotopy equivalence $\omega: CP^\infty \longrightarrow MU(1)$. In fact, $MU(1)$ is a quotient space formed from a disc-bundle over $BU(1)$ by identifying to one point a subbundle whose fibers are circles. This subbundle is the universal $U(1)$-bundle, so it is contractible, and the quotient map is a homotopy equivalence. The disc-bundle is clearly equivalent to $BU(1)$ under the projection.

We take $x^{MU} \in MU^2(CP^\infty)$ to be the class of ω .

Let us return to the general case. By using the projections of

$CP^\infty \times CP^\infty$ onto its two factors, we obtain two elements

$$x_1, x_2 \in \tilde{E}^*(CP^\infty \times CP^\infty).$$

LEMMA 2.5. (i) The spectral sequences

$$H^*(CP^n; \pi_*(E)) \Longrightarrow E^*(CP^n)$$

$$H^*(CP^\infty; \pi_*(E)) \Longrightarrow E^*(CP^\infty)$$

$$H^*(CP^n \times CP^m; \pi_*(E)) \Longrightarrow E^*(CP^n \times CP^m)$$

$$H^*(CP^\infty \times CP^\infty; \pi_*(E)) \Longrightarrow E^*(CP^\infty \times CP^\infty)$$

are trivial.

(ii) $E^*(CP^n)$ is the ring of polynomials $\pi_*(E)[x]$ modulo the ideal generated by x^{n+1}.

$E^*(CP^\infty)$ is the ring of formal power-series $\pi_*(E)[[x]]$.

$E^*(CP^n \times CP^m)$ is the ring of polynomials $\pi_*(E)[x_1, x_2]$ modulo the ideal generated by x_1^{n+1} and x_2^{m+1}.

$E^*(CP^\infty \times CP^\infty)$ is the ring of formal power-series $\pi_*(E)[[x_1, x_2]]$.

Proof. Consider each spectral sequence of part (i); the relevant powers x^i or $x_1^i x_2^j$ give a $\pi_*(E)$-base for the E_2-term on which all differentials d_r vanish. Since the differentials d_r are linear over $\pi_*(E)$, they vanish on everything.

We know that CP^∞ is an Eilenberg-Mac Lane space of type $(Z, 2)$; in particular it is an H-space, and its product map

$$m: CP^\infty \times CP^\infty \longrightarrow CP^\infty$$

is unique up to homotopy. One way to decribe m is to say that it is the classifying map for the tensor-product $\xi_1 \xi_2$ of the two line-bundles over $CP^\infty \times CP^\infty$; in other words, $m^* \xi = \xi_1 \xi_2$.

In general, we can form $m^* x$, and by Lemma 2.5 it is a formal power-series in two variables:

(2.6) $$m^* x = \mu(x_1, x_2) = \sum_{i,j} a_{ij} x_1^i x_2^j \qquad (a_{ij} \in \pi_*(E)) .$$

LEMMA 2.7. This formal power-series is a commutative formal product, in the sense of §1, over the ring $\pi_*(E)$.

The proof is easy.

If we have more than one spectrum E in sight, we write μ^E for E and a_{ij}^E for the coefficients in $\pi_*(E)$.

Examples. (2.8). $E = H$. We have

$$m^* x^H = x_1^H + x_2^H .$$

We get the "additive formal product" of (1.4).

(2.9). $E = K$. We have

$$m^* \xi = \xi_1 \xi_2 ,$$

that is,

$$m^*(1 + x) = (1 + x_1)(1 + x_2)$$

or

$$m^* x = x_1 + x_2 + x_1 x_2 .$$

We get the "multiplicative formal product" of (1.5).

(2.10). We see that there is a formal product defined over $\pi_*(MU)$, with

$$a_{ij} \in \pi_{2(i+j-1)}(MU) .$$

In this way we get a lot of useful elements in $\pi_*(MU)$.

(2.11). Let $n : CP^\infty \to CP^\infty$ be the map which classifies the line bundle ξ^{-1} inverse to ξ in the sense of the tensor-product. (Alternatively, n is the map of classifying spaces induced by the homomorphism $z \to z^{-1} = \bar{z} : U(1) \to U(1)$.) Then we have

$$n^* x^{MU} = \sum_{j \geq 0} a'_j (x^{MU})^j ,$$

where $\sum_{j \geq 1} a'_j x^j$ is the "formal inverse" corresponding to the formal product μ^{MU} (see (1.8)-(1.10)).

Next a remark on naturality. Suppose given a homomorphism $f: E \longrightarrow F$ of ring-spectra. If x^E is as above, then $i^* x^E = u^E \gamma^E$, so $i^*(f_* x^E) = (f_* u^E) \gamma^E$; here $f_* u^E$ is invertible in $\pi_*(F)$, so we can take $f_* x^E$ as a generator x^F. With this choice of generator we have $a^F_{ij} = f_* a^E_{ij}$, or in other words $\mu^F = f_* \mu^E$.

More usually, however, both E and F have given generators x^E, x^F. In this case we have

$$f_* x^E = \sum_{i \geq 1} c_i (x^F)^i ,$$

where the c_i are coefficients in $\pi_*(F)$ and

(2.12) $$f_* u^E = c_1 u^F .$$

Let us set

$$\sum_{i \geq 1} c_i (x^F)^i = g(x^F) ;$$

then we have the following result.

LEMMA 2.13. $\quad g(\mu^F(x_1^F, x_2^F)) = (f_* \mu^E)(g(x_1^F), g(x_2^F)).$

The proof is immediate, by naturality.

This lemma states that the power-series g is an isomorphism from the formal group with product μ^F to the formal group with product $f_* \mu^E$

Examples: (i) We will see in §4 that we have a map $f: MU \longrightarrow H$ such that $f_* x^{MU} = x^H$. Then $f_* a_{ij} = 0$ if $i \geq 1$ and $j \geq 1$.

(ii) We will see in §4 that we have a map $g: MU \longrightarrow K$ such that

$g_* x^{MU} = u^{-1} x^K$. Then $g_* a_{11} = u$ and $g_*(a_{ij}) = 0$ if $i > 1$ or $j > 1$.

Many calculations which are familiar for ordinary homology and cohomology can be carried over to E.

LEMMA 2.14. (i) The spectral sequences

$$H_*(CP^n; \pi_*(E)) \Longrightarrow E_*(CP^n)$$

$$H_*(CP^\infty; \pi_*(E)) \Longrightarrow E_*(CP^\infty)$$

$$H_*(CP^n \times CP^m; \pi_*(E)) \Longrightarrow E_*(CP^n \times CP^m)$$

$$H_*(CP^\infty \times CP^\infty; \pi_*(E)) \Longrightarrow E_*(CP^\infty \times CP^\infty)$$

are trivial.

(ii) $E^*(CP^n)$ and $E_*(CP^n)$ are dual finitely-generated free modules over $\pi_*(E)$.

(iii) There is a unique element $\beta_n \in E_*(CP^n)$ such that

$$\langle x^i, \beta_n \rangle = \begin{cases} 1 & (i = n) \\ 0 & (i \neq n) \end{cases} .$$

We can then consider the image of β_n in $E_*(CP^m)$ for $m \geq n$ and in $E_*(CP^\infty)$; these images we also write β_n.

(iv) $E_*(CP^n)$ is free over $\pi_*(E)$ on generators $\beta_0, \beta_1, \ldots, \beta_n$.

$E_*(CP^\infty)$ is free over $\pi_*(E)$ on generators $\beta_0, \beta_1, \ldots, \beta_n, \ldots$

$E_*(CP^n \times CP^m)$ is free over $\pi_*(E)$ with a base consisting of the external products $\beta_i \beta_j$ for $0 \leq i \leq n$, $0 \leq j \leq m$.

$E_*(CP^\infty \times CP^\infty)$ is free over $\pi_*(E)$ with a base consisting of the external products $\beta_i \beta_j$.

(v) The external product

$$E_*(CP^\infty) \otimes_{\pi_*(E)} E_*(CP^\infty) \longrightarrow E_*(CP^\infty \times CP^\infty)$$

is an isomorphism.

The proof of part (i) is easy, by considering the pairing of these spectral sequences with those of (2.5)(i). (Compare [3, p. 21], where however one is arguing in the opposite direction.) This leads immediately to parts (ii) and (iii). We see that in part (i), the E^2-term of each spectral sequence has a $\pi_*(E)$-base consisting of the appropriate elements β_i or $\beta_i\beta_j$. This leads to parts (iv) and (v).

If we have more than one spectrum E in sight, we write β_i^E for the generators in $E_*(CP^\infty)$. If we have a homomorphism $f: E \longrightarrow F$ of ring-spectra, and if we choose $x^F = f_* x^E$ (as above), then we have $\beta_i^F = f_* \beta_i^E$. More usually, however, both E and F have given generators x^E, x^F. In this case we have

$$f_* x^E = \sum_{i \geq 1} c_i (x^F)^i = g(x^F) ,$$

where the c_i are coefficients in $\pi_*(F)$ and $f_* u^E = c_1 u^F$, as above. In this case the appropriate move is to invert the power-series and get

$$x^F = g^{-1}(f_* x^E) = \sum_i d_i (f_* x^E)^i ;$$

passing to powers, we get

$$(x^F)^j = \sum_i d_i^j (f_* x^E)^i$$

for some coefficients $d_i^j \in \pi_*(F)$. Then we have:

LEMMA 2.15. $f_* \beta_i^E = \sum_j d_i^j \beta_j^F$.

The proof is immediate, by exploiting the pairing between generalized homology and cohomology.

Examples. (2.16). We will see in §4 that we have a map $f: MU \longrightarrow H$ such that $f_* x^{MU} = x^H$. Thus we have $f_* \beta_i^{MU} = \beta_i^H$.

(2.17). We will see in §4 that we have a map $g: MU \longrightarrow K$ such that $g_* x^{MU} = u^{-1} x^K$. Thus we have $g_* \beta_i^{MU} = u^i \beta_i^K$.

COROLLARY 2.18. The diagonal map

$$\Delta : CP^{\infty} \to CP^{\infty} \times CP^{\infty}$$

gives $E_*(CP^{\infty})$ the structure of a coalgebra, whose coproduct map is given by

$$\psi \beta_k = \sum_{i+j=k} \beta_i \otimes \beta_j .$$

This follows immediately from (2.14). It suggests that we regard $E_*(CP^{\infty})$ as a Hopf algebra, with product induced by

$$m: CP^{\infty} \times CP^{\infty} \to CP^{\infty}$$

and coproduct as in (2.18). We note that if we do this we shall have

$$m_*(\beta_i \otimes \beta_j) = \sum_k a^k_{ij} \beta_k ,$$

where the sum runs over $k \leq i+j$; for by cellular approximation we can suppose that m maps $CP^i \times CP^j$ into CP^{i+j}. Of course, the formulae which hold here can be written down in the general abstract case, and we will now indicate this.

3. REFORMULATION

In this section we will interpret a formal group over R as a group in the category of coalgebras over R.

The results of the previous section suggest that the algebra of formal power series $R[[x]]$, which arose in §1, is actually the dual of the object which should be considered. Let F be an R-module which is free on generators $\beta_0, \beta_1, \ldots, \beta_n, \ldots$. We make F into a coalgebra over R by setting

(3.1) $$\psi \beta_k = \sum_{i+j=k} \beta_i \otimes \beta_j .$$

The dual of F, given by $F^* = \mathrm{Hom}_R(F, R)$, is then an algebra over R, and it can be identified with $R[[x]]$; the pairing between $R[[x]]$ and F is given by

$$(3.2)\qquad \langle \sum_{i\geq 0} c_i x^i, \beta_n \rangle = c_n .$$

(Here the coefficients c_i lie in R.)

The analogy with the case of a Lie group confirms that this procedure is reasonable. Instead of looking at analytic functions $\sum_{i\geq 0} c_i x^i$ on G, we look at differential operators, because functions are contravariant and differential operators are covariant. More precisely, we interpret β_n as the differential operator $\frac{1}{n!}\frac{d^n}{dx^n}$, evaluated at $x = 0$. The result of applying this operator to the analytic function $\sum_{i\geq 0} c_i x^i$ is indeed c_n. The coproduct in F corresponds to Leibniz' formula

$$\frac{1}{k!}\frac{d^k}{dx^k}(fg) = \sum_{i+j=k} \left(\frac{1}{i!}\frac{d^i}{dx^i}f\right)\left(\frac{1}{j!}\frac{d^j}{dx^j}g\right) .$$

Since differential operators are covariant, it is reasonable that the product in G should induce a product of differential operators.

To continue, let F be as above; then we can form $F \otimes_R F$, and its dual, $\mathrm{Hom}_R(F \otimes_R F, R)$, may be identified with the algebra $R[[x_1, x_2]]$. The pairing between $R[[x_1, x_2]]$ and $F \otimes_R F$ is given by

$$(3.3)\qquad \langle \sum_{i,j} c_{ij} x_1^i x_2^j, \beta_p \otimes \beta_q \rangle = c_{pq} .$$

Each R-map

$$m_*: F \otimes_R F \longrightarrow F$$

induces a dual map

$$m^*: R[[x]] \longrightarrow R[[x_1, x_2]] .$$

This induces a 1-1 correspondence between maps m_* which are filtration-preserving (in a suitable sense) and maps m^* which are filtration-preserving; corresponding maps are given by the following formulae.

(3.4) $$m_*(\beta_i \otimes \beta_j) = \sum_{k \leq i+j} a^k_{ij}\beta_k$$

(3.5) $$m^* x^k = \sum_{i+j \geq k} a^k_{ij} x_1^i x_2^j .$$

(Here the coefficients a^k_{ij} lie in our ring R. The coefficients a^1_{ij} are the coefficients a_{ij} of §1.) The map m^* is a map of algebras if and only if the map m_* is a map of coalgebras. It is now easy to check that the relevant conditions on m_* (such as associativity and commutativity) are equivalent to the corresponding conditions on m^*. The map $e:R[[x]] \longrightarrow R$, which was introduced as a unit map in §1 and defined by $e(\sum_{i \geq 0} c_i x^i) = c_0$, now has the alternative name β_0; we take β_0 as our unit in F.

It is clear, of course, that if m^* is a map of algebras, then $m^* x^k$ is determined by $m^* x$. So in this case, the coefficients a^k_{ij} are determined by the $a^1_{ij} = a_{ij}$. For example, we easily obtain the following formula.

(3.6) $$a^k_{1j} = k a_{1,j+1-k} .$$

<u>Exercise.</u> Obtain a formula for a^k_{22}.

We conclude that there is a precise equivalence between group-object structures on $R[[x]]$, in the sense of §1, and suitable Hopf-algebra structures on F. A formal group is therefore also a group-object in a suitable category of coalgebras.

4. CALCULATIONS IN E-HOMOLOGY AND COHOMOLOGY

In this section we continue the programme of taking results which are familiar for ordinary homology and cohomology, and carrying them over to E. First we compute the E-homology of the spaces BU(n) and BU. The space BU is an H-space; its product corresponds to addition in K-cohomology; in particular, we have the following homotopy-commutative diagram, in which the upper arrow is the Whitney sum map.

$$\begin{array}{ccc} BU(n) \times BU(m) & \longrightarrow & BU(n+m) \\ \downarrow & & \downarrow \\ BU \times BU & \longrightarrow & BU \end{array}$$

This diagram gives rise to the following diagram of products.

$$\begin{array}{ccc} E_*(BU(n)) \otimes_{\pi_*(E)} E_*(BU(m)) & \longrightarrow & E_*(BU(n+m)) \\ \downarrow & & \downarrow \\ E_*(BU) \otimes_{\pi_*(E)} E_*(BU) & \longrightarrow & E_*(BU) \end{array}$$

By using the injection $BU(1) \rightarrow BU$, the classes $\beta_i \in E_*(CP^\infty)$ give classes in $E_*(BU)$; we write β_i for these classes also. The element β_0 acts as a unit for the products.

LEMMA 4.1. (i) The spectral sequences

$$H_*(BU(n); \pi_*(E)) \Longrightarrow E_*(BU(n))$$

$$H_*(BU; \pi_*(E)) \Longrightarrow E_*(BU)$$

are trivial.

(ii) $E_*(BU(n))$ is free over $\pi_*(E)$, with a base consisting of the monomials

$$\beta_{i_1} \beta_{i_2} \cdots \beta_{i_r}$$

such that $i_1 > 0$, $i_2 > 0$, ..., $i_r > 0$, $0 \leq r \leq n$. (The monomial with $r = 0$ is interpreted as 1.)

$E_*(BU)$ is a polynomial algebra

$$\pi_*(E)[\beta_1, \beta_2, \ldots, \beta_i, \ldots].$$

(iii) The coproduct in $E_*(BU(n))$ and $E_*(BU)$ is given by

$$\psi\beta_k = \sum_{i+j=k} \beta_i \otimes \beta_j,$$

where $\beta_0 = 1$.

The proof of parts (i) and (ii) is easy, because the monomials

$$\beta_{i_1}\beta_{i_2}\cdots\beta_{i_r}$$

give a $\pi_*(E)$-base for the E^2-term on which all differentials d_r vanish. Since the differentials are linear over $\pi_*(E)$, they vanish on everything. Part (iii) comes from (2.18).

We now introduce a general lemma.

LEMMA 4.2. Let X be a space (or a spectrum provided that $\pi_r(X) = 0$ for $r < -N$, some N.). Suppose that $H_*(X; \pi_*(E))$ is free over $\pi_*(E)$ and that the spectral sequence $H_*(X; \pi_*(E) \Longrightarrow E_*(X)$ is trivial. Let F be a module-spectrum over the ring-spectrum E. Then the spectral sequences

$$H_*(X; \pi_*(F)) \Longrightarrow F_*(X)$$
$$H^*(X; \pi_*(F)) \Longrightarrow F^*(X)$$

are trivial, and the maps

$$E_*(X) \otimes_{\pi_*(E)} \pi_*(F) \longrightarrow F_*(X)$$
$$F^*(X) \longrightarrow \mathrm{Hom}_{\pi_*(E)}(E_*(X), \pi_*(F))$$

are isomorphisms.

The proof is a routine exercise on pairings and spectral sequences (compare [3], p. 20, Proposition 17).

In particular, if E is as in §2, the lemma applies to $X = CP^\infty$, $BU(n)$ and BU. We will also see that it applies to $X = MU$ --see (4.5).

Although it is quite unnecessary for our main purposes, we pause to observe that Chern classes behave as expected in E-cohomology.

LEMMA 4.3. (i) $E^*(BU)$ contains a unique element c_i such that

$$\langle c_i, (\beta_1)^i \rangle = 1$$

and

$$\langle c_i, m \rangle = 0$$

when m is any monomial $\beta_1^{i_1}\beta_2^{i_2} \dots \beta_r^{i_r}$ distinct from $(\beta_1)^i$. We have $c_o = 1$.

(ii) The restriction of c_1 to $BU(1)$ is x^E, the generator given in §2.

(iii) The restriction of c_i to $BU(n)$ is zero for $i > n$. (Otherwise, the image of c_i in $E^*(BU(n))$ will also be written c_i.)

(iv) $E^*(BU(n))$ is the ring of formal power-series

$$\pi_*(E)[[c_1, c_2, \dots, c_n]];$$

and $E^*(BU)$ is the ring of formal power-series

$$\pi_*(E)[[c_1, c_2, \dots, c_i, \dots]].$$

(v) We have

$$\psi c_k = \sum_{i+j=k} c_i \otimes c_j.$$

Proof. The definition of c_i in (i) is legitimate by (4.2) applied to $X = BU$, $F = E$. We easily check that the unit $1 \in E_*(BU)$ plays the role laid down for c_o. Part (ii) is trivial; part (iii) follows easily from (4.1)(ii) plus (4.2) applied to $X = BU(n)$. We turn to part (iv).

Let m be a monomial

$$m = \beta_1^{i_1}\beta_2^{i_2} \dots \beta_r^{i_r} \quad \text{in } E_*(BU);$$

let the image of m under the iterated diagonal, which is determined by (4.1)(iii), be

$$\sum_\alpha m_{1\alpha} \otimes m_{2\alpha} \otimes \dots \otimes m_{s\alpha} .$$

Then

$$\langle c_{j_1} c_{j_2} \ldots c_{j_s}, m\rangle$$

$$= \sum_{\alpha} \langle c_{j_1}, m_{1\alpha}\rangle \langle c_{j_2}, m_{2\alpha}\rangle \ldots \langle c_{j_s}, m_{s\alpha}\rangle ;$$

and this is a well-determined integer independent of the spectrum E. In particular this integer is the same as in the case $E = H$. We conclude that that in the spectral sequence

$$H^*(BU(n); \pi_*(E)) \Longrightarrow E^*(BU(n)), \quad \text{or}$$

$$H^*(BU; \pi_*(E)) \Longrightarrow E^*(BU)$$

the E_2 term has a $\pi_*(E)$-base consisting of the appropriate monomials

$$c_{j_1} c_{j_2} \ldots c_{j_s} .$$

This leads to part (iv). Part (v) follows by duality from the definition in part (i).

The classes c_i are of course the generalized Chern classes in E-cohomology. If required they may be constructed as characteristic classes for $U(n)$-bundles over appropriate spaces by the method of Grothendieck, and then pulled back to $BU(n)$ and BU by limiting arguments. (Compare [2, pp. 8-9].) In the case $E = MU$ we get the Conner-Floyd Chern classes.

If we have more than one spectrum in sight we write β_i^E, c_i^E. If we are given a map $f: E \longrightarrow F$ of ring-spectra, and choose $x^F = f_* x^E$, as in § 2, then we have

$$c_i^F = f_* c_i^E .$$

The reader may carry over (2.15) to cohomology, but it is not necessary for our purposes.

For the next lemma, we note that $E_*(MU)$ is a ring, and that the "inclusion" of MU(1) in MU induces a homomorphism

$$\tilde{E}_p(MU(1)) \longrightarrow E_{p-2}(MU).$$

Following the analogy of ordinary homology, we take the element

$$u^E \beta^E_{i+1} \in \tilde{E}_*(MU(1)) \quad (i \geq 0)$$

and write b_i^E for its image in $E_*(MU)$. The factor u^E (see § 2) is introduced in order to ensure that $b_o^E = 1$ in $E_o(MU)$.

Suppose given a map $f: E \to F$ of ring-spectra. Then it is clear that Lemma 2.15 carries over; with the notation of (2.15), we have the following result.

$$f_* b_i^E = c_1 \sum_j d_{i+1}^{j+1} b_j^F . \tag{4.4}$$

In particular, as soon as we obtain the canonical map $f: MU \to H$, it will send b_i^{MU} to b_i^H; as soon as we obtain the canonical map $g: MU \to K$, it will send b_i^{MU} to $u^i b_i^K$, where $u = u^K \in \pi_2(K)$.

With an eye to later applications (§15) we include a little spare generality in the next two lemmas. Let R be a subring of the rational numbers Q; the reader interested only in the immediate applications may take R = Z. We recall from § 2 that MUR is the representing spectrum for complex bordism and cobordism with coefficients in R.

We assume that for each integer d invertible in R, the groups $\pi_*(E)$ have no d-torsion. This assumption is certainly vacuous if R = Z.

LEMMA 4.5. (i) The spectral sequences

$$H_*(MUR; \pi_*(E)) \Rightarrow E_*(MUR)$$

$$H_*(MUR \wedge MUR; \pi_*(E)) \Rightarrow E_*(MUR \wedge MUR)$$

are trivial, so that Lemma 4.2 applies.

(ii) $E_*(MUR)$ is the polynomial ring

$$(\pi_*(E) \otimes R)[b_1, b_2, \ldots, b_n, \ldots] .$$

Proof. For (i), in the case of MUR we note that the monomials in the b_i form a $\pi_*(E) \otimes R$-base for the E_2-term on which all differentials d_r vanish. The differentials d_r are linear over $\pi_*(E)$, and by using the assumption on $\pi_*(E)$ we see they are linear over R. So the differentials d_r vanish on everything. Similarly for $MUR \wedge MUR$, using exterior products of such monomials. This proves (i) and (ii).

For the next lemma, let R be again a subring of the rational numbers Q, and let E be a ring-spectrum, with x^E as in §2, such that

$$\pi_*(E) \longrightarrow \pi_*(E) \otimes R$$

is iso. (For example, we might have $E = FR$.)

LEMMA 4.6. Suppose given a formal power-series

$$f(x^E) = \sum_{i \geq 0} d_i (x^E)^{i+1} \in \tilde{E}^2(CP^\infty)$$

with $u^E d_o = 1$. Then there is one and (up to homotopy) only one map of ring-spectra

$$g: MUR \longrightarrow E$$

such that $g_* x^{MU} = f(x^E)$.

Notes. (i) By abuse of language, we have written x^{MU} also for the image of $x^{MU} \in \widetilde{MU}^2(CP^\infty)$ in $\widetilde{MUR}^2(CP^\infty)$.

(ii) The necessity of the condition $u^E d_o = 1$ is shown by (2.12).

Proof. We check that the conditions of Lemma 4.2 apply to $X = MUR$, $F = E$. We certainly have

$$H_*(MUR; \pi_*(E)) \cong H_*(MU; \pi_*(E) \otimes R) \cong H_*(MU; \pi_*(E))$$

(by the assumption on E), so $H_*(MUR; \pi_*(E))$ is free over $\pi_*(E)$. Similarly

$$E_*(MUR) = (\pi_*(E) \otimes R)[b_1, b_2, \ldots, b_n, \ldots] = \pi_*(E)[b_1, b_2, \ldots, b_n, \ldots].$$

If $\pi_*(E) \longrightarrow \pi_*(E) \otimes R$ is iso, then $\pi_*(E)$ has no d-torsion for any integer d invertible in R, and Lemma 4.5 shows that the spectral sequence

$$H_*(MUR;\ \pi_*(E)) \Longrightarrow E_*(MUR)$$

is trivial. So Lemma 4.2 shows that there is a (1-1) correspondence between homotopy classes of maps

$$g: MUR \longrightarrow E$$

and maps

$$\theta: E_*(MUR) \longrightarrow \pi_*(E)$$

which are linear over $\pi_*(E)$, and of degree zero. Similarly for maps

$$h: MUR \wedge MUR \longrightarrow E\ ;$$

and this allows us to check whether a map $g: MUR \longrightarrow E$ makes the following diagram homotopy-commutative.

$$\begin{array}{ccc} MUR \wedge MUR & \xrightarrow{\quad g \wedge g \quad} & E \wedge E \\ \downarrow & & \downarrow \\ MUR & \xrightarrow{\quad g \quad} & E \end{array} .$$

We find by diagram-chasing that for this, it is necessary and sufficient that the map θ corresponding to g should be a map of algebras over $\pi_*(E)$. Now the condition

$$g_* x^{MU} = \sum_{i \geq 0} d_i (x^E)^{i+1}$$

is equivalent to

$$\theta(b_i) = u^E d_i \qquad (i \geq 0) \quad .$$

Provided that $u^E d_o = 1$, there is one and only one map θ of $\pi_*(E)$-algebras which satisfies this condition. This proves Lemma 4.6.

<u>Examples.</u> (4.7). There is one and only one map $f: MU \longrightarrow H$ of ring-spectra such that

$$f_* x^{MU} = x^H \quad .$$

This is of course a trivial example.

(4.8). There is one and only one map $g: MU \longrightarrow K$ of ring-spectra such that

$$g_* x^{MU} = (u^K)^{-1} x^K \; .$$

This map is, of course, the usual one, which provides a K-orientation for complex bundles.

We can also use Lemma 4.6 to construct multiplicative cohomology operations from MU^* to MU^*, following Novikov [13].

We can also use Lemma 4.6 to obtain Hirzebruch's theory of "multiplicative sequences of polynomials" in the (ordinary) Chern classes. If we think for a moment about the gradings in Hirzebruch's formulae, we see that for this purpose we need to take E to be a product of Eilenberg-Mac Lane spectra, having homotopy groups

$$\pi_r(E) = \begin{cases} Q & \text{for } r \text{ even} \\ 0 & \text{for } r \text{ odd.} \end{cases}$$

A suitable candidate is the spectrum $H \wedge K$, which has the required properties.

Some readers may perhaps be used to thinking of "multiplicative sequences of polynomials" as elements of the cohomology of the space BU (elements of $(H \wedge K)^*(BU)$, in fact); and they may perhaps be surprised to see them treated as maps of MU. On this point several comments are in order.

(a) Lemma 4.6 provides us with all the Thom classes we need, so we have a Thom isomorphism

$$(H \wedge K)^*(BU) \cong (H \wedge K)^*(MU).$$

(b) "Multiplicative sequences of polynomials" carry the Whitney sum in BU into the product in cohomology. The Whitney sum in BU corresponds to the product in MU, so it is more convenient to describe the behavior on products by saying that we have a map of ring-spectra defined in the spectrum MU.

(c) "Multiplicative sequences of polynomials" are intended for use on manifolds, so that we actually require their values on elements of $\pi_*(MU)$. For this reason, their definition in terms of MU may be more transparent than their expression in terms of ordinary Chern classes in BU. For example, consider the map of ring-spectra

$$MU \xrightarrow{g} K \cong S^0 \wedge K \longrightarrow H \wedge K,$$

where the map $g: MU \longrightarrow K$ is that mentioned above.

Exercise. Follow up these hints.

Lemma 4.6 shows that if we consider pairs (E, x^E), as above, and such that $u^E = 1$, then among them the pair (MU, x^{MU}) has a universal property; for any other pair (E, x^E), there is a map $g: MU \longrightarrow E$ such that $g_* x^{MU} = x^E$. In particular, for any such (E, x^E) we have a homomorphism of rings $g_*: \pi_*(MU) \longrightarrow \pi_*(E)$ such that $g_* \mu^{MU} = \mu^E$ (see §2); that is, g_* carries the one formal product into the other. We will see in the next section that there is a ring L, with a formal product defined over it, which enjoys a similar universal property in a purely algebraic setting. It is known that $\pi_*(MU)$ is a polynomial algebra, over Z, on generators of dimension 2, 4, 6, 8, ... The ring L can be made into a graded ring, and it is known that it is then a polynomial algebra, over Z, on generators of dimension 2, 4, 6, 8, ... Following Quillen, we regard these as plausibility arguments, to introduce the theorem that the canonical map from L to $\pi_*(MU)$ is an isomorphism.

5. LAZARD'S UNIVERSAL RING

In this section we introduce Lazard's universal ring. Following Fröhlich [8], we call this ring L (for Lazard).

THEOREM 5.1. There is a commutative ring L with unit, and a commutative formal product μ^L defined over L, such that for any commutative ring R with unit and any commutative formal product μ^R defined over R there is one and only one homomorphism $\theta: L \longrightarrow R$ such that

$$\theta_* \mu^L = \mu^R .$$

<u>Proof.</u> We define L by generators and relations; that is, we define L as the quotient of a polynomial ring P by an ideal I. Take formal symbols a_{ij} for $i \geq 1$, $j \geq 1$, and set

$$P = Z[a_{11}, a_{12}, a_{21}, \ldots, a_{ij}, \ldots].$$

Form the formal power-series

$$\mu(x, y) = x + y + \sum_{i,j \geq 1} a_{ij} x^i y^j \tag{5.2}$$

and set

$$\mu(x, \mu(y, z)) - \mu(\mu(x, y), z) = \sum_{i,j,k} b_{ijk} x^i y^j z^k . \tag{5.3}$$

Then each coefficient b_{ijk} is a well-defined polynomial in the a_{ij}. Take I to be the ideal in P generated by the elements b_{ijk} and $a_{ij} - a_{ji}$. It is trivial to check that $L = P/I$ has the required properties.

We note that we can make L into a graded ring if we wish. In fact, we assign to x, y and $\mu(x, y)$ the degree -2; then a_{ij} has degree

$2(i+j-1)$, and b_{ijk} is a homogeneous polynomial of degree $2(i+j+k-1)$. It follows that I is a graded ideal and P/I is a graded ring.

We note that the structure of L is in principle computable. For example,

$L_o \cong Z$, generated by 1,

$L_2 \cong Z$, generated by a_{11},

$L_4 \cong Z \oplus Z$, generated by a_{11}^2 and a_{12},

$L_6 \cong Z \oplus Z \oplus Z$, generated by $a_{11}^3, a_{11}a_{12}$ and $a_{22}-a_{13}$.

(Exercise: obtain the relation which allows one to write a_{22} and a_{31} in terms of the three generators given.)

The structure of L will be described in more detail in our next algebraic section, §7.

In order to obtain the structure of L, we use algebraic arguments which are openly obtained by analogy with the situation in algebraic topology.

6. MORE CALCULATIONS IN E-HOMOLOGY

The element a_{ij} in $\pi_{2(i+j-1)}(MU)$ can be represented by a weakly almost-complex manifold; we might well be asked to compute the (normal) characteristic numbers of this manifold. It is equivalent to ask for the image of a_{ij} under the Hurewicz homomorphism

$$\pi_*(MU) \longrightarrow H_*(MU).$$

It is the object of this section to answer this question.

To do so we introduce the Boardman homomorphism, which is slightly more general than the Hurewicz homomorphism. Let E be a (commutative) ring-spectrum; then for any (space or spectrum) Y we can consider the map

$$Y \simeq S^o \wedge Y \xrightarrow{i \wedge 1} E \wedge Y;$$

composition with this map induces a homomorphism

$$[X, Y]_* \xrightarrow{B} [X, E \wedge Y]_*.$$

We recover the Hurewicz homomorphism by setting $X = S^o$, $E = H$.

The Boardman homomorphism is more or less guaranteed to be useful when $E = H$, because of the following lemma.

LEMMA 6.1. $H \wedge Y$ is equivalent (though not canonically, in general) to a product of Eilenberg-Mac Lane spectra, whose homotopy groups are the groups

$$\pi_n(H \wedge Y) = H_n(Y).$$

It follows that

$$[X, H \wedge Y]_r \cong \prod_n H^{n-r}(X; H_n(Y))$$

(not canonically); so the groups $[X, E \wedge Y]_r$ are computable for $E = H$.

Proof of (6.1): For each n, we can construct a Moore spectrum $M(G_n, n)$ for the group $G_n = \pi_n(H \wedge Y)$ in dimension n, and construct a map

$$f_n : M(G_n, n) \longrightarrow H \wedge Y$$

which induces an isomorphism

$$(f_n)_* : \pi_n(M(G_n, n)) \longrightarrow \pi_n(H \wedge Y).$$

We can then construct the map

$$H \wedge M(G_n, n) \xrightarrow{1 \wedge f_n} H \wedge H \wedge Y \xrightarrow{\mu \wedge 1} H \wedge Y,$$

where $H \wedge M(G_n, n)$ is an Eilenberg-Mac Lane spectrum for the group G_n in dimension n. We can then form the map

$$\bigvee_n H \wedge M(G_n, n) \longrightarrow H \wedge Y$$

whose n-th component is the map $(\mu \wedge 1)(1 \wedge f_n)$ just constructed; we observe that it is a homotopy equivalence by Whitehead's Theorem (in the category of spectra). Let $\prod_n H \wedge M(G_n, n)$ be the product in the categorical sense; then there is a map

$$\bigvee_n H \wedge M(G_n, n) \longrightarrow \prod_n H \wedge M(G_n, n),$$

and this too is a homotopy equivalence by Whitehead's Theorem. This proves Lemma 6.1.

Returning to the general case, since $E \wedge Y$ is at least a module-spectrum over the ring-spectrum E, we may hope to obtain information about $[X, E \wedge Y]_r = (E \wedge Y)^{-r}(X)$ from $E_*(X)$; for example, we may have available a universal coefficient theorem.

LEMMA 6.2. We have the following commutative diagram.

$$\begin{array}{ccc} [X, Y]_* & \xrightarrow{B} & [X, E \wedge Y]_* \\ & {\scriptstyle \alpha} \searrow \quad \swarrow {\scriptstyle p} & \\ & \mathrm{Hom}_{\pi_*(E)}(E_*(X), E_*(Y)) & \end{array}$$

Here α is defined by

$$\alpha(f) = f_*: E_*(X) \longrightarrow E_*(Y) ,$$

while p is the homomorphism of the universal coefficient theorem, defined by

$$(p(h))(k) = \langle h, k \rangle \in \pi_*(E \wedge Y).$$

In the last formula we have $h \in (E \wedge Y)^*(X)$, $k \in E_*(X)$, and the Kronecker product $\langle h, k \rangle$ is defined using the obvious pairing of $E \wedge Y$ and E to $E \wedge Y$.

The proof of the lemma from the definitions is easy diagram-chasing. The lemma is of course mainly useful when p is an isomorphism; but since $E \wedge Y$ is a module-spectrum over E, Lemma 4.2 shows that p is an isomorphism when E is as in §2, and $X = CP^\infty$, BU, MU, etc.

Let E be a ring-spectrum which satisfies the assumptions made in §2. Then we can consider the following two maps.

$$E \simeq E \wedge S^o \longrightarrow E \wedge MU$$

$$MU \simeq S^o \wedge E \longrightarrow E \wedge MU .$$

Both are of course maps of ring-spectra. The generators x^E and x^{MU} will yield two generators in $(E \wedge MU)^*(CP^\infty)$, and these generators may well be different. In order to remember which is which, we call them x^E and x^{MU} also (abusing notation to avoid complicating it). Our next task is to compare x^E and x^{MU}.

LEMMA 6.3. In $(E \wedge MU)^*(CP^\infty)$ we have

$$x^{MU} = \sum_{i \geq 0} (u^E)^{-1} b_i^E (x^E)^{i+1} .$$

Note that the ooefficients $(u^E)^{-1} b_i^E$ lie in $\pi_*(E \wedge MU)$.

Proof. Apply Lemma 6.2 to the case $X = CP^\infty$, $Y = MU$. Since x^{MU} is a reduced class, so is Bx^{MU}. By definition, we have

$$(\sigma x^{MU})(u^E \beta_{i+1}^E) = b_i^E .$$

But we also have

$$(p(x^E)^j)(b_i^E) = \begin{cases} 1 & (i = j) \\ 0 & (i \neq j) \end{cases} .$$

The result follows by comparing these formulae, since p is an isomorphism.

In order to exploit this result, let $g(x^E)$ be the formal power-series

$$(6.4) \qquad g(x^E) = \sum_{i \geq 0} (u^E)^{-1} b_i^E (x^E)^{i+1} ,$$

with coefficients in $\pi_*(E \wedge MU)$, and let g^{-1} be the inverse power-series, so that

$$x^E = g^{-1} x^{MU} .$$

COROLLARY 6.5. After applying the homomorphisms

$$\pi_*(E) \longrightarrow \pi_*(E \wedge MU)$$

$$\pi_*(MU) \longrightarrow \pi_*(E \wedge MU)$$

the formal products μ^E, μ^{MU} are related by

$$\mu^{MU}(x_1^{MU}, x_2^{MU}) = g(\mu^E(g^{-1}x_1^{MU}, g^{-1}x_2^{MU})).$$

The proof is immediate from Lemma 2.13; or directly, the map $m: CP^\infty \times CP^\infty \longrightarrow CP^\infty$ yields an induced homomorphism m^* which commutes with products and limits, so that

$$m^* g(x^E) = g(m^* x^E).$$

One just rewrites this equation.

COROLLARY 6.6. Take $E = H$. Then after applying the homomorphism $\pi_*(MU) \longrightarrow \pi_*(H \wedge MU)$ we have

$$\mu^{MU}(x_1, x_2) = \exp^H(\log^H x_1 + \log^H x_2),$$

where

$$\exp^H(x) = \sum_{i \geq 0} b_i x^{i+1},$$

$b_i \in H_{2i}(MU)$ are the usual generators coming from $H_{2i+2}(MU(1))$, and $\log^H$ is the formal power-series inverse to $\exp^H$.

This is immediate from (6.5), using (2.8).

This corollary yields a method of calculating the image of a_{ij} in $H_{2(i+j-1)}(MU)$, in terms of the usual base in $H_*(MU)$. For example, we have

$$a_{11} \longrightarrow 2b_1$$

$$a_{12} \longrightarrow 3b_2 - 2b_1^2$$

$$a_{13} \longrightarrow 4b_3 - 8b_1b_2 + 4b_1^3$$

$$a_{22} \longrightarrow 6b_3 - 6b_1b_2 + 2b_1^3 \quad \text{etc.}$$

COROLLARY 6.7. Take $E = K$. Then after applying the homomorphism $\pi_*(MU) \longrightarrow \pi_*(K \wedge MU)$, we have

$$\mu^{MU}(x_1, x_2) = g(g^{-1}x_1 + g^{-1}x_2 + (g^{-1}x_1)(g^{-1}x_2)),$$

where

$$g(x) = \sum_{i \geq 0} u^{-1} b_i x^{i+1},$$

$u \in \pi_2(K)$, $b_i \in K_0(MU)$ are the generators defined above, and g^{-1} is the formal power-series inverse to g.

This is immediate from (6.5), using (2.9).

This corollary yields a method of calculating the image of a_{ij} in $K_{2(i+j-1)}(MU)$, in terms of the base in $K_*(MU)$. For example, we have

$$a_{11} \longrightarrow u(1 + 2b_1)$$

$$a_{12} \longrightarrow u^2(b_1 + 3b_2 - 2b_1^2)$$

$$a_{13} \longrightarrow u^3(2b_2 - 2b_1^2 + 4b_3 - 8b_1b_2 + 4b_1^3)$$

$$a_{22} \longrightarrow u^3(b_1 + 6b_2 - 3b_1^2 + 6b_3 - 6b_1b_2 + 2b_1^3) \quad \text{etc.}$$

We can also use the same method to calculate the Hurewicz homomorphism

$$\pi_*(MU) \longrightarrow MU_*(MU).$$

For this purpose we need to distinguish between the two copies of MU. We borrow the notation of [3], and write

$$\eta_L, \eta_R : \pi_*(MU) \longrightarrow MU_*(MU)$$

for the homomorphisms induced by the maps

$$MU \simeq MU \wedge S^0 \xrightarrow{1 \wedge i} MU \wedge MU$$

$$MU \simeq S^0 \wedge MU \xrightarrow{i \wedge 1} MU \wedge MU.$$

The Hurewicz homomorphism is η_R. The usual action of $\pi_*(MU)$ on

$MU_*(X)$, which works for any X, is given for $X = MU$ by η_L.

COROLLARY 6.8. The value of η_R on the generators a_{ij} is given by

$$\mu^R(x_1, x_2) = g\mu^L(g^{-1}x_1, g^{-1}x_2)$$

where

$$\mu^R(x_1, x_2) = \sum_{i,j} (\eta_R a_{ij}) x_1^i x_2^j ,$$

$$\mu^L(x_1, x_2) = \sum_{i,j} (\eta_L a_{ij}) x_1^i x_2^j ,$$

$$g(x) = \sum_{i \geq 0} b_i^{MU} x^{i+1} ,$$

$b_i^{MU} \in MU_{2i}(MU)$ is the generator described in §4, and g^{-1} is the power-series inverse to g.

This corollary is strictly on the same footing as the preceding two.

This yields a method of calculating $\eta_R(a_{ij})$. For example, we find

$$\eta_R(a_{11}) = 2b_1 + a_{11}$$

$$\eta_R(a_{12}) = (3b_2 - 2b_1^2) + a_{11}b_1 + a_{12}$$

$$\eta_R(a_{13}) = (4b_3 - 8b_1b_2 + 4b_1^3) + a_{11}(2b_2 - 2b_1^2) + a_{13}$$

$$\eta_R(a_{22}) = (6b_3 - 6b_1b_2 + 2b_1^3) + a_{11}(6b_2 - 3b_1^2) + (2a_{11} + a_{11}^2)b_1 + a_{22}.$$

From these formulae for the images of the a_{ij} under the Hurewicz homomorphism

$$\pi_*(MU) \longrightarrow MU_*(MU)$$

one can of course deduce the formulae for the images of the a_{ij} under the Hurewicz homomorphisms

$$\pi_*(MU) \longrightarrow H_*(MU)$$

$$\pi_*(MU) \longrightarrow K_*(MU).$$

One just applies the maps $MU \longrightarrow H$, $MU \longrightarrow K$. In fact, the map

$MU \to H$ sends b_i^{MU} to b_i^H, and sends a_{ij} to 0 for $i \geq 1$, $j \geq 1$. The map $MU \to K$ sends b_i^{MU} to $u^i b_i^K$, a_{11} to u, and a_{ij} to 0 if $i > 1$ or $j > 1$.

7. THE STRUCTURE OF LAZARD'S UNIVERSAL RING L

We propose to prove:

THEOREM 7.1. The graded ring L is a polynomial algebra over Z on generators of dimension $2, 4, 6, 8, \ldots$.

In order to prove this, we will use a faithful representation of L. Its construction is suggested by the results of the last section. As a matter of pure algebra, we define a (graded) commutative ring R by

$$R = Z[b_1, b_2, \ldots, b_n, \ldots]$$

where b_i is assigned degree $2i$; b_0 is interpreted as 1 if it arises. The generator b_i is to be distinguished from the generator β_i in §3.

We define a formal power-series

$$\exp(y) \in R[[y]]$$

by

$$\exp(y) = \sum_{i \geq 0} b_i y^{i+1}, \tag{7.2}$$

and we define $\log(x)$ to be the power-series inverse to exp, so that

$$\begin{aligned} \exp \log(x) &= x \\ \log \exp(y) &= y . \end{aligned} \tag{7.3}$$

For later use, we make the log series more explicit. Let its coefficients be

$$m_i \in Z[b_1, b_2, \ldots, b_n, \ldots]],$$

so that

$$\log x = \sum_{i \geq 0} m_i x^{i+1} . \tag{7.4}$$

If S is an inhomogeneous sum, let us write S_i for the component of S of dimension $2i$. Then we have:

PROPOSITION 7.5. $$m_n = \frac{1}{n+1}\left(\sum_{i=0}^{\infty} b_i\right)_n^{-n-1},$$

$$b_n = \frac{1}{n+1}\left(\sum_{i=0}^{\infty} m_i\right)_n^{-n-1}.$$

<u>Examples.</u> $$m_1 = -b_1$$
$$m_2 = 2b_1^2 - b_2$$
$$m_3 = -5b_1^3 + 5b_1 b_2 - b_3 \quad \text{etc.}$$

<u>Proof.</u> If

$$\omega = \sum_{i \geq -N} c_i y^i dy ,$$

define $\operatorname{res} \omega$ to be c_{-1}, the residue of ω at $y = 0$. This definition of the residue is purely algebraic, and the property of the residue which we shall use can be established purely algebraically. Set

$$x = \sum_{i \geq 0} b_i y^{i+1}$$
$$y = \sum_{j \geq 0} m_j x^{j+1} .$$

Then $(\sum_{i \geq 0} b_i)_n^{-n-1}$ is the coefficient of y^n in $(\sum_{i \geq 0} b_i y^i)^{-n-1}$, that is, the coefficient of y^{-1} in $(\sum_{i \geq 0} b_i y^{i+1})^{-n-1}$. So we have

$$\begin{aligned}
\left(\sum_{i \geq 0} b_i\right)_n^{-n-1} &= \operatorname{res}(x^{-n-1} dy) \\
&= \operatorname{res}\left(x^{-n-1} \frac{dy}{dx} dx\right) \\
&= \operatorname{res}\left[x^{-n-1}\left(\sum_{j \geq 0} m_j (j+1) x^j\right) dx\right] \\
&= (n+1) m_n .
\end{aligned}$$

Of course, the relation between the coefficients b_i and m_i of the two inverse series is symmetric.

In the future, whenever symbols b_i and m_i appear in various contexts, they will be related as in (7.5).

Remark 7.6. Suppose that instead of Z we have in sight a ring U, that we replace R by

$$U[b_1, b_2, \ldots, b_n, \ldots],$$

and that we replace our series exp by

$$x = u^{-1} \sum_{i \geq 0} b_i y^{i+1}$$

where u is invertible in U. (An application is given in (6.3), (6.7).) Then we have

$$y = \sum_{j \geq 0} m_j u^{j+1} x^{j+1},$$

by substituting ux for x in our previous work.

Let us return to formal groups. We define a formal product over $R = Z[b_1, b_2, \ldots, b_n, \ldots]$ by

$$\mu^R(x_1, x_2) = \exp(\log x_1 + \log x_2). \tag{7.7}$$

It is easy to check that this does define a formal product. We have simply taken the additive formal product, (1.4), and made a change of variables; but the change of variables is of a fairly general nature. The topologist who has read §6 knows that this piece of pure algebra is read off from the structure of $H_*(MU)$; the algebraist doesn't have to worry.

According to §5, there is one and only one homomorphism

$$\theta : L \longrightarrow R$$

which carries the formal product μ^L into μ^R. We propose to prove:

THEOREM 7.8. The map θ is monomorphic.

This theorem shows that we have made the ring R big enough to provide a faithful representation of L. The proof will require various intermediate results.

We first recall that the augmentation ideal of a connected graded ring A is defined by

$$I = \sum_{n>0} A_n .$$

The elements of I^2 are often called "decomposable elements". The "indecomposable quotient" $Q_*(A)$ is defined by

$$Q_*(A) = I/I^2.$$

We can often use $Q_*(A)$ to get a hold on A.

It is clear that $Q_m(L)$ and $Q_m(R)$ are both zero unless $m = 2n$, $n > 0$. In this case we have $Q_{2n}(R) \cong Z$, generated by the coset $[b_n]$.

LEMMA 7.9. (i) $\log(x) = \sum_{i \geq 0} m_i x^{i+1}$, where $m_o = 1$ and $m_i \equiv -b_i \mod I^2$ for $i \geq 1$.

(ii) $\theta(a_{ij}) \equiv \frac{(i+j)!}{i!\, j!} b_{i+j-1} \mod I^2$ for $i \geq 1$, $j \geq 1$.

(iii) The image of $Q_{2n}(\theta): Q_{2n}(L) \longrightarrow Q_{2n}(R)$ consists of the multiples of $d[b_n]$, where

$$d = \begin{cases} p & \text{if } n+1 = p^f, \ p \text{ prime}, \ f \geq 1 \\ 1 & \text{otherwise} . \end{cases}$$

<u>Proof.</u> Part (i) is immediate. Part (ii) follows from (7.7) by an easy calculation, ignoring coefficients in I^2. Since L is generated as a ring by the a_{ij}, $Q_{2n}(L)$ is certainly generated as an abelian group by the a_{ij} with $i+j = n+1$, $i \geq 1$, $j \geq 1$. To prove part (iii) we need only show that the highest common factor of the binomial coefficients

$$\frac{(i+j)!}{i!\, j!} \quad (i + j = n+1, \ i \geq 1, \ j \geq 1)$$

is the integer d defined in the enunciation.

It is well known, and easy to see, that if $n+1 = p^f$ all these binomial coefficients are divisible by p, and that if $n+1 \neq p^f$ at least one of them is not divisible by p. One has only to add that if $n+1 = p^f$, then the binomial coefficient with $i = \lambda p^{f-1}$, $j = \mu p^{f-1}$ is

$$\frac{p!}{\lambda! \mu!} \bmod p^2 ,$$

and it is divisible by p but not by p^2.

Topologists will note that this calculation is exactly the same as one which Milnor made in the topological case. He was, of course, computing the image of

$$Q_{2n}(\pi_*(MU)) \longrightarrow Q_{2n}(H_*(MU)).$$

The "Milnor genus" may be regarded as the projection

$$H_{2n}(MU) \longrightarrow Q_{2n}(H_*(MU)) ,$$

and the "hypersurfaces of type (1, 1) in $CP^i \times CP^j$" are related to the elements $a_{ij} \in \pi_*(MU)$ (see Corollary 10.9).

In order to obtain the structure of $Q_*(L)$, we propose to consider formal groups defined over graded rings S of a particular form. Given an abelian group A, and an integer $n > 0$, we can make $Z \oplus A$ into a graded ring so that

$$S_o = Z$$

$$S_{2n} = A$$

$$S_r = 0 \quad \text{for } r \neq 0, 2n .$$

LEMMA 7.10. Among formal groups defined over such rings S, the obvious formal group defined over $Z \oplus Q_{2n}(L)$ is universal.

The proof is immediate; any homomorphism of rings

$$L \longrightarrow Z \oplus A$$

factors to give the following diagram.

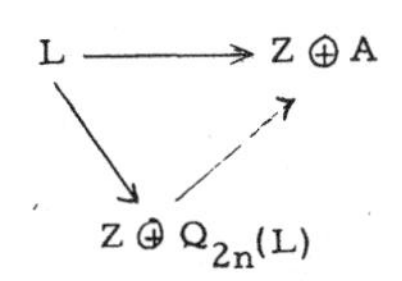

We can now reformulate the main lemma used by Lazard and by Fröhlich. Let T_n be the image of $Q_{2n}(\theta): Q_{2n}(L) \longrightarrow Q_{2n}(R)$, described in (7.9).

LEMMA 7.11. (After Lazard and Fröhlich). For any (commutative) formal group defined over a ring $Z \oplus A$, the homomorphism

$$Z \oplus Q_{2n}(L) \longrightarrow Z \oplus A$$

factors through the quotient map

$$Z \oplus Q_{2n}(L) \longrightarrow Z \oplus T_n \, .$$

The main results of this section follow very easily from this lemma; but we will defer the proofs until we have proved Lemma 7.11.

<u>Proof.</u> We recall the reformulation of §3. A formal group defined over $Z \oplus A$ is a Hopf algebra structure on a certain coalgebra F; the coalgebra F is free over $Z \oplus A$ on generators $\beta_0, \beta_1, \ldots, \beta_i, \ldots$ and the coaction is given by

$$\psi \beta_k = \sum_{i+j=k} \beta_i \otimes \beta_j \, .$$

Inspecting the formulae in §3 again, we see that now our rings are graded, F can be graded so that β_i has degree $2i$.

In our case, part of the product structure is determined by the coproduct structure; we must have

$$\beta_i \beta_j = \frac{(i+j)!}{i!\, j!} \beta_{i+j} + \sum_{k=i+j-n>0} a^k_{ij} \beta_k \, . \tag{7.12}$$

Here the a^k_{ij} are coefficients in A, which have to be determined, and we are interested in their values for $k = 1$. More precisely, let d be the highest common factor of the binomial coefficients $\frac{(i+j)!}{i!\,j!}$ over $i+j = n+1$, $i \geq 1$, $j \geq 1$, as in (7.9); we wish to show that

$$a^1_{ij} = \frac{1}{d}\frac{(i+j)!}{i!\,j!}\,a \tag{7.13}$$

for some fixed element $a \in A$; for then the required map φ from T_n to A will be defined by $\varphi(d\,\mathfrak{b}_n) = a$.

We emphasize that the product $\beta_i\beta_j$ is known, from (7.12), if $i+j < n+1$. We now divide cases.

Case (i). $A \cong Z$; let us write as if $A = Z$. We have

$$(\beta_1)^{n+1} = (n+1)!(\beta_{n+1} + \frac{a}{d}\beta_1)$$

for some $a \in Q$. When $i+j = n+1$, we have

$$\begin{aligned}(i!\beta_i)(j!\beta_j) &= (\beta_1)^i(\beta_1)^i = (\beta_1)^{n+1} \\ &= (i+j)!(\beta_{i+j} + \frac{a}{d}\beta_1).\end{aligned}$$

Comparing this with (7.9) we have

$$a^1_{ij} = \frac{(i+j)!}{i!\,j!}\frac{a}{d}\,.$$

Here a is a rational number such that $\frac{(i+j)!}{i!\,j!}\frac{a}{d}$ is an integer for $i+j = n+1$, $i \geq 1$, $j \geq 1$. The highest common factor of the numbers $\frac{(i+j)!}{i!\,j!}\frac{1}{d}$ is 1, so a is an integer, and we have obtained the required result (7.13) in this case.

Case (ii). $A \cong Z_p$. Take i, j such that $i+j = n+1$, $i \geq 1$, $j \geq 1$ and write

$$i = \lambda_o + \lambda_1 p + \lambda_2 p^2 + \dots + \lambda_r p^r,$$

$$j = \mu_o + \mu_1 p + \mu_2 p^2 + \dots + \mu_r p^r,$$

where $0 \leq \lambda_i < p$, $0 \leq \mu_i < p$ for each i. Then

$$\beta_1^{\lambda_o}\beta_p^{\lambda_1}\beta_{p^2}^{\lambda_2}\dots\beta_{p^r}^{\lambda_r} = c'\beta_i$$

$$\beta_1^{\mu_0} \beta_p^{\mu_1} \beta_{p^2}^{\mu_2} \ldots \beta_{p^r}^{\mu_r} = c''\beta_j$$

where the coefficients c' and c'' are non-zero mod p; in fact,

$$c' = \lambda_0!\lambda_1!\lambda_2! \ldots \lambda_r! \quad \text{mod } p$$

$$c'' = \mu_0!\mu_1!\mu_2! \ldots \mu_r! \quad \text{mod } p.$$

Then we have

$$\frac{(i+j)!}{i!\,j!} \beta_{i+j} + a^1_{ij}\beta_1 = \beta_i\beta_j = \frac{1}{c'c''} \beta_1^{\lambda_0+\mu_0} \beta_p^{\lambda_1+\mu_1} \ldots \beta_{p^r}^{\lambda_r+\mu_r} .$$

At this point we separate cases further.

Case (a): Suppose that $n+1 \neq p^f$ and $\lambda_i + u_i \geq p$ for some i. Then we have $p^{i+1} \leq n+1$, and since $n+1 \neq p^f$ we actually have $p^{i+1} < n+1$; so $(\beta_{p^i})^p = 0$ by (7.12) and $a^1_{ij} = 0$. Since $\frac{(i+j)!}{i!\,j!}$ is also 0 mod p, the required formula (7.13) will be true in this case whatever choice of a we make later.

Case (b). Suppose that $n+1 = p^f$ and $\lambda_i + u_i \geq p$ for some $i \leq f-2$. Then the same argument applies, except that we have to remark that $\frac{1}{p}\frac{(i+j)!}{i!\,j!}$ is 0 mod p. (I am willing to assume that the reader knows or can work out all the required results on binomial coefficients.)

Case (c). Suppose $n+1 \neq p^f$ and $\lambda_i + \mu_i < p$ for all i. If we write

$$n+1 = \nu_0 + \nu_1 p + \nu_2 p^2 + \ldots + \nu_r p^r,$$

with $0 \leq \nu_i < p$ for each i, we must have

$$\lambda_i + \mu_i = \nu_i .$$

But we can set, once for all,

$$\beta_1^{\nu_0} \beta_p^{\nu_1} \beta_{p^2}^{\nu_2} \ldots \beta_{p^r}^{\beta_r} = c(\beta_{n+1} + a\beta_1)$$

where the coefficient c is non-zero mod p; in fact,

$$c = \nu_0!\nu_1! \dots \nu_r! \bmod p.$$

Then

$$a^1_{ij} = \frac{c}{c'c''} a$$

$$= \frac{(i+j)!}{i!\,j!} a .$$

So the required formula (7.13) holds if $n+1 \neq p^f$.

Case (d). The only remaining case is that in which $n+1 = p^f$, $i = \lambda_{f-1} p^{f-1}$, $j = \mu_{f-1} p^{f-1}$. In this case we can set, once for all,

$$(\beta_{p^{f-1}})^p = e\beta_{n+1} + (p-1)!a\beta_1$$

where $a \in A$. Then we have

$$a^1_{ij} = \frac{(p-1)!}{\lambda_{f-1}!\mu_{f-1}!} a$$

$$= \frac{1}{p} \frac{(i+j)!}{i!\,j!} a .$$

So the required formula (7.13) holds if $n+1 = p^f$. This completes case (ii).

Case (iii). $A \cong Z_{p^f}$. We first remark that a homomorphism of graded rings $Z \oplus A \longrightarrow Z \oplus A'$ is equivalent to a homomorphism of abelian groups $A \longrightarrow A'$. We now proceed by induction over f, and assume the result true for $f-1$. Suppose given a homomorphism

$$Q_{2n}(L) \xrightarrow{\theta} Z_{p^f} ;$$

and form the following diagram.

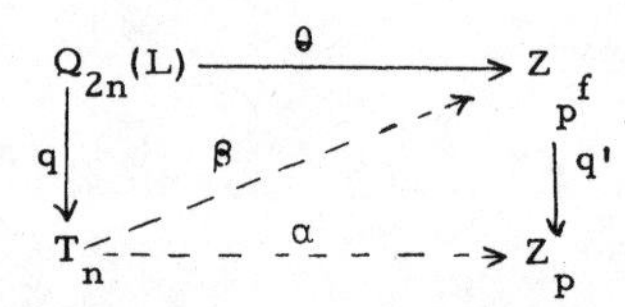

By case (ii) the homomorphism $q'\theta$ factors in the form αq. Since T_n is free, we can factor α in the form $q'\beta$. Then $q'(\theta - \beta q) = 0$, and so $\theta - \beta q$ maps into $Z_{p^{f-1}}$. By the inductive hypothesis, $\theta - \beta q$ factors in the form γq. Therefore $\theta = (\beta + \gamma)q$. This completes the induction, and finishes case (iii).

Case (iv). A is any finitely-generated abelian group. Then A can be written as a direct sum of groups Z and Z_{p^f}. The result follows from cases (i) and (iii).

Case (v). A is any abelian group. Let $\theta : Q_{2n}(L) \longrightarrow A$ be a homomorphism. Since $Q_{2n}(L)$ is finitely-generated, so is the image of θ. The result follows from case (iv).

This completes the proof of Lemma 7.11.

COROLLARY 7.14. The quotient map

$$Q_{2n}(\theta) : Q_{2n}(L) \longrightarrow T_n$$

of (7.9) and (7.11) is an isomorphism.

<u>Proof.</u> Of course, the quotient map is an epimorphism. Consider the following diagram.

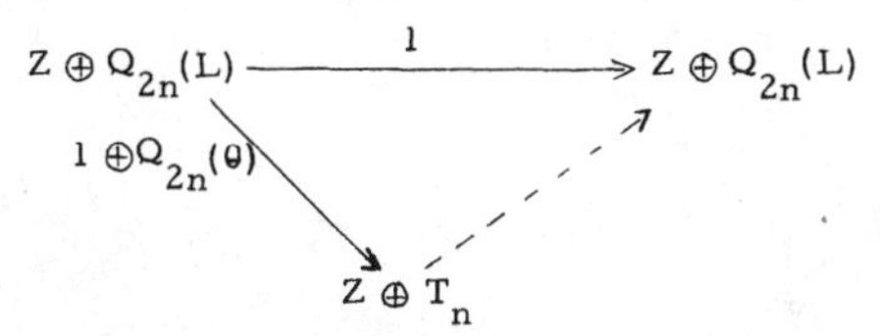

By Lemma 7.11, the identity map of $Q_{2n}(L)$ factors through $Q_{2n}(\theta)$. Therefore $Q_{2n}(\theta)$ is monomorphic.

We now prove Theorems 7.1 and 7.8. Choose in L_{2n} an element t_n which projects to the generator of T_n. We immediately obtain a map

$$Z[t_1, t_2, \ldots, t_n, \ldots] \xrightarrow{\alpha} L$$

By Corollary 7.14, $Q_{2n}(\alpha)$ is an isomorphism for each n, and therefore α is an epimorphism. But it is obvious that the composite map

$$Z[t_1, t_2, \ldots, t_n, \ldots] \xrightarrow{\alpha} L \xrightarrow{\theta} R = Z[b_1, b_2, \ldots, b_n, \ldots]$$

is monomorphic, since $\theta\alpha t_n$ is a non-zero multiple of b_n, modulo decomposables. Therefore α is an isomorphism and θ is a monomorphism. This proves (7.1) and (7.8).

COROLLARY 7.15. Let μ^S be any formal product defined over a ring S containing the rational numbers Q. Then the formal group with formal product μ^S is isomorphic to the additive formal group (1.4).

Proof. We have a homomorphism $\theta: L \longrightarrow S$ carrying μ^L into μ. Since $S \supset Q$, θ extends to give a homomorphism $\theta: L \otimes Q \longrightarrow S$. Let R be as above; then we may identify $L \otimes Q$ with $R \otimes Q$. Then the power-series

$$\exp(y) = \sum_{i \geq 0} (\theta b_i) y^{i+1}$$

$$\log(x) = \sum_{i \geq 0} (\theta m_i) x^{i+1}$$

give the required isomorphism.

Of course, this result is much easier than the proof we have given of it; and it does not need the hypothesis that the formal product μ^S is commutative (as we are always assuming.) We have given the result to stress that in what follows, log and exp will always be as in the proof of (7.15).

8. QUILLEN'S THEOREM

By Theorem 5.1 we have a map

$$\theta' : L \longrightarrow \pi_*(MU).$$

The object of this section is to prove the following results.

THEOREM 8.1. (Milnor) $\pi_*(MU)$ is a polynomial algebra over Z on generators of dimension $2, 4, 6, 8, \ldots$.

THEOREM 8.2. (Quillen) The map

$$\theta' : L \longrightarrow \pi_*(MU)$$

is an isomorphism.

Following Milnor, we base our calculation of $\pi_*(MU)$ on the spectral sequence

$$\mathrm{Ext}_A^{s,t}(H^*(MU;Z_p), Z_p) \overset{s}{\Longrightarrow} \pi_{t-s}(MU). \tag{8.3}$$

Here A is the mod p Steenrod algebra.

LEMMA 8.4. $H^*(MU;Z_p)$ is a free module over $A/(A\beta A)$, where $A\beta A$ is the two-sided ideal generated by the Bokštein boundary $\beta = \beta_p$.

This lemma is an absolutely standard consequence of the following facts. (i) $A/(A\beta A)$ acts freely on the Thom class $u \in H^0(MU;Z_p)$. (ii) $H^*(MU;Z_p)$ is a coalgebra over $A/(A\beta A)$.

Unfortunately, we do not only need to know that $H^*(MU;Z_p)$ is free over $A/(A\beta A)$; we need to know about its base; or more precisely, we need the following result.

LEMMA 8.5. $\mathrm{Hom}_A(H^*(MU;Z_p), Z_p)$, which can be identified with the set of primitive elements in the comodule $H_*(MU;Z_p)$, is a polynomial algebra on generators of dimension $2n$ for $n > 0$, $n \neq p^f - 1$.

We prove (8.4) and (8.5) together, following Liulevicius. More precisely, let A_* be the dual of $A/(A\beta A)$; it is a polynomial algebra

$$Z_p[\xi_1, \xi_2, \ldots, \xi_f, \ldots]$$

with ξ_f of dimension $2(p^f-1)$. Let N_* be a polynomial algebra

$$Z_p[x_1, x_2, \ldots, x_{p-2}, x_p, \ldots]$$

with one generator x_i of dimension $2i$ whenver $i \neq p^f-1$. Define a homomorphism

$$\alpha: H_*(MU; Z_p) \longrightarrow N_*$$

by

$$\alpha(b_i) = \begin{cases} x_i & (i \neq p^f-1) \\ 0 & (i = p^f-1) \end{cases} .$$

Define a homomorphism from $H_*(MU; Z_p)$ to $A_* \otimes N_*$ by

$$H_*(MU; Z_p) \xrightarrow{\psi} A_* \otimes H_*(MU; Z_p) \xrightarrow{1 \otimes \alpha} A_* \otimes N_* ,$$

where ψ is the coproduct map. Make $A_* \otimes N_*$ into a comodule over A_* by giving it the coproduct map

$$A_* \otimes N_* \xrightarrow{\psi \otimes 1} A_* \otimes (A_* \otimes N_*).$$

Then $(1 \otimes \alpha)\psi$ is a homomorphism of rings and a homomorphism of comodules over A_*.

Now, in $BU(1)$ we have

$$\psi \beta_{p^f} = \xi_f \otimes \beta_1 + \xi_{f-1}^p \otimes \beta_p + \ldots + 1 \otimes \beta_{p^f}.$$

So in MU we have

$$\psi b_{p^f-1} = \xi_f \otimes 1 + \xi_{f-1}^p \otimes b_{p-1} + \ldots + 1 \otimes b_{p^f-1} .$$

We see that the map

$$Q((1 \otimes \alpha)\psi) : Q(H_*(MU; Z_p)) \longrightarrow Q(A_* \otimes N_*)$$

is given by

$$Q((1 \otimes \alpha)\psi) b_i = \begin{cases} 1 \otimes x_i \mod I^2 & (i \neq p^f-1) \\ \xi_f \otimes 1 \mod I^2 & (i = p^f-1). \end{cases}$$

So $Q((1 \otimes \alpha)\psi)$ is an isomorphism, and $(1 \otimes \alpha)\psi$ is an epimorphism. By counting dimensions, $(1 \otimes \alpha)\psi$ must be an isomorphism.

Since the dual of $A_* \otimes N_*$ is free, we have proved (8.4). Since the set of primitive elements in $A_* \otimes N_*$ is precisely N_*, we have proved (8.5) too.

COROLLARY 8.6. In the spectral sequence (8.3), the E_2-term

$$\mathrm{Ext}_A^{s,t}(H^*(MU;Z_p), Z_p)$$

is a polynomial algebra on generators x_n, $n = 0, 1, 2, 3, \ldots$ of bidegree

$$s = 0, \quad t = 2n \qquad (n \neq p^f - 1)$$

$$s = 1, \quad t = 2n+1 \quad (n = p^f - 1).$$

This follows from (8.4), (8.5) by standard methods; see [12].

It follows from (8.6) that the spectral sequence (8.3) has non-zero groups only in even dimensions; so the spectral sequence is trivial.

In order to deduce the required results on $\pi_*(MU)$, we need a technical lemma on the convergence of the spectral sequence (8.3).

LEMMA 8.7. Suppose given a connected spectrum X, such that $\pi_r(X)$ is finitely generated for each r and zero for $r < 0$. Suppose given integers m, e. Then there exists $s = s(m, e)$ such that any element in $\pi_m(X)$ of filtration $\geq s$ in the spectral sequence

$$\mathrm{Ext}_A^{s,t}(H^*(X;Z_p), Z_p) \xrightarrow{s} \pi_{t-s}(X)$$

is divisible by p^e in $\pi_n(X)$.

This may be proved by the method given in my original paper [1].

COROLLARY 8.8.

$$Q_m(\pi_*(MU)) \otimes Z_p \cong \begin{cases} Z_p & \text{for } m = 2n, \ n > 0 \\ 0 & \text{otherwise}. \end{cases}$$

<u>Proof.</u> When $m \neq 2n$ $(n > 0)$ the result is trivial, so we

assume $m = 2n$, $n > 0$. There are of course many ways of seeing that $Q_{2n}(\pi_*(MU)) \otimes Z_p$ has dimension at least one over Z_p; for example,

$$Q_{2n}(\pi_*(MU)) \otimes Q \cong Q_{2n}(H_*(MU)) \otimes Q \cong Q .$$

We need to prove that $Q_{2n}(\pi_*(MU)) \otimes Z_p$ has dimension at most 1.

Let $t_i \in \pi_{2i}(MU)$ be an element whose class in the E_2-term is the generator x_i of (8.6). I claim that $Q_{2n}(\pi_*(MU)) \otimes Z_p$ is generated by the image of t_n. In fact, let y be any element in $\pi_{2n}(MU)$, and let s be as in (8.7), taking $m = 2n$, $e = 1$; then (by induction over the filtration) we can find a polynomial $q(t_o, t_1, \ldots, t_n)$ such that $y - q(t_o, t_1, \ldots, t_n)$ has filtration $\geq s$, and so

$$y = q(t_o, t_1, \ldots, t_n) + pz.$$

Since $\pi_o(MU) = Z$, the coefficient of t_n (which a priori is a polynomial in t_o) must be an integer c. We deduce that

$$y = ct_n \bmod I^2 + p\pi_*(MU),$$

where $I = \sum_{i \geq 0} \pi_i(MU)$. That is, $Q_{2n}(\pi_*(MU)) \otimes Z_p$ is generated by the image of t_n. This proves (8.8).

COROLLARY 8.9.

$$Q_m(\pi_*(MU)) \cong \begin{cases} Z & \text{for } m = 2n,\ n > 0 \\ 0 & \text{otherwise} . \end{cases}$$

Proof. $Q_m(\pi_*(MU))$ is a finitely generated abelian group; use the structure theorem for finitely generated abelian groups, plus (8.8).

We now consider the following diagram.

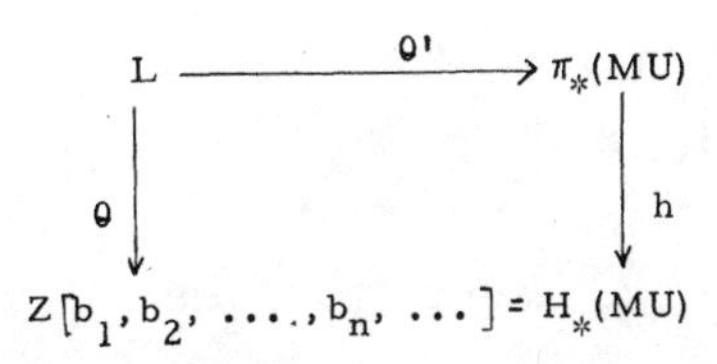

Here θ has been carefully defined so that the diagram is commutative, as we see by comparing (6.6) with (7.2), (7.7). The behavior of θ has been studied in §7.

LEMMA 8.10. The image of

$$Q_{2n}(h): Q_{2n}(\pi_*(MU)) \longrightarrow Q_{2n}(H_*(MU))$$

is the same as the image of $Q_{2n}(\theta)$ (which has been described in (7.9)).

Proof. It is clear that $\mathrm{Im}\, Q_{2n}(\theta) \subset \mathrm{Im}\, Q_{2n}(h)$; we have to prove $\mathrm{Im}\, Q_{2n}(h) \subset \mathrm{Im}\, Q_{2n}(\theta)$. If $n+1 \neq p^f$ there is nothing to prove. If $n+1 = p^f$, consider the canonical map

$$MU \longrightarrow H \longrightarrow HZ_p;$$

call it g. The induced homomorphism

$$q_*: H_*(MU) \longrightarrow (HZ_p)_*(HZ_p)$$

clearly annihilates the image of $\pi_{2n}(MU)$. On the other hand, it carries b_n into the Milnor generator ξ_f in $(HZ_p)_*(HZ_p)$ (since both come from $MU(1) = BU(1)$). The class ξ_f remains non-zero when we pass to $Q_{2n}(HZ_p)_*(HZ_p) \cong Z_p$. So the image of $Q_{2n}(h)$ consists at most of the multiples of $p[b_n]$. This proves (8.10).

Exercise. See if you can refrain from translating this proof into cohomology.

We proceed to prove (8.1) and (8.2). Recall our diagram.

$$\begin{array}{ccc} L & \xrightarrow{\theta'} & \pi_*(MU) \\ \downarrow{\scriptstyle\theta} & & \downarrow{\scriptstyle h} \\ Z[b_1, b_2, \ldots, b_n, \ldots] & = & H_*(MU) \end{array}$$

It follows from (8.9) and (8.10) that

$$Q_{2n}(h):Q_{2n}(\pi_*(MU)) \longrightarrow \mathrm{Im}\ Q_{2n}(\theta))$$

is iso. Using (7.14), we see that

$$Q_{2n}(\theta'):Q_{2n}(L) \longrightarrow Q_{2n}(\pi_*(MU))$$

is iso. Therefore

$$\theta':L \longrightarrow \pi_*(MU)$$

is epi. But by (7.8), the map $\theta = h\theta'$ is mono; so θ' is mono, and θ is an isomorphism. This proves (8.2), and (8.1) follows from (7.1).

Taking a last look at our diagram, we conclude that the homomorphism θ studied in §7 was, up to isomorphism, the Hurewicz homomorphism

$$h: \pi_*(MU) \longrightarrow H_*(MU).$$

COROLLARY 8.11. The Hurewicz homomorphism

$$h: \pi_*(MU) \longrightarrow H_*(MU)$$

is a monomorphism.

Exercise. Deduce (8.1) directly from (8.6).

9. COROLLARIES

In this section we will record various results which follow from the results in §8, or supplement them, and are needed later.

Recall that the complex manifold CP^n defines an element $[CP^n] \in \pi_{2n}(MU)$.

LEMMA 9.1. With the notation of §7, the image of $[CP^n]$ in $H_{2n}(MU)$ is $(n+1)m_n$.

Proof. Algebraic topologists will instantly recognize the formula

$$\left(\sum_{i=0}^{\infty} b_i\right)_n^{-n-1}$$

of (7.5) as giving the normal Chern numbers of CP^n.

We know from §8 that the map

$$\pi_*(MU) \longrightarrow \pi_*(MU) \otimes Q$$

is an injection, and we may identify $\pi_*(MU) \otimes Q$ with $H_*(MU) \otimes Q$. It is often convenient to work in $\pi_*(MU) \otimes Q$, and we now know that we lose nothing by doing so. In what follows, then, we will often regard $m_n = m_n^H \in H_{2n}(MU)$ as the element $\frac{[CP^n]}{n+1}$ of $\pi_*(MU) \otimes Q$. If we do so, we have the following result.

COROLLARY 9.2. (Miščenko [13, Appendix 1]) The logarithmic series for the formal product μ^{MU} may be written

$$\log^H x^{MU} = \sum_{n \geq 0} \frac{[CP^n]}{n+1} (x^{MU})^{n+1} .$$

LEMMA 9.3. Suppose that $R \subset S$ are subrings of the rationals. Then a map

$$f: MUR \rightarrow MUS$$

is determined up to homotopy by

$$f_*: \pi_*(MUR) \longrightarrow \pi_*(MUS).$$

Proof. There are many variants possible; we argue as follows. Applying (4.2) as in the proof of (4.6), we see that f is determined up to homotopy by

$$f_*: MUR_*(MUR) \longrightarrow MUR_*(MUS).$$

Since $\pi_*(MU)$ is torsion-free by (8.1), we see that the vertical arrows of the following commutative diagram are monomorphisms.

$$\begin{array}{ccc} MUR_*(MUR) & \xrightarrow{f_*} & MUR_*(MUS) \\ \downarrow & & \downarrow \\ MUR_*(MUR) \otimes Q & \longrightarrow & MUR_*(MUS) \otimes Q \\ \| \wr & & \| \wr \\ \pi_*(MUQ) \otimes \pi_*(MUR) & \xrightarrow{1 \otimes f_*} & \pi_*(MUQ) \otimes \pi_*(MUS) \end{array}$$

So f is determined by $f_*: \pi_*(MUR) \longrightarrow \pi_*(MUS)$. This proves the lemma.

Next we go back to the work of (6.8). We now know that the Hurewicz homomorphism

$$\eta_R: \pi_*(MU) \longrightarrow MU_*(MU)$$

is adequately described by giving

$$\eta_R \otimes 1: \pi_*(MU) \otimes Q \longrightarrow MU_*(MU) \otimes Q,$$

and this can be done by giving its effect on the generators $m_i = m_i^H \in \pi_{2i}(MU) \otimes Q$. For this purpose we propose the following formula. We write M_j for the generator $m_j^{MU} \in MU_{2j}(MU)$, to distinguish it from $m_j = m_j^H$.

PROPOSITION 9.4.

$$\sum_{i \geq 0} (\eta_R m_i) x^{i+1} = \sum_{i \geq 0} m_i \Big(\sum_{j \geq 0} M_j x^{j+1} \Big)^{i+1}.$$

Proof. Consider again the two maps

$$\eta_L: MU \simeq MU \wedge S^0 \xrightarrow{1 \wedge i} MU \wedge MU$$

$$\eta_L: MU \simeq S^0 \wedge MU \xrightarrow{i \wedge 1} MU \wedge MU$$

of (6.8). Applying them to x^{MU}, we obtain two generators in $(MU \wedge MU)^2(CP^\infty)$; we call these generators x^L, x^R. (We no longer need L for the Lazard ring.) Applying Lemma 6.3, we find

$$x^R = \sum_{i \geq 0} b_i^{MU} (x^L)^{i+1}. \tag{9.5}$$

Passing to the inverse power-series, we find

$$(9.6) \qquad x^L = \sum_{i\geq 0} m_i^{MU}(x^R)^{i+1} = \sum_{j\geq 0} M_j(x^R)^{j+1}.$$

Now our log series are

$$x^H = \log^L x^L = \sum_{i\geq 0} (\eta_L m_i)(x^L)^{i+1}$$

$$x^H = \log^R x^R = \sum_{i\geq 0} (\eta_R m_i)(x^R)^{i+1}.$$

So we obtain

$$\sum_{i\geq 0} (\eta_R m_i)(x^R)^{i+1} = \sum_{i\geq 0} (\eta_L m_i)\Big(\sum_{j\geq 0} M_j(x^R)^{j+1}\Big)^{i+1}.$$

This proves the proposition.

10. VARIOUS FORMULAE IN $\pi_*(MU)$

In this section we will derive various relations between different elements lying in $\pi_*(MU)$ or $\pi_*(MU) \otimes Q$. In particular, we will give the relationship between the coefficients a_{ij} and Milnor's hypersurfaces of type (1, 1) in $CP^i \times CP^j$ (10.6).

To begin with we try to answer various questions that might arise in practical calculations.

(i) To write the coefficients m_i in the $\log^H$ series in terms of the coefficients b_i in the $\exp^H$ series. See (7.5).

(ii) To write the coefficients b_i in the $\exp^H$ series in terms of the coefficients m_i in the $\log^H$ series. See (7.5).

(iii) To write the coefficients a_{ij} in terms of the b_i or m_i, regarded as elements of $\pi_*(MU) \otimes Q$. See (6.6).

(iv) To write the b_i or m_i, regarded as elements of $\pi_*(MU) \otimes Q$, in terms of the a_{ij}. The most convenient formula is the following.

(10.1) $$[CP^n] = (n+1)m_n = (\sum_{i\geq 0} a_{i1})_n^{-1}.$$

COROLLARY 10.2. If $n \geq 1$ we have $[CP^n] \equiv -a_{n1}$ mod decomposables in $\pi_*(MU)$.

Proof of (10.1): Take the equation

$$\log(x_1 + \sum_{i\geq 0, j>1} a_{ij} x_1^i x_2^j) = \log x_1 + \log x_2$$

and equate coefficients of x_2. We obtain

$$(\sum_{n\geq 0} (n+1)m_n x_1^n)(\sum_{i\geq 0} a_{i1} x_1^i) = 1.$$

Following Lemma 9.1, it is plausible to observe that the injection $i_n: CP^n \longrightarrow CP^\infty$ defines an element $[i_n] \in MU_{2n}(CP^\infty)$, and to relate this element to those we have already studied. The element $[i_n]$ is not equal to β_n^{MU}, because the constant map $c: CP^\infty \longrightarrow pt$ sends $[i_n]$ to $[CP^n]$ and β_n^{MU} to 0. The required relation will be given in (10.5).

LEMMA 10.3. If $n \geq 1$, we have

$$x^{MU} \cap [i_n] = [i_{n-1}] \text{ in } MU_{2(n-1)}(CP^\infty).$$

This is the sort of result which should obviously be proved geometrically. However, since we are proceeding homologically and not assuming much familiarity with the geometric approach, we check the result by applying the homomorphism

$$MU_*(CP^\infty) \longrightarrow (H \wedge MU)_*(CP^\infty),$$

which we know to be monomorphic by (2.14), (8.11).

The image of $[i_n]$ in $(H \wedge MU)_{2n}(CP^\infty)$ is

$$\sum_{p+q=n} (\sum_{k=0}^{\infty} b_k)_p^{-n-1} \otimes \beta_q$$

where $b_k = b_k^H$, $\beta_q = \beta_q^H$. The image of x^{MU} in $(H \wedge MU)^2(CP^\infty)$ is

$$\sum_r b_r (x^H)^{r+1},$$

by (6.3). The cap product of these two classes is

$$\sum_{\substack{p+q=n\\ r}} \left(\sum_{k=0}^{\infty} b_k\right)_p^{-n-1} b_r \otimes \beta_{q-r-1} .$$

Set $q-r-1 = s$; we obtain

$$\sum_{p+r+s=n-1} \left(\sum_{k=0}^{\infty} b_k\right)_p^{-n-1} b_r \otimes \beta_s = \sum_{t+s=n-1} \left(\sum_{k=0}^{\infty} b_k\right)_t^{-n} \otimes \beta_s .$$

This is the same as the image of $[i_{n-1}]$.

COROLLARY 10.4.

$$(x^{MU})^r \cap [i_n] = \begin{cases} [i_{n-r}] & (r \leq n) \\ 0 & (r > n). \end{cases}$$

This follows immediately, by induction over r.

COROLLARY 10.5.

$$[i_n] = \sum_{r+s=n} [CP^r] \beta_s^{MU} \quad \text{in } MU_{2n}(CP^{\infty}).$$

Proof.

$$\langle (x^{MU})^s, [i_n] \rangle = c_*((x^{MU})^s \cap [i_n])).$$

If $s > n$ we obtain 0; if $s \leq n$ we obtain $c_*[i_{n-s}] = [CP^{n-s}]$.

We are now ready to explain the connection between the coefficients a_{ij} of (2.10) and Milnor's hypersurfaces $H_{i,j}$ of type (1, 1) in $CP^i \times CP^j$.

PROPOSITION 10.6. $[H_{p,q}] = \sum_{\substack{r+u=p\\ s+v=q}} a_{r,s}[CP^u][CP^v]$.

(I understand this formula was also obtained by Boardman.)

COROLLARY 10.7. If $p > 1$ and $q > 1$, we have

$$[H_{p,q}] = a_{p,q} \quad \text{mod decomposables in } \pi_*(MU).$$

Proof of (10.6). The construction of $H_{p,q}$ yields the following formula.

$$[H_{p,q}] = c_*((m^* x^{MU}) \cap ([i_p] \times [i_q])) .$$

Here $c: CP^{\infty} \times CP^{\infty} \longrightarrow pt$ is the constant map, and $m: CP^{\infty} \times CP^{\infty} \longrightarrow CP^{\infty}$

is the product map of §2; we have

$$m^* x^{MU} \in MU^2(CP^\infty \times CP^\infty)$$

and

$$[i_p] \times [i_q] \in MU_{2(p+q)}(CP^\infty \times CP^\infty).$$

This yields

$$[H_{p,q}] = \langle m^* x^{MU}, [i_p] \times [i_q] \rangle .$$

But here we have

$$m^* x^{MU} = \sum_{r,s} a_{rs} (x_1^{MU})^r (x_2^{MU})^s ,$$

$$[i_p] = \sum_{r+u=p} [CP^u] \beta_r^{MU} ,$$

$$[i_q] = \sum_{s+v=q} [CP^v] \beta_s^{MU} .$$

The result follows immediately.

COROLLARY 10.8. $\pi_*(MU)$ is generated by the elements $[CP^n]$ for $n \geq 1$ together with the elements $[H_{p,q}]$ for $p > 1$, $q > 1$.

Proof. By (8.2), $\pi_*(MU)$ is generated by the a_{ij}; but by (10.2) and (10.7) these coincide with the $[CP^n]$ and $[H_{p,q}]$ modulo decomposables.

11. $MU_*(MU)$

It is shown in [3, Lecture 3, pp. 56-76] that $MU_*(MU)$ may be considered as a Hopf algebra. We may think of $MU^*(MU)$, the Novikov algebra of operations on MU-cohomology, as analogous to the Steenrod algebra; if we do so, we should think of $MU_*(MU)$ as analogous to the dual of the Steenrod algebra, which was studied by Milnor [11]. There is only one point at which we need to take care in generalizing from the classical case to the case of generalized homology; the Hopf algebra

$MU_*(MU) = \pi_*(MU \wedge MU)$ is a bimodule over the ring of coefficients $\pi_*(MU)$, because we can act either on the left-hand factor of $MU \wedge MU$ or on the right-hand factor. On this point, see [3, Lecture 3, pp. 59-60].

I would now advance the thesis that instead of considering $MU^*(X)$ as a (topologised) module over the (topologised) ring $MU^*(MU)$, we should consider $MU_*(X)$ as a comodule with respect to the Hopf algebra $MU_*(MU)$. For this purpose I propose to record the structure of $MU_*(MU)$ as a Hopf algebra. I would like to regard this account as superseding, to a large extent, the account which I gave in my earlier Chicago notes [2].

At this point I pause to insert various remarks intended to make the spectrum $MU \wedge MU$ seem more familiar. Some may like to think of it as the representing spectrum for $U \times U$-bordism; that is, we consider manifolds M^n, which are given embedded in a sphere $S^{n+2p+2q}$, and whose normal bundle is given the structure of a $U(p) \times U(q)$-bundle -- say as $\nu = \nu_1 \oplus \nu_2$. With this interpretation, some of the structure maps to be considered are obvious ones. For example, we shall consider a conjugation map or canonical anti-automorphism

$$c: MU_*(MU) \longrightarrow MU_*(MU);$$

this is induced by the usual switch map

$$\tau: MU \wedge MU \longrightarrow MU \wedge MU,$$

which interchanges the two factors. The effect of c on M^n is to leave the manifold alone and take the new ν_1 to be the old ν_2 and vice versa. We can easily construct $U \times U$-manifolds, for example, by taking CP^n and taking the stable classes of ν_1, ν_2 to be $p\xi, q\xi$, where $p+q = -(n+1)$. However, we will make no further use of this approach.

I also remark that $MU \wedge MU$ is homotopy-equivalent to a wedge-sum of suitable suspensions of MU. This follows from the following lemma, plus (4.5).

LEMMA 11.1. Let E be a ring-spectrum. In order that $E \wedge X$ be equivalent, as a module-spectrum over E, to a wedge-sum $\bigvee_\alpha E \wedge S^{n(\alpha)}$, it is necessary and sufficient that $\pi_*(E \wedge X)$ should be a free module over $\pi_*(E)$.

Proof. $\pi_*(\bigvee_\alpha E \wedge S^{n(\alpha)}) \cong \sum_\alpha \pi_*(E \wedge S^{n(\alpha)})$ is indeed a free module over $\pi_*(E)$. So if $E \wedge X$ is equivalent, as a module-spectrum over E, to $\bigvee_\alpha E \wedge S^{n(\alpha)}$, then $\pi_*(E \wedge X)$ is also free.

Conversely, assume that $\pi_*(E \wedge X)$ is free over $\pi_*(E)$, with a base of elements $b_\alpha \in \pi_{n(\alpha)}(E \wedge X)$. Represent b_α by a map

$$f_\alpha : S^{n(\alpha)} \longrightarrow E \wedge X,$$

and consider the map

$$f : \bigvee_\alpha E \wedge S^{n(\alpha)} \longrightarrow E \wedge X$$

whose α-th component is

$$E \wedge S^{n(\alpha)} \xrightarrow{1 \wedge f_\alpha} E \wedge E \wedge X \xrightarrow{\mu \wedge 1} E \wedge X.$$

Then f is clearly a map of module-spectra over E, and f induces an isomorphism of homotopy groups; so f is a homotopy equivalence, by Whitehead's Theorem (in the category of spectra.)

Let us return to the structure of $MU_*(MU)$. Recall from (4.5) that $MU_*(MU)$ is free as a left module over $\pi_*(MU)$, with a base consisting of the monomials in the generators $b_i = b_i^{MU} \in MU_{2i}(MU)$.

Recall also from [3, p. 61] that the structure maps to be considered are as follows.

(i) A product map

$$\varphi: MU_*(MU) \otimes MU_*(MU) \longrightarrow MU_*(MU).$$

This is the same product in $MU_*(MU)$ that we have been using all along, and we do not need to give any formulae for it, because $MU_*(MU)$ is described in terms of this product.

(ii) Two unit maps

$$\eta_L, \eta_R: \pi_*(MU) \longrightarrow MU_*(MU).$$

These are induced by the maps

$$MU \simeq MU \wedge S^0 \xrightarrow{1 \wedge i} MU \wedge MU,$$

$$MU \simeq S^0 \wedge MU \xrightarrow{i \wedge 1} MU \wedge MU$$

respectively. They are introduced so that left multiplication by $a \in \pi_*(MU)$ is multiplication by $\eta_L(a)$, and right multiplication by $a \in \pi_*(MU)$ is multiplication by $\eta_R(a)$. The map η_L sends $a \in \pi_*(MU)$ to $a.1$, and we do not need to give any other formula for it. The map η_R is essentially the Hurewicz homomorphism

$$\pi_*(MU) \longrightarrow MU_*(MU).$$

It figures in the next result; to motivate it, we recall that one should describe the action of cohomology operations $h \in MU^*(MU)$ on the ring of coefficients $\pi_*(MU)$; compare [2, p. 19; Theorem 8.1, p. 23].

PROPOSITION 11.2. Let E be as in [3, Lecture 3], and let $h \in E^*(E)$. Then the effect of the cohomology operation h on the element $\lambda \in \pi_*(E)$ is given by

$$h\lambda = \langle h, \eta_R \lambda \rangle.$$

This may be proved either directly from the definitions by diagram-chasing, or by substituting $X = S^0$, $\psi\lambda = (\eta_L \lambda) \otimes 1$ in [3, Proposition 2, p. 72].

We return to listing the structure maps to be considered.

(iii) A counit map

$$\varepsilon: MU_*(MU) \rightarrow \pi_*(MU).$$

This is induced by the product map

$$\mu: MU \wedge MU \longrightarrow MU.$$

(iv) A canonical anti-automorphism, or conjugation map

$$c: MU_*(MU) \rightarrow MU_*(MU).$$

This is induced by the switch map

$$\tau: MU \wedge MU \rightarrow MU \wedge MU,$$

as remarked above.

(v) A diagonal or coproduct map

$$\psi: MU_*(MU) \longrightarrow MU_*(MU) \otimes_{\pi_*(MU)} MU_*(MU).$$

The maps which have not been discussed already are given by the following result.

THEOREM 11.3. (i) The homomorphism η_R is calculated in §6, §9.

(ii) The map ε is a map of algebras which are bimodules over $\pi_*(MU)$; it satisfies

$$\varepsilon(1) = 1$$

$$\varepsilon(b_i) = 0 \text{ for } i \geq 1.$$

(iii) The map c is a map of rings; it satisfies

$$\left.\begin{aligned} c(\eta_L a) &= \eta_R a \\ c(\eta_R a) &= \eta_L a \end{aligned}\right\} \quad (a \in \pi_*(MU))$$

and

$$c(b_i) = m_i$$

where b_i and m_i are related as in (7.5).

(iv) The coproduct map ψ is a map of bimodules over $\pi_*(MU)$.

It is given by

$$\psi b_k = \sum_{i+j=k} \left(\sum_{h \geq 0} b_h \right)_i^{j+1} \otimes b_j .$$

(Compare [2, p. 20, Theorem 6.3].)

Proof. We begin with part (ii). The formal properties of ε are given in [3]. Instead of saying that ε is induced by $\mu: MU \wedge MU \longrightarrow MU$, we may proceed as follows. Let $x \in MU_*(MU)$, let $1 \in MU^0(MU)$ be the class of the identity map $1: MU \longrightarrow MU$, and let $<1, x> \in \pi_*(MU)$ be their Kronecker product; then

$$\varepsilon(x) = <1, x>.$$

Applying the naturality of the Kronecker product to the map $MU(1) \longrightarrow MU$, we find that

$$\begin{aligned} <1, b_i> &= <x^{MU}, \beta_{i+1}> \\ &= 0 \text{ for } i > 0. \end{aligned}$$

We turn to part (iii) of (11.3). The formal properties of c are given in [3]. By (9.5) we have

$$x^R = \sum_{i \geq 0} b_i^{MU} (x^L)^{i+1} .$$

Applying c, we find

$$x^L = \sum_{i \geq 0} (c b_i^{MU})(x^R)^{i+1} .$$

So $c b_i^{MU} = m_i^{MU}$.

We turn to part (iv) of (11.3). The formal properties of ψ are given in [3]. We begin work by determining the coproduct map

$$\psi: MU_*(CP^\infty) \longrightarrow MU_*(MU) \otimes_{\pi_*(MU)} MU_*(CP^\infty).$$

By definition, this coproduct map is the following composite:

$$\begin{array}{ccc} MU_*(CP^\infty) & \longrightarrow & (MU \wedge MU)_*(CP^\infty) \\ & & \uparrow \cong \\ & & MU_*(MU) \otimes_{\pi_*(MU)} MU_*(CP^\infty). \end{array}$$

Here the first factor can be described by adopting the notation of the proof of (9.4); it maps $\beta_i \in MU_{2i}(CP^\infty)$ into $\beta_i^L \in (MU \wedge MU)_{2i}(CP^\infty)$. The isomorphism maps the element $1 \otimes \beta_i$ in the tensor-product into $\beta_i^R \in (MU \wedge MU)_{2i}(CP^\infty)$. By (9.5) we have

$$x^R = \sum_{i \geq 0} b_i^{MU} (x^L)^{i+1}$$

and therefore

$$(x^R)^j = \sum_k \left(\sum_{i \geq 0} b_i^{MU} \right)_k^j (x^L)^{j+k} .$$

Dualizing, we find

$$\beta_i^L = \sum_{0 \leq j \leq i} \left(\sum_{k \geq 0} b_k^{MU} \right)_{i-j}^j \otimes \beta_j^R ;$$

that is,

$$\psi\beta_i = \sum_{0 \leq j \leq i} \left(\sum_k b_k^{MU} \right)_{i-j}^j \otimes \beta_j . \tag{11.4}$$

(Note that this formula determines the coaction map

$$\psi : MU_*(BU) \longrightarrow MU_*(MU) \otimes_{\pi_*(MU)} MU_*(BU)$$

for the space BU.) Transferring (11.4) to MU by the "inclusion" $CP^\infty = MU(1) \longrightarrow MU$, we find

$$\psi b_{i-1} = \sum_{0 \leq j-1 \leq i-1} \left(\sum_{\ell \geq 0} b_\ell \right)_{i-j}^j \otimes b_{j-1} ,$$

which is equivalent to the result given. This completes the proof of (11.3).

<u>Note.</u> Consider the subalgebra

$$S_* = Z[b_1, b_2, \ldots, b_n, \ldots]$$

(compare [2, p. 20, Theorem 6.3).) The product map φ, diagonal map ψ and conjugation c all carry this subalgebra to itself; the counit restricts to give a map

$$\varepsilon : S_* \longrightarrow Z$$

such that $\varepsilon(1) = 1$, $\varepsilon(b_i) = 0$ for $i \geq 1$. We conclude that the restriction of c to this subalgebra must coincide with the conjugation it would have if considered in its own right as a Hopf algebra over Z.

12. BEHAVIOUR OF THE BOTT MAP

We recall that in the spectrum K, every even term is the space BU, and the maps between them are all the same; each is the map

$$B : S^2 \wedge BU \longrightarrow BU$$

adjoint to the Bott equivalence

$$B' : BU \simeq \Omega_0^2 BU.$$

(Here Ω_0^2 means the component of the base-point in the double loop-space Ω^2.)

In order to compute $E_*(BU)$, it is therefore desirable to compute

$$B_* : \tilde{E}_n(BU) \longrightarrow \tilde{E}_{n+2}(BU).$$

This will be done in (12.5), (12.6).

We first describe the primitive elements in $E_*(BU)$.

We have seen that

$$E_*(BU) = \pi_*(E)[\beta_1, \beta_2, \ldots, \beta_n, \ldots]$$

with coproduct

$$\psi \beta_k = \sum_{i+j=k} \beta_i \otimes \beta_j .$$

As usual, we define the Newton polynomial Q_n^k so that

$$x_1^k + x_2^k + \ldots + x_n^k = Q_n^k(\sigma_1, \sigma_2, \ldots, \sigma_k),$$

where σ_i is the i-th elementary symmetric function of $x_1, x_2, \ldots, x_n$. Q_n^k is independent of n for $n \geq k$, and then we write Q^k for Q_n^k.

We define elements $s_k \in E_*(BU)$ for $k \geq 1$ by

$$s_k = Q^k(\beta_1, \beta_2, \ldots, \beta_k).$$

Examples.

$$s_1 = \beta_1$$

$$s_2 = \beta_1^2 - 2\beta_2$$

$$s_3 = \beta_1^3 - 3\beta_1\beta_2 + 3\beta_3 .$$

PROPOSITION 12.1. The primitive elements in $E_*(BU)$ form a free module over $\pi_*(E)$, with a base consisting of the elements $s_1, s_2, s_3, \ldots$.

The proof goes precisely as in ordinary homology.

We need two formulas about the s_i.

(12.2) $$s_n - \beta_1 s_{n-1} + \beta_2 s_{n-2} + \ldots + (-1)^{n-1}\beta_{n-1}s_1 + (-1)^n n\beta_n = 0.$$

This is well-known.

(12.3) $$\left(\sum_{n=1}^{\infty} (-1)^{n-1} s_n\right) = \left(\sum_{s=1}^{\infty} s\beta_s\right)\left(\sum_{t=0}^{\infty} \beta_t\right)^{-1}.$$

Proof. Write (12.2) in the form

$$(-1)^{n-1}s_n + b_1(-1)^{n-2}s_{n-1} + \ldots + b_{n-1}s_1 = nb_n$$

and add over $n \geq 1$; we find

$$\left(\sum_{n=1}^{\infty} (-1)^{n-1} s_n\right)\left(\sum_{t=0}^{\infty} \beta_t\right) = \left(\sum_{s=1}^{\infty} s\beta_s\right).$$

This yields (12.3).

We next consider the tensor product map. We recall that the map

$$BU(n) \times BU(m) \longrightarrow BU(nm)$$

which classifies the ordinary tensor product of bundles does not behave well under the inclusion of $BU(n)$ in $BU(n+1)$; it is necessary to consider the product on reduced K-theory defined by the "tensor product of virtual bundles of virtual dimension zero"; this is represented by a map

$$t: BU \wedge BU \longrightarrow BU.$$

We calculate

$$t_*: \tilde{E}_*(BU) \otimes \tilde{E}_*(BU) \longrightarrow \tilde{E}_*(BU)$$

at least on the elements $\beta_i \otimes \beta_j$.

PROPOSITION 12.4. If $i > 0$, $j > 0$ we have

$$t_*(\beta_i \otimes \beta_j) = \sum_{\substack{p \leq i \\ q \leq j \\ k \leq p+q}} a^k_{pq} \beta_k \left(\sum_{\ell=0}^{\infty} \beta_\ell \right)^{-1}_{i-p} \left(\sum_{\ell=0}^{\infty} \beta_\ell \right)^{-1}_{j-q}$$

Proof. The restriction of t to $BU(1) \wedge BU(1)$ corresponds to the element

$$(\xi_1 - 1)(\xi_2 - 1) = \xi_1\xi_2 - \xi_1 - \xi_2 + 1$$

in $BU^0(BU(1) \times BU(1))$. We therefore introduce the following maps.

$BU(1) \times BU(1) \xrightarrow{m} BU(1) \longrightarrow BU$, corresponding to $\xi_1\xi_2$

$BU(1) \times BU(1) \xrightarrow{\pi_1} BU(1) \longrightarrow BU$, corresponding to ξ_1

$BU(1) \times BU(1) \xrightarrow{\pi_2} BU(1) \longrightarrow BU$, corresponding to ξ_2

$BU(1) \times BU(1) \xrightarrow{c} BU(1) \longrightarrow BU$, corresponding to 1.

Here π_1 is projection onto the first factor, π_2 is projection onto the second factor, and c is the constant map. The required element of $BU^0(BU(1) \times BU(1))$ can therefore be represented in the following form.

$$\begin{array}{ccc} (BU(1) \times BU(1)) & \xrightarrow{\Delta} & (BU(1) \times BU(1))^4 \\ & & \downarrow f \\ & & BU^4 \\ & & \downarrow g \\ & & BU^4 \xrightarrow{\mu} BU . \end{array}$$

Here Δ is the iterated diagonal map; f is the map whose four components are the four maps given above; g is a map whose four components represent 1, -1, -1 and 1; and μ is the iterated product map.

We have

$$\Delta_*(\beta_i \otimes \beta_j) = \sum_{\substack{i_1+i_2+i_3+i_4 = i \\ j_1+j_2+j_3+j_4 = j}} \beta_{i_1} \otimes \beta_{j_1} \otimes \beta_{i_2} \otimes \beta_{j_2} \otimes \beta_{i_3} \otimes \beta_{j_3} \otimes \beta_{i_4} \otimes \beta_{j_4}$$

$$m_*(\beta_{i_1} \otimes \beta_{j_1}) = \sum_{k \leq i_1+j_1} a^k_{i_1 j_1} \beta_k$$

$$(\pi_1)_*(\beta_{i_2} \otimes \beta_{j_2}) = \begin{cases} \beta_{i_2} & (j_2 = 0) \\ 0 & (j_2 > 0) \end{cases}$$

$$(\pi_2)_*(\beta_{i_3} \otimes \beta_{j_3}) = \begin{cases} \beta_{j_3} & (i_3 = 0) \\ 0 & (i_3 > 0) \end{cases}$$

$$c_*(\beta_{i_4} \otimes \beta_{j_4}) = \begin{cases} 1 & (i_4 = j_4 = 0) \\ 0 & (\text{otherwise}). \end{cases}$$

and

$$(-1)_*\left(\sum_{\ell=0}^{\infty} \beta_\ell\right) = \left(\sum_{\ell=0}^{\infty} \beta_\ell\right)^{-1}.$$

So we obtain

$$t_*(\beta_i \otimes \beta_j) = \sum_{\substack{i_1+i_2 = i \\ j_1+j_3 = j \\ k \leq i_1+j_1}} a^k_{i_1 j_1} b_k \left(\sum_{\ell=0}^{\infty} \beta_\ell\right)^{-1}_{i_2} \left(\sum_{\ell=0}^{\infty} \beta_\ell\right)^{-1}_{j_3}.$$

This proves (12.4).

PROPOSITION 12.5. The map

$$B_*: \tilde{E}_n(BU) \longrightarrow \tilde{E}_{n+2}(BU)$$

annihilates decomposable elements.

Proof. We have the following commutative diagram.

$$\begin{array}{ccc} \tilde{E}_n(BU) & \cong & \tilde{E}_{n+2}(S^2 \wedge BU) \\ \downarrow B'_* & & \downarrow B_* \\ \tilde{E}_n(\Omega^2_0 BU) & \xrightarrow{\sigma^2} & \tilde{E}_{n+2}(BU) \end{array}$$

Here the bottom horizontal map σ^2 is the appropriate double suspension;

and it is well-known that it annihilates products, providing the products in $\tilde{E}_n(\Omega_o^2 BU)$ are those induced by the loop-space product; the proof for ordinary homology goes over. But BU is an H-space, so the loop-space product μ_Ω on $\Omega_o^2(BU)$ is homotopic to the product μ_H induced from the H-space product in BU. Now the periodicity isomorphism

$$\widetilde{BU}^o(X) \cong \widetilde{BU}^o(S^2 \wedge X)$$

is an isomorphism of additive groups; this says that under $B':BU \longrightarrow \Omega_o^2 BU$ the H-space product in BU corresponds to the product μ_H in $\Omega_o^2 BU$. So $\sigma^2 \beta_*^!$ annihilates elements which are decomposable in the usual sense.

PROPOSITION 12.6. If $j > 0$ we have

$$B_*(\beta_j) = \sum_{\substack{r+t=j+1 \\ t>0}} u^E a_{1r} (-1)^{t-1} s_t$$

$$\equiv \sum_{\substack{r+t=j+1 \\ t>0}} u^E a_{1r} t\beta_t \quad \text{mod decomposables.}$$

Proof. The second line follows from the first by (12.2), so we need only prove the first.

Recall that β_1 is not the canonical generator in $\tilde{E}_*(S^2)$; the latter is given by $u^E \beta_1 \in \tilde{E}_2(S^2)$. Since the Bott map B is the restriction of t to $S^2 \wedge BU$, we have

$$B_*(\beta_j) = t_*(u^E \beta_1 \otimes \beta_j) \quad \text{for } j > 0.$$

We apply (12.4), and find that the sum in (12.4) can be divided into two parts, one with $p = 1$ and one with $p = 0$. In the latter, we use the fact that

$$a_{0q}^k = \begin{cases} 1 & \text{if } k = q \\ 0 & \text{if } k \neq q \end{cases}.$$

We find

$$B_*(\beta_j) = u^E \sum_{\substack{q+s=j \\ k}} a^k_{1q}\, \beta_k \Big(\sum_{\ell=0}^{\infty} \beta_\ell \Big)^{-1}_s$$

$$+ u^E \sum_{q+s=j} \beta_q(-\beta_1)\Big(\sum_{\ell=0}^{\infty} \beta_\ell \Big)^{-1}_s .$$

The second sum is zero unless $j = 0$, so we can forget it. In the first sum, we have

$$a^k_{1q} = k a_{1\, q+1-k}$$

by (3.6). Writing r for $q+1-k$, we find

$$B_*(\beta_j) = u^E \sum_{r+s+k=j+1} a_{1r}(k\beta_k)\Big(\sum_{\ell=0}^{\infty} \beta_\ell \Big)^{-1}_s .$$

Using (12.3), we find

$$B_*(\beta_j) = u^E \sum_{\substack{r+t=j+1 \\ t>0}} a_{1r}(-1)^{t-1} s_t .$$

This proves (12.6).

13. $K_*(K)$

In this section we compute the Hopf algebra $K_*(K)$. The results represent joint work with Mr. A.S. Harris.

We recall from [4] that $\pi_*(K)$ is the ring of finite Laurent series $Z[u, u^{-1}]$, where $u \in \pi_2(K)$ is the element introduced in §2. By (4.1), $K_*(BU)$ is torsion-free. Passing to the limit along the BU-spectrum K, we see that $K_*(K)$ is torsion-free. Therefore the map

$$K_*(K) \longrightarrow K_*(K) \otimes Q$$

is a monomorphism. But $K_*(K) \otimes Q$ is the ring of finite Laurent series $Q[u, u^{-1}, v, v^{-1}]$, where we have written u for $\eta_L u$, v for $\eta_R u$. We propose to describe $K_*(K)$ as a subring of $Q[u, u^{-1}, v, v^{-1}]$. It is sufficient to describe $K_0(K)$ as a subring of $K_0(K) \otimes Q = Q[u^{-1}v, uv^{-1}]$, but we will work in full generality.

We first observe that the operation ψ^k was originally introduced as an unstable operation; to make it a stable operation we need to introduce coefficients $Z[\frac{1}{k}]$. (Here $Z[\frac{1}{k}]$ is the ring of rational numbers of the form n/k^m.) Crudely speaking, we cannot define a map of spectra $K \longrightarrow K$ by taking each component map to be $\psi^k : BU \longrightarrow BU$, because the following diagram does not commute.

$$\begin{array}{ccc} S^2 \wedge BU & \xrightarrow{B} & BU \\ {\scriptstyle 1\wedge\psi^k}\downarrow & & \downarrow{\scriptstyle \psi^k} \\ S^2 \wedge BU & \xrightarrow{B} & BU \end{array}$$

We have to take the (2n)-th component of our map to be

$$\frac{1}{k^n}\psi^k : BU \longrightarrow BUZ[\tfrac{1}{k}] .$$

Here the space $BUZ[\frac{1}{k}]$ is constructed by taking the spectrum $KZ[\frac{1}{k}]$ representing K-theory with coefficients in $Z[\frac{1}{k}]$ (see §2), converting it into an Ω-spectrum, and taking the (2n)-th space of this Ω-spectrum.

For any element $h \in K_*(K)$ we can form

$$<\psi^k, h> \in \pi_*(K) \otimes Z[\tfrac{1}{k}] \qquad (k \neq 0).$$

But if we identify h with a finite Laurent series $f(u, v)$, as above, then we have

$$(13.1) \qquad <\psi^k, h> = f(u, ku).$$

COROLLARY 13.2. A necessary condition that a finite Laurent series $f(u, v)$ lie in the image of $K_*(K)$ is

$$(13.3) \qquad f(u, ku) \in \pi_*(K) \otimes Z[\tfrac{1}{k}] \quad \text{for } k > 0.$$

THEOREM 13.4. (i) $K_*(K)$ may be identified with the set of finite Laurent series $f(u, v)$ which satisfy (13.3).

(ii) The product in $K_*(K)$ is the product of Laurent series.

(iii) The unit maps are given by

$$\eta_L(u) = u$$

$$\eta_R(u) = v.$$

(iv) The counit map is given by

$$\varepsilon(u) = u$$

$$\varepsilon(v) = u$$

$$\varepsilon(u^{-1}v) = 1$$

$$\varepsilon(uv^{-1}) = 1.$$

(v) The conjugation map is given by

$$c(u) = v$$

$$c(v) = u$$

$$c(u^{-1}v) = uv^{-1}$$

$$c(uv^{-1}) = u^{-1}v .$$

(vi) The coproduct map is given by

$$\psi(u) = u \otimes 1$$

$$\psi(v) = 1 \otimes v$$

$$\psi(u^{-1}v) = u^{-1}v \otimes u^{-1}v$$

$$\psi(uv^{-1}) = uv^{-1} \otimes uv^{-1} .$$

The proof of (13.4) will be built up in stages.

LEMMA 13.5. The Bott map

$$B_*: \tilde{K}_n(BU) \longrightarrow \tilde{K}_{n+2}(BU)$$

annihilates decomposables, and is given by

$$B_*\beta_j = u((j+1)\beta_{j+1} + j\beta_j) \text{ mod decomposables.}$$

Proof. Immediate from (12.5) and (12.6); the values of the coefficients a_{1r} come from (2.9).

We observe that the generator in $\pi_{2n}(BU)$ gives an element in

$K_{2n}(BU)$; we write the latter element w^n (noting that the multiplication involved is in the sense of the tensor-product map $t: BU \wedge BU \longrightarrow BU$, and is not to be confused with our usual multiplication, which comes from the Whitney sum map $BU \times BU \longrightarrow BU$.) If we regard BU as the 2m-th term of the spectrum K, then the image of w^n in $K_{2(n-m)}(BU)$ is v^{n-m} (assuming $n \geq 1$).

LEMMA 13.6. In $K_{2n}(BU) \otimes Q$ we have

$$\beta_n = \frac{u^{-1}w(u^{-1}w-1)\dots(u^{-1}w-n+1)}{1\cdot 2\cdot \dots \cdot n}$$

modulo decomposables in the sense of Whitney sum, where the product is taken in the sense of the tensor-product.

<u>Proof.</u> By induction over n; for $n = 1$ we have $\beta_1 = u^{-1}w$. Suppose the result true for n. Since $B_* w^r = w^{r+1}$, we have

$$B_*(\beta_n) = \frac{u^{-1}w(u^{-1}w-1)\dots(u^{-1}w-n+1)w}{1\cdot 2\cdot \dots \cdot n}$$

By (13.5) we have

$$\begin{aligned}\beta_{n+1} &= \frac{1}{n+1}(u^{-1}B_*\beta_n - n\beta_n) \quad \text{mod decomposables}\\ &= \frac{u^{-1}w(u^{-1}w-1)\dots(u^{-1}w-n+1)(u^{-1}w-n)}{1\cdot 2\cdot \dots \cdot n\cdot(n+1)}\\ &\qquad\qquad \text{mod decomposables.}\end{aligned}$$

This completes the induction and proves (13.6).

LEMMA 13.7. The image of $K_*(K)$ in $K_*(K) \otimes Q$ is generated over $Z[u, u^{-1}, v, v^{-1}]$ by the elements

$$\frac{u^{-1}v(u^{-1}v-1)\dots(u^{-1}v-n+1)}{1\cdot 2\cdot \dots \cdot n} \qquad (n = 1, 2, 3, \dots).$$

<u>Proof.</u> Immediate, since it is generated over $Z[u, u^{-1}]$ by the images of the elements β_n in the 2m-th term of the spectrum K $(n = 1, 2, 3, \dots;\ m = 0, 1, 2, \dots)$.

LEMMA 13.8. A polynomial $f(x) \in Q[x]$ can be written as an

integral linear combination of the binomial polynomials

$$\frac{x(x-1)\ \dots\ (x-n+1)}{1\cdot 2\cdot\ \dots\ \cdot n} \quad (n = 0, 1, 2, \dots)$$

if and only if it takes integer values for $x = 1, 2, 3, \dots$.

The proof is a piece of standard algebra, which can be left to the reader.

Proof of Theorem 13.4. The substantial part is part (i). First, take an element of $K_*(K)$; its image in $K_*(K) \otimes Q$ is a finite Laurent series of the type described in (13.7), and $f(u, ku) \in Z[u, u^{-1}, 1/k]$ by (13.8).

Conversely, take a finite Laurent series $f(u, v)$ which satisfies (13.3); without loss of generality we may suppose that f is homogeneous, say $f(u, v) = u^d g(u^{-1}v)$, where $g(k) \in Z[\frac{1}{k}]$ for $k = 1, 2, 3, \dots$. The power to which z^{-1} occurs in $g(z)$ is bounded, say by N. Also $g(z)$ contains only a finite number of coefficients in Q; their denominators contain only a finite number of prime factors p, and each prime p occurs to a power which is bounded, say by M (independent of p). Then

$$h(z) = z^{N+M} g(z)$$

has the property that $h(k) \in Z$ for $k = 1, 2, 3, \dots$. In fact, each prime p dividing k cannot occur in the denominator of $h(k)$, by construction; nor can any other prime, by assumption. By Lemma 13.8, $h(u^{-1}v)$ is an integral linear combination of binomial polynomials

$$\frac{u^{-1}v(u^{-1}v-1)\ \dots\ (u^{-1}v-n+1)}{1\cdot 2\cdot\ \dots\ \cdot n} \quad (n \geq 0).$$

So $f(u, v) = u^d (uv^{-1})^{N+M} h(u^{-1}v)$ is a linear combination over $Z[u, u^{-1}, v^{-1}]$ of these polynomials. We do not need the polynomial for $n = 0$ (namely 1) since it is a multiple over $Z[u, v^{-1}]$ of the polynomial for $n = 1$

(namely $u^{-1}v$). By Lemma 13.7, $f(u, v)$ lies in the image of $K_*(K)$. This proves (13.4)(i).

The remaining parts of (13.4) are easy. It is only necessary to comment on one point. In (vi), the fact that ψ is a map of bimodules gives

$$\psi(u^{-1}v) = u^{-1} \otimes v \; ;$$

but in $K_*(K) \otimes_{\pi_*(K)} K_*(K)$ we have

$$u^{-1} \otimes v = u^{-1}v \otimes u^{-1}v \; ,$$

since the tensor product is taken over $\pi_*(K)$ and $v = \eta_R u$. Similarly for $\psi(uv^{-1})$.

14. THE HATTORI-STONG THEOREM

In this section I will present a slight reformulation of the result of Hattori and Stong. (Stong proved it first, but his name creeps to the back for reasons of euphony--it brings a phrase or sentence to such a resounding end.) This reformulation has been used by L. Smith [15].

Recall from [3, Lecture 3] that for suitable spectra E, such as $E = K$, $E_*(X)$ is a comodule over the Hopf algebra $E_*(E)$. We say that an element x in a comodule is <u>primitive</u> if $\psi x = 1 \otimes x$; we write $PE_*(X)$ for the subgroup of primitive elements in $E_*(X)$. One can see directly from the definition of ψ that the Hurewicz homomorphism in E-homology,

$$h \colon \pi_*(X) \longrightarrow E_*(X),$$

maps into $PE_*(X)$.

THEOREM 14.1 (after Stong [16] and Hattori [9]). The Hurewicz homomorphism in K-homology gives an isomorphism

$$h: \pi_*(MU) \cong PK_*(MU).$$

Remark. As soon as one knows that $\pi_*(MU)$ is torsion-free, it is easy to show that this Hurewicz homomorphism is a monomorphism. For example, consider the following commutative diagram.

$$\begin{array}{ccc} \pi_*(MU) & \xrightarrow{h} & K_*(MU) \\ \downarrow & & \downarrow \\ \pi_*(MU) \otimes Q & \xrightarrow{h \otimes 1} & K_*(MU) \otimes Q \end{array}$$

We have $K_*(MU) \otimes Q \cong \pi_*(K) \otimes \pi_*(MU) \otimes Q$; so the bottom horizontal map and the left-hand vertical map are both monomorphisms.

The essential content of the theorem, then, is that it identifies the images of h.

Proof of (14.1). For lack of time in writing out these notes to work out a direct proof, I will deduce this result from the formulation given by Hattori. (After all, Hattori's proof is very elegant.) Hattori proves precisely that if $x \in K_*(MU)$ and $nx \in \text{Im } h$ for some integer $n \neq 0$, then $x \in \text{Im } h$. It is rather easy to see that any primitive in $K_*(MU) \otimes Q$ lies in the image of $h \otimes 1$. So suppose $x \in PK_*(MU)$; then by preceding sentence, x lies in $\text{Im}(h \otimes 1)$; that is, for some integer $n \neq 0$ we have $nx \in \text{Im } h$. So by Hattori's form of the result, $x \in \text{Im } h$. This proves (14.1).

Exercise. Deduce Hattori's form of the result from (14.1).

15. QUILLEN'S IDEMPOTENT COHOMOLOGY OPERATIONS

Suppose given a spectrum E and an abelian group G. It may happen that when we form the spectrum EG, as in §2, it splits as a sum or product. Examples are given in [3, Lecture 4]. In such cases, it is highly desirable to have a splitting which is canonical and doesn't depend on any choices. I have developed this point in [3, Lecture 4]. In particular, I have made the rather obvious point that one should look for canonical idempotent cohomology operations, that is, idempotent maps $\varepsilon : EG \longrightarrow EG$.

An important special case is that in which $E = MU$ and $G = Q_p$, the integers localized at p (that is, the ring of rational numbers n/m with m prime to p.) In this case the possibility of splitting MUQ_p was proved by Brown and Peterson [7], and again by Novikov [13]; but both methods involved choice.

Quillen has succeeded in giving canonical idempotents $\varepsilon : MUQ_p \longrightarrow MUQ_p$ (one for each p). This is profitable in two ways. Firstly, it means that we no longer have to construct the Brown-Peterson spectrum by synthesis, building it up from its homotopy groups and k-invariants; we can construct it by taking MUQ_p and splitting off the piece we want. Secondly, we obtain a very precise hold on the Brown-Peterson spectrum, and can obtain information about it by passing to the quotient from MUQ_p. This process yields good, explicit formulae.

THEOREM 15.1. Let $d > 1$ be an integer, and let $R \subset Q$ be a subring of the rationals containing d^{-1}. Then there is a unique map of ring-spectra

$$e = e_d : MUR \longrightarrow MUR$$

satisfying the following conditions.

(i) e is idempotent: $e^2 = e$.

(ii) e has the following effect on $\pi_*(MUR)$.

$$e[CP^n] = \begin{cases} 0 & \text{if } n \equiv -1 \bmod d \\ [CP^n] & \text{if } n \not\equiv -1 \bmod d. \end{cases}$$

Two such idempotents e_d, e_δ commute.

THEOREM 15.2. (D. Quillen [14]) Let p be a prime. Then there is a unique map of ring-spectra

$$\mathcal{E} = \mathcal{E}_p : MUQ_p \longrightarrow MUQ_p$$

satisfying the following conditions.

(i) $\mathcal{E}$ is idempotent: $\mathcal{E}^2 = \mathcal{E}$.

(ii) $\mathcal{E}$ has the following effect on $\pi_*(MUQ_p)$.

$$\mathcal{E}[CP^n] = \begin{cases} [CP^n] & \text{if } n = p^f - 1 \text{ for some } f \\ 0 & \text{otherwise.} \end{cases}$$

Proof of (15.2) from (15.1). Take

$$\mathcal{E} = \prod_q e_q$$

where the product ranges over all primes $q \neq p$, observing that the product is convergent in the filtration topology on $MUQ_p^*(MUQ_p)$, which is complete and Hausdorff.

We turn to consider the proof of (15.1). We know from Lemma 4.5 that so long as $\pi_*(E) \longrightarrow \pi_*(E) \otimes R$ is iso (which is certainly true for $E = MUR$), maps of ring-spectra $g: MUR \longrightarrow E$ are in (1-1) correspondence with power-series

$$g_*(x^{MU}) = f(x^E) = \sum_{i \geq 0} d_i (x^E)^{i+1}$$

with $u^E d_0 = 1$, $d_i \in \pi_*(E)$. Assume for simplicity that $u^E = 1$, which is the case in the applications. All we have to do is pick the right power-series. Let us consider how the choice of f will affect

$$g_*: \pi_*(MUR) \longrightarrow \pi_*(E).$$

Let us take the primitive elements

$$\log^{MU} x^{MU} = \sum_{i \geq 0} m_i (x^{MU})^{i+1}, \quad m_i = \frac{[CP^i]}{i+1} \in \pi_*(MU) \otimes Q$$

$$\log^E x^E = \sum_{i \geq 0} n_i (x^E)^{i+1}, \quad \text{say,} \quad n_i \in \pi_*(E) \otimes Q, \quad n_0 = 1.$$

Let us define the modified $\log^{MU}$ series by

$$\text{mog}\, x^{MU} = \sum_{i \geq 0} (g_* m_i)(x^{MU})^{i+1},$$

so that it serves to store the coefficients $g_* m_i$. Let $\exp^E$ be the series inverse to $\log^E$.

PROPOSITION 15.3. The elements $g_* m_i \in \pi_*(E) \otimes Q$ are given by

$$\text{mog}(fx^E) = \log^E x^E$$

or equivalently

$$\text{mog}\, z = \log^E (f^{-1} z).$$

For our applications we need to know how to construct f given the coefficients $g_* m_i$, and the appropriate formula is as follows.

COROLLARY 15.4. $f^{-1} z = \exp^E \text{mog}\, z.$

Proof of (15.3), (15.4). The element

$$\log^{MU} x^{MU} = \sum_{i \geq 0} m_i (x^{MU})^{i+1}$$

is primitive. Therefore

$$g_* \log^{MU} x^{MU} = \sum_{i \geq 0} (g_* m_i)(fx^E)^{i+1} = \text{mog}(fx^E)$$

is primitive. But the primitive elements in $\widetilde{EQ}^*(CP^\infty)$ form a free module over $\pi_*(EQ)$, with one generator $\log^E x^E$; and we check that $\text{mog}(fx^E)$ has first term x^E; so

$$\text{mog}(fx^E) = \log^E x^E.$$

This proves (15.3) and (15.4).

Next suppose given a formal product μ, over a ring R, and consider formal power-series, with zero constant term, over R. We can make these formal power-series into an abelian group by defining

$$\sigma +_\mu \tau = \mu(\sigma, \tau).$$

Subtraction in this abelian group will be written $-_\mu$. If our ring R also contains d^{-1}, we can divide by d in this abelian group; we write

$$\sigma = (\frac{1}{d})_\mu \tau$$

for the solution of

$$\tau = \sigma +_\mu \sigma +_\mu \cdots +_\mu \sigma \qquad \text{(d summands)}.$$

If our ring R contains Q, we can write

$$\sigma +_\mu \tau = \exp(\log \sigma + \log \tau).$$

where exp and log are as in §7.

<u>Proof of (15.1).</u> Our proposal is to take

$$\text{(15.5)} \qquad \operatorname{mog} z = \log z - \frac{1}{d}(\log \zeta_1 z + \log \zeta_2 z + \cdots + \log \zeta_d z).$$

Here $\zeta_1, \zeta_2, \ldots, \zeta_d$ are the complex d-th roots of 1, and

$$\log z = \sum_{i \geq 0} m_i z^{i+1} \quad , \quad m_i = \frac{[CP^i]}{i+1}$$

as it should be for MU or MUR. It is easy to see that this power-series (15.5) has the coefficients $g_*(m_i)$ given in (15.1)(ii). A priori the coefficients of mog z lie in $\pi_*(MU) \otimes Q[\exp 2\pi i/d]$.

Applying exp to (15.5), we get

$$\text{(15.6)} \qquad f^{-1}z = z -_\mu (\frac{1}{d})_\mu (\zeta_1 z +_\mu \zeta_2 z +_\mu \cdots +_\mu \zeta_d z).$$

For any $\zeta_1, \zeta_2, \ldots, \zeta_d$ we can consider

$$\zeta_1 z +_\mu \zeta_2 z +_\mu \cdots +_\mu \zeta_d z$$

as a formal power-series with coefficients in

$$\pi_*(MU) \times Z[\zeta_1, \zeta_2, \ldots, \zeta_d].$$

The coefficients are clearly polynomials symmetric in $\zeta_1, \zeta_2, \ldots, \zeta_d$, so we can write them in terms of the elementary symmetric functions $\sigma_1, \sigma_2, \ldots, \sigma_d$. When we substitute for $\zeta_1, \ldots, \zeta_d$ the complex d-th roots of 1, we have

$$\sigma_1 = 0, \ \ldots, \ \sigma_{d-1} = 0, \quad \sigma_d = (-1)^{d-1}.$$

We obtain a power-series with coefficients in $\pi_*(MU)$.

So (15.6) shows that $f^{-1}z$, and hence fz, has coefficients in $\pi_*(MU) \otimes Z[\frac{1}{d}]$. This proves the existence of a map $e: MUR \longrightarrow MUR$ of ring-spectra satisfying (15.1)(ii).

The fact that e_d is idempotent follows from the fact that its effect on $\pi_*(MUR)$ is obviously idempotent, by Lemma 9.3. The fact that two such idempotents e_d, d_δ commute is proved in the same way.

16. THE BROWN-PETERSON SPECTRUM

In this section we introduce the Brown-Peterson spectrum, and discuss its properties. In particular, we prove the homology analogue of Quillen's result on the algebra $BP^*(BP)$ of cohomology operations.

We keep a prime p fixed throughout. For any X, consider

$$\varepsilon_*: MUQ_p^*(X) \longrightarrow MUQ_p^*(X),$$

where $\varepsilon = \varepsilon_p$ is as in §15. The image of ε_* is a natural direct summand of $MUQ_p^*(X)$, so it is a functor turning cofibrations into exact sequences. It also satisfies the wedge axiom, so (by Brown's theorem in the category of spectra) it is a representable functor. We write BP for its representing spectrum, after Brown and Peterson [7]. The map ε is a map of ring-spectra, so the image of ε_* is a cohomology functor with (external) products. Therefore BP is a ring-spectrum. We have canonical maps of ring-spectra which make up the following commutative diagram.

$$\begin{array}{ccc} MUQ_p & \xrightarrow{\varepsilon} & MUQ_p \\ & {\scriptstyle\pi}\searrow \quad \nearrow{\scriptstyle\iota} & \\ & BP & \end{array}$$

We have $\pi\iota = 1 : BP \longrightarrow BP$.

If we were to follow Quillen's line [14], we would now copy the work of §15, taking $E = BP$, to construct a whole family of cohomology operations from MUQ_p to BP, and prove that they factor through the canonical projections $\pi : MUQ_p \longrightarrow BP$.

To construct the different operations of the family, Quillen introduces into his work formal variables $t_1, t_2, \ldots, t_n, \ldots$ and constructs an operation

$$r_t : MUQ_p \longrightarrow BP(Z[t_1, t_2, \ldots, t_n, \ldots]).$$

He then takes the components of this operation; for any sequence $\alpha = (\alpha_1, \alpha_2, \ldots, \alpha_n, \ldots)$ such that $\alpha_i = 0$ for all but a finite number of i, he takes the operation r_α to be the coefficient of $t_1^{\alpha_1} t_2^{\alpha_2} \ldots t_n^{\alpha_n}$ in r_t.

It would not really give us any trouble to afflict BP with coefficients $Z[t_1, t_2, \ldots, t_n, \ldots]$; we could construct a Moore space M for the graded ring $Z[t_1, t_2, \ldots, t_n, \ldots]$ by taking a wedge of spheres of suitable dimensions, and giving it a suitable product; and then we could form $BP \wedge M$. But since we are only trying to explain the direction of Quillen's work, we won't labor these details.

We give BP a class x^{BP} by using the canonical maps $MU \longrightarrow MUQ_p \xrightarrow{\pi} BP$. The log function for BP is obtained by naturality from that for MU. Let us recall that

$$m_i = \frac{[CP^i]}{i+1} \in \pi_*(MU) \otimes Q,$$

and that $\pi : MUQ_p \longrightarrow BP$ annihilates m_i unless $i = p^f - 1$. Let us write

$$m_{p-1}, m_{p^2-1}, m_{p^3-1}, \text{ etc.}$$

for the images of these surviving generators in $\pi_*(BP) \otimes Q$. Then we have

$$\log^{BP} x = x + m_{p-1} x^p + m_{p^2-1} x^{p^2} + m_{p^3-1} x^{p^3} \ldots$$

In our present language, Quillen's method is to construct r_t by taking its modified log series to be

$$\begin{aligned} \operatorname{mog} z = z &+ m_{p-1} z^p + m_{p^2-1} z^{p^2} + m_{p^3-1} z^{p^3} \\ &+ t_1 z^p + m_{p-1} t_1^p z^{p^2} + m_{p^2-1} t_1^{p^2} z^{p^3} \ldots \\ &+ t_2 z^{p^2} + m_{p-1} t_2^p z^{p^3} \ldots \\ &+ t_3 z^{p^3} \end{aligned}$$

(Note how one can read off the effect of r_t on $\pi_*(BP) \otimes Q$ from this display.) The reason that the coefficients in the display are introduced is that they represent the cheapest way to get the corresponding formal power-series defined over $\pi_*(BP)$; for we have

$$\begin{aligned} f^{-1} z &= \exp^{BP} \operatorname{mog} z \\ &= z +_\mu t_1 z^p +_\mu t_2 z^{p^2} +_\mu t_3 z^{p^3} +_\mu \cdots . \end{aligned}$$

Here μ means μ^{BP}, the formal product defined over $\pi_*(BP)$.

From our present point of view, however, Quillen's formal variables t_i are crying out to be located in $BP_*(BP)$. That is: for any element

$$u \in \operatorname{Hom}^*_Z(Z[t_1, t_2, \ldots], \pi_*(BP))$$

(say assigning the value u_α to $t_1^{\alpha_1} t_2^{\alpha_2} \ldots t_n^{\alpha_n}$) Quillen constructs a cohomology operation

$$\sum_\alpha u_\alpha r_\alpha$$

He then obtains each operation once and once only [14, Theorem 5(i)], so he is asserting $BP^*(BP) = \mathrm{Hom}_Z^*(Z[t_1, t_2, \ldots], \pi_*(BP))$

$$= \mathrm{Hom}^*_{\pi_*(BP)}(\pi_*(BP)[t_1, t_2, \ldots], \pi_*(BP)).$$

But we know we should have

$$BP^*(BP) = \mathrm{Hom}^*_{\pi_*(BP)}(BP_*(BP), \pi_*(BP)).$$

We therefore try to copy Quillen's work in homology.

THEOREM 16.1. (i) There is a unique system of classes

$$t_i \in BP_{2(p^i-1)}(BP)$$

such that $t_0 = 1$ and in $BPQ_*(BP)$ we have

$$\eta_R(m_{p^k-1}) = \sum_{i+j=k} m_{p^i-1}(t_j)^{p^i}.$$

(ii) We have

$$BP_*(BP) = \pi_*(BP)[t_1, t_2, \ldots].$$

(This describes the product map φ and the map η_L, or the structure as a left module over $\pi_*(BP)$; the map η_R, or the structure as a right module over $\pi_*(BP)$, is given by (i).)

(iii) The counit map is given by

$$\varepsilon(1) = 1$$

$$\varepsilon(t_i) = 0 \quad \text{for } i > 0.$$

(iv) The conjugation is given by the following inductive formula.

$$\sum_{h+i+j=k} m_{p^h-1}(t_i)^{p^h}(ct_j)^{p^{h+i}} = m_{p^k-1}.$$

(v) The coproduct is given by the following inductive formula.

$$\sum_{i+j=k} m_{p^i-1} (\psi t_j)^{p^i} = \sum_{h+i+j=k} m_{p^h-1} (t_i)^{p^h} \otimes (t_j)^{p^{h+i}}$$

The formula in part (i) restates that in Quillen's Theorem 5(iii) [14], and the formula in part (v) restates that in Quillen's Theorem 5 (iv) [14].

As for the formulae which are claimed as "inductive", we note that (iv) does indeed contain the leading term ct_k (take $h = 0$, $i = 0$) and otherwise contains terms in ct_j with $j < k$; and similarly, (v) contains the leading term ψt_k (take $i = 0$) and otherwise contains terms in ψt_j with $j < k$.

Proof of (16.1). We first prove the uniqueness clause of part (i). The formula

$$\eta_R(m_{p^k-1}) = \sum_{i+j=k} m_{p^i-1} (t_j)^{p^i}$$

contains the leading term t_k (take $i = 0$) and otherwise contains terms in t_j with $j < k$; so by induction, it determines the image of t_k in $BPQ_*(BP)$. But the map

$$BP_*(BP) \longrightarrow BPQ_*(BP)$$

is monomorphic, so the formula of part (i) characterises the t_k.

The essential part is the existence clause of part (i). We first recall the following equation from the proof of (9.4):

$$\sum_i \eta_R(m_i)(x^R)^{i+1} = \sum_i m_i \Big(\sum_{j \geq 0} M_j (x^R)^{j+1} \Big)^{i+1} . \tag{16.2}$$

Here $m_i = \dfrac{[CP^i]}{i+1} \in \pi_{2i}(MU) \otimes Q$,

$\eta_R(m_i) \in MU_{2i}(MU) \otimes Q$,

$M_j \in MU_{2j}(MU)$ is as in Proposition (9.4),

and the equation takes place in $(MU \wedge MUQ)_*(CP^\infty)$.

To this equation we apply the homomorphism induced by the map $\pi \wedge \pi : MU \wedge MU \longrightarrow BP \wedge BP$. If, for the moment, we write N_j for the image of M_j in $BP_{2i}(BP)$, we obtain the following equation in $(BP \wedge BPQ)_*(CP^\infty)$.

(16.3)
$$\sum_i \eta_R(m_{p^f-1})(x^R)^{p^f} = \sum_f (m_{p^f-1})(\sum_{j\geq 0} N_j(x^R)^{j+1})^{p^f}.$$

Use the equation of (16.1)(i), namely

$$\eta_R m_{p^k-1} = \sum_{i+j=k} m_{p^i-1}(t_j)^{p^i}$$

to define t_k (inductively) as an element of $BPQ_*(BP)$. Substituting in (16.3) we get

$$\sum_{i,j} m_{p^i-1}(t_j)^{p^i}(x^R)^{p^{i+j}} = \sum_f (m_{p^f-1})(\sum_{j\geq 0} N_j(x^R)^{j+1})^{p^f}.$$

That is,

$$\sum_i \log^{BP}(t_j(x^R)^{p^j}) = \log^{BP}(\sum_{j\geq 0} N_j(x^R)^{j+1}.$$

Apply $\exp^{BP}$. We get

(16.4)
$$x^R +_\mu t_1(x^R)^p +_\mu t_2(x^R)^{p^2} +_\mu t_3(x^R)^{p^3} \ldots$$
$$= \sum_{j\geq 0} N_j(x^R)^{j+1}.$$

Here μ means μ^{BP}, the formal product defined over $\pi_*(BP)$.

Suppose, as an inductive hypothesis, that we have shown $t_i \in BP_*(BP)$ for $i > k$. (The induction starts, since $t_o = 1$.) Extract from (16.4) the coefficient of $(x^R)^{p^k}$. We obtain

(16.5)
$$t_k + f(t_1, t_2, \ldots, t_{k-1}) = N_{p^k-1}.$$

Here N_{p^k-1} lies in $BP_*(BP)$; and $f(t_1, t_2, \ldots, t_{k-1})$ is a polynomial in $t_1, t_2, \ldots, t_{k-1}$, with coefficients in $\pi_*(BP)$, so it lies in $BP_*(BP)$ by the inductive hypothesis. Therefore t_k lies in $BP_*(BP)$. This completes the induction, and proves part (i).

We notice that (16.4) answers the obvious question: how do the homology generators in $MU_*(MU)$ map into $BP_*(BP)$? That is, the image N_j of M_j in $BP_*(BP)$ is the coefficient of $(x^R)^{j+1}$ in the left-hand side of (16.4), and this coefficient is a definite polynomial in $t_1, t_2, \ldots$.

We turn to part (ii). It is clear that $BP_*(BP)$ is the image under $(\pi \wedge \pi)_*$ of $MU_*(MU)$; so it is generated, over $\pi_*(BP)$, by the classes N_j. Using the last paragraph, this means that it is generated by the classes t_k. Similarly, $H_*(BP)$ is the image under π_* of $H_*(MU)$, and so it is

$$Q_p[m_{p-1}, m_{p^2-1}, m_{p^3-1}, \ldots].$$

Consider the spectral sequence

$$H_*(BP; \pi_*(BP)) \Longrightarrow BP_*(BP).$$

It is trivial, because it is a direct summand of the corresponding sequence for $MUQ_p^*(MUQ_p)$; and in the E_2-term, t_k is equal to m_{p^k-1} modulo decomposables, by (16.5). Therefore

$$BP_*(BP) = \pi_*(BP)[t_1, t_2, \ldots].$$

This proves part (ii).

We turn to part (iii). It is one of the formal properties of the counit that $\varepsilon 1 = 1$. Suppose, as an inductive hypothesis, that we have proved $\varepsilon t_i = 0$ for $0 < i < k$. Apply the counit ε to the formula in (16.1)(i). Using the fact that $\varepsilon \eta_R = 1$, and the inductive hypothesis, we find that

$$m_{p^k-1} = m_{p^k-1} + \varepsilon t_k .$$

So $\varepsilon t_k = 0$. This completes the induction and proves part (iii).

We turn to part (iv). Apply the conjugation map c to the formula in (16.1)(i). Since $c\eta_R = \eta_L$ and $c\eta_L = \eta_R$, we obtain the following result.

$$m_{p^k-1} = \sum_{f+j=k} (\eta_R m_{p^f-1})(ct_j)^{p^i} .$$

Substituting for $\eta_R m_{p^f-1}$ from (16.1)(i), we find

$$m_{p^k-1} = \sum_{h+i+j=k} m_{p^h-1}(t_i)^{p^h}(ct_j)^{p^{h+i}} .$$

This proves part (iv).

We turn to part (v). Take the formula in (16.1)(i), and apply the coproduct map ψ. Taking the right-hand side first, we have

$$\sum_{i+j=k} m_{p^i-1}(\psi t_j)^{p^i} = 1 \otimes \eta_R(m_{p^h-1}).$$

Substituting for $\eta_R(m_{p^h-1})$ from (16.1)(i), we have

$$\sum_{i+j=k} m_{p^i-1}(\psi t_j)^{p^i} = 1 \otimes \sum_{f+j=k} m_{p^f-1}(t_j)^{p^f}.$$

Since the tensor-product is taken over $\pi_*(BP)$, acting on the left of the right-hand factor and on the right of the left-hand factor, this gives

$$\sum_{i+j=k} m_{p^i-1}(\psi t_j)^{p^i} = \sum_{f+j=k} (\eta_R m_{p^f-1}) \otimes (t_j)^{p^f} .$$

Substituting for $\eta_R m_{p^f-1}$ from (16.1)(i), we find

$$\sum_{i+j=k} m_{p^i-1}(\psi t_j)^{p^i} = \sum_{h+i+j=k} m_{p^h-1}(t_i)^{p^h} \otimes (t_j)^{p^{h+i}} .$$

This proves part (v), and completes the proof of Theorem 16,1.

17. $KO_*(KO)$ (Added May 1970)

The results of §13 carry over to real K-theory. The material which follows represents joint work with R.M. Switzer.

We write KO for the BO-spectrum. The groups $KO_{4n}(KO)$ are torsion-free, so the map

$$KO_{4n}(KO) \longrightarrow KO_{4n}(KO) \otimes Q$$

is a monomorphism. By means of the complexification map

$$KO \longrightarrow K$$

we can identify $\sum_k KO_{4n}(KO) \otimes Q$ with a subalgebra of $K_*(K) \otimes Q$, namely (with the notation of §13) $Q[u^2, u^{-2}, v^2, v^{-2}]$.

THEOREM 17.1. The map

$$\sum_n KO_{4n}(KO) \longrightarrow K_*(K) \otimes Q$$

gives an isomorphism between $\sum_n KO_{4n}(KO)$ and the set of finite Laurent series $f(u, v)$ which satisfy the following conditions.

(17.2) $$f(-u, v) = f(u, v), \quad f(u, -v) = f(u, v).$$

(17.3) For any pair of non-zero integers h, k we have

$$f(ht, kt) \in Z[t^4, t^{-4}, 2t^2, \frac{1}{hk}].$$

Notes. (17.4). It is clear from the above that any f in the image of $\sum_n KO_{4n}(KO)$ satisfies (17.2).

(17.5). By using the operation ψ^k, as in §13, one easily proves that such an f satisfies (17.3).

(17.6). Condition (17.3) has been written with two integers h, k in order to emphasize that it is invariant under the switch map $\tau: KO \wedge KO \longrightarrow KO \wedge KO$, which interchanges u and v. It would actually be sufficient to use the special case of (17.3) in which $h = 1$. Similarly, in §13 we could replace (13.3) by

$$f(ht, kt) \in Z[t, t^{-1}, \frac{1}{hk}].$$

The proof of Theorem 17.1 is similar to that in §13.

Since $KO_*(X)$ is a left module over $\pi_*(KO)$, we have a product map

$$\pi_m(KO) \otimes_Z KO_0(KO) \longrightarrow KO_m(KO).$$

THEOREM 17.7. This map

$$\pi_m(KO) \otimes_Z KO_o(KO) \longrightarrow KO_m(KO)$$

is an isomorphism.

Thus we have

$$KO_m(KO) \cong \begin{cases} Z_2 \otimes_Z KO_o(KO) & (m \equiv 1, 2 \mod 8) \\ 0 & (m \equiv 3, 5, 6, 7 \mod 8) \end{cases}$$

At the risk of laboring the obvious, we make the following result explicit.

PROPOSITION 17.8. An element of $KO_o(KO)$ lies in the kernel of

$$KO_o(KO) \longrightarrow Z_2 \otimes_Z KO_o(KO)$$

if and only if the corresponding Laurent series $f(u, v)$ satisfies the following condition.

(17.9) For any pair of odd integers h, k we have

$$f(h, k) \in 2Z[\frac{1}{hk}] .$$

By (17.1), this is the condition for $\frac{1}{2}f$ to lie in the image of $KO_o(KO)$.)

As in §13, the structure of $KO_*(KO)$ as a Hopf algebra is determined by this representation. We need the following extra informatio

PROPOSITION 17.10. The generator $g \in \pi_1(KO)$ satisfies

$$\eta_L(g) = \eta_R(g) .$$

This is immediate, since g lies in the image of

$$i_*: \pi_1(S^o) \longrightarrow \pi_1(KO).$$

REFERENCES

[1] J.F. Adams, On the structure and applications of the Steenrod algebra, Comm. Math. Helv. 32 (1958), 180-214.

[2] __________, S.P. Novikov's work on operations on complex cobordism, Lecture Notes, University of Chicago, 1967.

[3] __________, Lectures on generalized cohomology, in Lecture Notes in Mathematics, vol. 99 (1969), Springer-Verlag Berlin-Heidelberg-New York.

[4] M.F. Atiyah and R. Bott, On the periodicity theorem for complex vector bundles, Acta Math. 112 (1964), 229-247.

[5] J.M. Boardman, Thesis, Cambridge, 1964.

[6] __________, Stable homotopy theory, mimeographed notes, University of Warwick, 1965 onward.

[7] E.H. Brown, Jr. and F.P. Peterson, A spectrum whose Z_p cohomology is the algebra of reduced p-th powers, Topology 5 (1966), 149-154.

[8] A. Fröhlich, Formal Groups, Lecture Notes in Mathematics, vol. 74 (1968), Springer-Verlag Berlin-Heidelberg-New York.

[9] A. Hattori, Integral characteristic numbers for weakly almost complex manifolds, Topology 5 (1966), 259-280.

[10] M. Lazard, Sur les groupes de Lie formels à un paramètre, Bull. Soc. Math. France 83 (1955), 251-274.

[11] J.W. Milnor, The Steenrod algebra and its dual, Annals of Math. 67 (1958), 150-171.

[12] __________, On the cobordism ring Ω^* and a complex analogue, Amer. Jour. Math. 82 (1960), 505-521.

[13] S.P. Novikov, The methods of algebraic topology from the viewpoint of cobordism theories (Russian), Izvestija Akademii Nauk SSSR, Serija Matematičeskaja 31 (1967), 855-951.

[14] D. Quillen, On the formal group laws of unoriented and complex cobordism theory, Bull. Amer. Math. Soc. 75 (1969), 1293-1298.

[15] L. Smith, (Private communication).

[16] R.E. Stong, Relations among characteristic numbers I, Topology 4 (1965), 267-281.

[17] G.W. Whitehead, Generalized homology theories, Trans. Amer. Math. Soc. 102 (1962), 227-283.

PART III

STABLE HOMOTOPY AND GENERALISED HOMOLOGY

1. INTRODUCTION

These notes, prepared by R. Ming, are based on a course I gave at the University of Chicago in the spring of 1971. I propose to construct a stable homotopy category equivalent to Boardman's, but whose construction will be accessible to those without a specialized knowledge of category theory. I will then formulate a number of classical topics in this framework, and finally present some new applications.

First I have to explain the meaning of the word "stable" in algebraic topology. We say that some phenomenon is <u>stable</u>, if it can occur in any dimension, or in any sufficiently large dimension, and if it occurs in essentially the same way independent of dimension, provided perhaps that the dimension is sufficiently large.

<u>Example (a).</u> We can consider the homotopy groups of spheres, $\pi_{n+r}(S^n)$. We have the suspension homomorphism

$$E: \pi_{n+r}(S^n) \longrightarrow \pi_{n+r+1}(S^{n+1}).$$

The Freudenthal suspension theorem says that this homomorphism is an isomorphism for $n > r+1$. For example, $\pi_{n+1}(S^n)$ is isomorphic to Z_2 for $n > 2$. The groups $\pi_{n+r}(S^n)$ $(n > r+1)$ are called the stable homotopy groups of spheres.

More generally, let X and Y be two CW-complexes with base-point. When we mention a CW-complex with base-point, we will always assume that the base-point is a 0-cell. By $[X, Y]$ we will mean the set of homotopy classes of maps from X to Y; here maps and homotopies are required to preserve the base-point. The product $W \times X$ of two CW-complexes will always be taken with the CW-topology. The smash-product $W \wedge X$ of two CW-complexes with base-point is defined, as usual, by

$$W \wedge X = W \times X / W \vee X.$$

The suspension SX of a CW-complex with base-point is to be the reduced suspension, either $S^1 \wedge X$ or $X \wedge S^1$, whichever suits our sign conventions better when we come to use it. Of course the two are homeomorphic. If $f: X \longrightarrow Y$ is map between CW-complexes with base-point, its suspension Sf is to be $1 \wedge f: S^1 \wedge X \longrightarrow S^1 \wedge Y$ (or $f \wedge 1: X \wedge S^1 \longrightarrow Y \wedge S^1$). Suspension defines a function

$$S: [X, Y] \longrightarrow [SX, SY].$$

Theorem 1.1. Suppose that Y is $(n-1)$-connected. Then S is onto if $\dim X \leq 2n-1$ and is a 1-1 correspondence if $\dim X < 2n-1$. ([14], p. 458).

Under these circumstances we call an element of $[X, Y]$ a stable homotopy class of maps.

Example (b). We consider the notion of a cohomology operation. Such an operation is a natural transformation

$$\varphi: H^n(X, Y; \pi) \longrightarrow H^m(X, Y; G).$$

Here n, m, π and G are fixed. In other words, φ is a function defined on $H^n(X,Y;\pi)$ and taking values in $H^m(X,Y;G)$, subject to one axiom only: if $f: X, Y \rightarrow X', Y'$ and $h \in H^n(X', Y'; \pi)$ then $\varphi(f^* h) = f^*(\varphi h)$.

By contrast, a stable cohomology operation is a collection of cohomology operations, say

$$\varphi_n : H^n(X,Y;\pi) \rightarrow H^{n+d}(X,Y;G).$$

Here n runs over Z, while d, π and G are fixed. Each φ_n is required to be a natural, as above. But also we require that the following diagram be commutative for each n.

$$\begin{array}{ccc} H^n(Y,Z;\pi) & \xrightarrow{\delta} & H^{n+1}(X,Y;\pi) \\ \downarrow \varphi_n & & \downarrow \varphi_{n+1} \\ H^{n+d}(Y,Z;G) & \xrightarrow{\delta} & H^{n+d+1}(X,Y;G) \end{array}$$

That is, we require φ to commute with δ as well as f^*.

For an example, take $\pi = G = Z_2$, and let φ_n be the Steenrod square Sq^d.

So a stable cohomology operation is something which can be applied in any dimension. Given a cohomology operation

$$\varphi : H^n(X,Y;\pi) \rightarrow H^m(X,Y;G)$$

it need not appear as the n-th term of any stable cohomology operation.

(For more on cohomology operations, see for example [11], [15] and [14] pp. 429-430.)

To do algebraic topology, it is rather important to be able to distinguish between unstable problems, which arise in some definite dimension, and stable problems, which arise in any sufficiently large dimension. We have actually come quite a long way since Eilenberg said, "We can distinguish two cases--the stable case and the interesting case."

Sometimes we solve an unstable problem first and use the result to solve a stable problem. For example, one might begin by proving $\pi_3(S^2) \cong Z$ (unstable) and then go on to deduce that $\pi_{n+1}(S^n) \cong Z_2$ for $n > 2$ (stable). More usually, however, we face some geometrical problem which looks like an unstable problem, but we reduce it to a stable problem and then solve the stable problem.

For example, we might consider the problem, "Is S^{n-1} an H-space?" Examples: for $n = 4$, S^3 is an H-space; for $n = 6$, S^5 is not. This problem is unstable. However, one way to solve the problem is to reduce it to the following one. "Assuming $m \geq n$, is there a complex $X = S^m \cup e^{m+n}$ in which

$$Sq^n : H^m(X;Z_2) \longrightarrow H^{m+n}(X;Z_2)$$

is non-zero?" This problem is stable; for a given n the answer is independent of m, provided $m \geq n$. But this problem is equivalent to the former one.

Another case arises in cobordism theory. Here, for example, one might want to take compact oriented smooth manifolds, of dimension n, without boundary, and classify them under a certain equivalence relation to get a group Ω_n. The problem would be to find the structure of Ω_n. The problem as stated is not yet in the form of a homotopy problem, but it appears to be unstable--there is one problem for each n. However, René Thom reduced the problem to a homotopy problem, and found it was a problem of stable homotopy theory. More precisely, he introduced the Thom complex $MSO(n)$, and he gave an important construction which yields an isomorphism

$$\Omega_r \cong \pi_{n+r}(MSO(n)) \qquad (n > r+1).$$

The computation of $\pi_{n+r}(MSO(n))$ is a stable problem, which was begun by Thom, continued by Milnor and completed by Wall. (A suitable reference on cobordism is Stong [16].)

Now of course to solve stable problems, or to compute groups such as $[X, Y]$ or $\pi_{n+r}(MSO(n))$, we need computable invariants. In the first instance this means homology and cohomology, but we could certainly agree to go as far as generalized homology and cohomology theories. I will suppose it known that a generalized homology or cohomology theory is a functor K_* or K^* which satisfies the first six axioms of Eilenberg-Steenrod [6], but not necessarily the seventh, the dimension axiom. I will suppose it known that the material in Eilenberg-Steenrod Chapter 1 carries over to this situation. For example, if X is a space with base-point one can define reduced groups $\tilde{K}_*(X)$, $\tilde{K}^*(X)$; and one can define a suspension isomorphism

$$\tilde{K}_n(X) \cong \tilde{K}_{n+1}(SX)$$
$$\tilde{K}^n(X) \cong \tilde{K}^{n+1}(SX).$$

This already tells us that the study of generalized homology and cohomology is part of stable homotopy theory. At least, what I said is true if you consider $\tilde{K}_*(X)$ or $\tilde{K}^*(X)$ as an additive group; if you started to use products, or unstable cohomology operations, you would get outside the realm of stable homotopy theory.

To go on with Eilenberg-Steenrod Chapter 1, we have Mayer-Vietoris sequences

$$\ldots \longrightarrow K_*(U \cap V) \longrightarrow K_*(U) \oplus K_*(V) \longrightarrow K_*(U \cup V) \longrightarrow \ldots$$
$$\ldots \longrightarrow K^*(U \cup V) \longrightarrow K^*(U) \oplus K^*(V) \longrightarrow K^*(U \cap V) \longrightarrow \ldots .$$

Also we have the Atiyah-Hirzebruch spectral sequence, which was really invented by G. W. Whitehead but not published by him:

$$H_*(X; K_*(pt.)) \Longrightarrow K_*(X)$$
$$H^*(X; K^*(pt.)) \Longrightarrow K^*(X) .$$

This spectral sequence replaces the Eilenberg-Steenrod uniqueness theorem when we go from the ordinary to the generalized case. The Atiyah-Hirzebruch spectral sequence emphasizes that before computing, we need to know the coefficient groups $K_*(pt.)$ and $K^*(pt.)$.

At this point I should give some motivation for some of the topics to be considered. One of these we will treat in some detail is that of products; they may not be part of stable homotopy theory, but they have numerous applications. For example, suppose we wanted to take the classical results on duality in manifolds, and carry them over to the generalized case. We would proceed like this.

"Let X be a topological manifold; I don't care whether it is compact or not, but let us assume it has no boundary." (If it starts with a boundary I add an open collar, which doesn't change the homology and gives a non-compact manifold without boundary.) "Suppose that X is orientable with respect to E, where E is a ring-spectrum. Let K, L be a compact pair in X, and assume that F is a module-spectrum over E. Then a certain homomorphism (which has to be described) is an isomorphism

$$F_r(\mathcal{C}L, \mathcal{C}K) \longrightarrow \check{F}^{n-r}(K, L),$$

where n is the dimension of the orientation class." (The homology on the left is the singular homology associated with F, the cohomology on the right is of the Čech type.)

Theorems of this sort were introduced by G. W. Whitehead in his well-known paper on generalized homology theories [17], but unfortunately he did not go quite as far as the result I have stated. To prove this result one follows a simple recipe: take the treatment in Spanier and do it all over again, with ordinary homology replaced by generalized homology.

For this purpose, of course, one needs products, as in the ordinary case. Indeed, the duality map is defined by a product. There are four basic external products: an external product in homology, an external product in cohomology, and two slant products. From this one gets two internal products, the cup product and the cap product. There is also the Kronecker product, which can be obtained as a special case of either slant product or the cap product.

Of course one needs to know the formal properties of the products. For example, the four external products satisfy eight associativity formulae. I do not know a good source in print where they are collected and numbered 1 to 8. Again, when you prove the duality theorem for manifolds, you need to know that the duality homomorphism commutes (up to sign) with boundary maps. So you need to know the properties of the products with respect to boundary maps. Again I know of no good source in print; Eilenberg-Steenrod volume II is not out yet.

Once you have all the material about duality in manifolds, you can have a certain amount of fun. For example, there is a formula for computing the index of a compact oriented manifold. It says that you take a certain characteristic class of the tangent bundle τ and evaluate it on the fundamental homology class. Now, you may think I mean

Hirzebruch's formula in ordinary homology, but I don't; I mean the analogue in complex K-theory. If M is an almost-complex manifold, it has a fundamental class $[M]_K$ in K-homology, and its tangent bundle τ has a characteristic class $\rho_2(\tau)$ in K-cohomology, and we can form their Kronecker product $<\rho_2(\tau), [M]_K>$. Then we have

$$\text{Index}(M) = <\rho_2(\tau), [M]_K> .$$

In ordinary cohomology, one uses not only products, but also cohomology operations. For example, suppose that X and Y are finite complexes, and that we want to study the stable groups

$$\lim_{n \to \infty} [S^{n+r}X, S^n Y] .$$

There is a recipe which goes as follows. Form $\tilde{H}^*(X; Z_p)$ and $\tilde{H}^*(Y; Z_p)$ and consider them as modules over the mod p Steenrod algebra A, that is, the algebra of stable operations on mod p cohomology. Form

$$\text{Ext}_A^{**}(\tilde{H}^*(Y; Z_p), \tilde{H}^*(X; Z_p)) .$$

Then there is a spectral sequence with this E_2-term and converging to the stable group above, at least if one ignores q-torsion for q prime to p. People seem to call this the Adams spectral sequence, so I suppose I had better do so too. This was the way Milnor computed $\pi_*(MU)$.

At one time I used to make the point that one ought to take this spectral sequence and replace mod p cohomology by a generalized cohomology theory; but the first person to do so successfully was Novikov, who took complex cobordism, MU^*. In these notes I have developed the spectral sequence in sufficient generality so as to include spectral sequences constructed from a number of commonly used theories, using homology instead of cohomology for reasons which will become apparent in § 16.

Recently Anderson has been considering the Adams spectral sequence (for computing stable homotopy groups of spheres) based on bu, the connective BU-spectrum, and Mahowald has proved various results, including one on the image of the J-homomorphism, by considering a similar construction based on bo, the connective BO-spectrum. I have reproved some of their results. The calculations to be given here give a sample application of the Adams spectral sequence, as well as giving some of the information needed to use these spectral sequences based on bu and bo.

2. SPECTRA

The notion of a spectrum is due to Lima [7]. It is generally supposed that G. W. Whitehead also had something to do with it, but the latter takes a modest attitude about that.

By definition, a <u>spectrum</u> E is a sequence of spaces E_n with base-point, provided with structure maps, either

$$\epsilon_n : SE_n \longrightarrow E_{n+1}$$

or

$$\epsilon'_n : E_n \longrightarrow \Omega E_{n+1} .$$

Of course giving a map ϵ_n is equivalent to giving a map ϵ'_n, as S and Ω are adjoint. There is one other variant; if we choose to work with connected spaces, then E_n will automatically map into $\Omega_o E_{n+1}$, where Ω_o is the component of the base-point in Ω; we might prefer to write

$$\epsilon'_n : E_n \longrightarrow \Omega_o E_{n+1} .$$

The index n may run over the integers or over $\{0, 1, 2, 3, \ldots\}$. Examples will appear in a moment.

The notion of a spectrum is very natural if one starts from cohomology theory. Let K^* be a generalized cohomology theory, defined on CW pairs. We have

$$K^n(X) = K^n(X, pt.) + K^n(pt.),$$

and so define $\widetilde{K}^n(X) = K^n(X, pt.)$. We assume that K^* satisfies the wedge axiom of Milnor and Brown. More precisely, let X_α $(\alpha \in A)$ be CW-complexes with base-point, and let $i_\alpha : X_\alpha \longrightarrow \bigvee_\alpha X_\alpha$ be the inclusion of one summand in the wedge-sum. This induces

$$i_\alpha^* : \widetilde{K}^n(\bigvee_\alpha X_\alpha) \longrightarrow K^n(X_\alpha).$$

Let

$$\theta : \widetilde{K}^n(\bigvee_\alpha X_\alpha) \longrightarrow \prod_{\alpha \in A} \widetilde{K}^n(X_\alpha)$$

be the homomorphism with components i_α^*. We assume that θ is an isomorphism (for all choices of $\{X_\alpha\}$ and n.)

We can now apply the representability theorem of E. H. Brown [4]. We see that there exist connected CW-complexes E_n with base-point and natural equivalences

$$\widetilde{K}^n(X) \cong [X, E_n].$$

(Here X runs over connected CW-complexes with base-point.) So we obtain a collection of spaces E_n $(n \in Z)$. However, a cohomology theory does not consist only of functors K^n; they are connected by coboundary maps. If we divert attention from the relative groups $K^n(X, Y)$ to reduced groups $\widetilde{K}^n(X)$, we should divert attention from the coboundary maps δ to the suspension isomorphisms

$$\sigma : \widetilde{K^n} \xrightarrow{\cong} \widetilde{K^{n+1}}(SX).$$

Here SX is considered as the union of two cones CX and C'X over the same copy of X. The suspension isomorphism is defined as

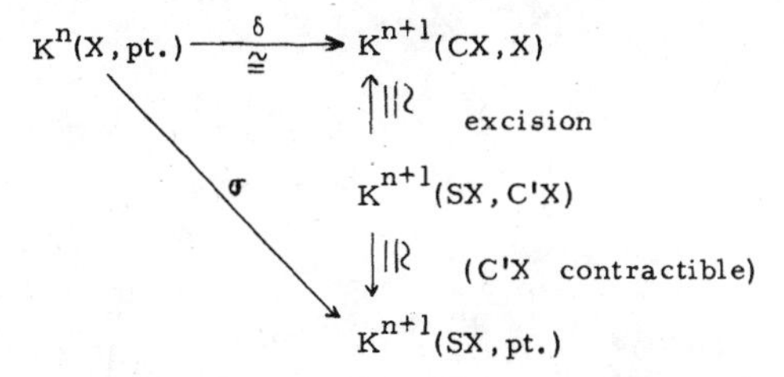

The map δ is the coboundary for the exact sequence of the triple (CX, X, pt.). The vertical isomorphism is also induced by the collapsing map $(CX, X) \longrightarrow (SX, pt.)$.

We now observe that we have the following natural equivalences, at least if X is connected.

$$[X, E_n] \cong \widetilde{K^n}(X) \cong \widetilde{K^{n+1}}(SX)$$
$$\cong [SX, E_{n+1}] \cong [X, \Omega_o, E_{n+1}].$$

This natural equivalence must be induced by a weak equivalence

$$\epsilon'_n : E_n \longrightarrow \Omega_o E_{n+1} .$$

So our sequence of spaces becomes a spectrum.

It is usual to make the following definition. A spectrum E is an <u>Ω-spectrum</u> (resp. <u>Ω_o-spectrum</u>) if $\epsilon'_n : E_n \longrightarrow \Omega E'_{n+1}$ (resp. $\Omega_o E'_{n+1}$ is a weak equivalence for each n. So we have constructed an Ω_o-spectrum.

These considerations also show us how to construct a CW-complex F_n (with base-point) and a natural equivalence $[X, F_n] \cong \widetilde{K^n}(X)$ valid whether X is connected or not. In fact, we have only to take F_n weakly equivalent to ΩE_{n+1}. Then we have

$$\widetilde{K}^n(X) \cong \widetilde{K}^{n+1}(SX) \cong [SX, E_{n+1}]$$
$$\cong [X, \Omega E_{n+1}] \cong [X, F_{n+1}] .$$

As before, we have the following natural equivalences.

$$[X, F_n] \cong \widetilde{K}^n(X) \cong \widetilde{K}^{n+1}(SX)$$
$$\cong [SX, F_{n+1}] \cong [X, \Omega F_{n+1}]$$

This time we conclude that this natural equivalence must be induced by a weak homotopy equivalence

$$\varphi_n : F_n \longrightarrow \Omega F_{n+1} .$$

We have constructed an Ω-spectrum.

Example 2.1. Take K^* to be ordinary cohomology; $K^n(X, Y) = H^n(X, Y; \pi)$. The corresponding spectrum E is the Eilenberg-MacLane spectrum for the group π; the nth space is the Eilenberg-MacLane space of type (π, n). That is, we have

$$\pi_r(E_n) = [S^r, E_n] \cong \widetilde{H}^n(S^r; \pi) = \begin{cases} \pi & (r = n) \\ 0 & (r \neq n). \end{cases}$$

Example 2.2. Take K^* to be complex K-theory. The corresponding spectrum is called the BU-spectrum. Each even term E_{2n} is the space BU, or $Z \times BU$, depending on whether you choose to work with connected spaces or not. Each odd term E_{2n+1} is the space U.

Similarly, we can take K^* to be real K-theory. The corresponding spectrum is called the BO-spectrum. Every eighth term E_{8n} is the space BO, or $Z \times BO$, depending on whether you choose to work with connected spaces or not. Each term E_{8n+4} is the space BSp.

Of course, not all spectra are Ω-spectra.

Example 2.3. Given a CW-complex X, let $E_n = \begin{cases} S^n X & (n \geq 0) \\ pt & (n < 0) \end{cases}$ with the obvious maps. We might define a spectrum F to be a suspension spectrum or S-spectrum if

$$\varphi_n : SF_n \longrightarrow F_{n+1}$$

is a weak homotopy equivalence for n sufficiently large. Then this spectrum E would be an S-spectrum, but usually not an Ω-spectrum. E is called the suspension spectrum of X. In particular, the sphere spectrum S is the suspension spectrum of S^o; it has n^{th} term S^n for $n \geq 0$.

Example 2.4. Let $MO(n)$ be the Thom complex of the universal n-plane bundle ξ_n over $BO(n)$. Then the Whitney sum $\xi_n \oplus 1$ admits a bundle map to ξ_{n+1}. (Here 1 means the trivial line bundle.) The Thom complex of $\xi_n \oplus 1$ is $MO(n) \wedge S^1$ and the Thom complex of ξ_{n+1} is $MO(n+1)$; so we get a map $MO(n) \wedge S^1 \longrightarrow MO(n+1)$. The Thom spectrum MO is the spectrum in which the nth space is $MO(n)$ and the maps are the ones just indicated.

Similar remarks apply to the Thom spectra MSO, MSpin, MU, MSU and MSp. However, $MU(n)$ is the $2n^{th}$ term of the spectrum MU, the $(2n+1)^{th}$ term being $MU(n) \wedge S^1$ (because in the complex case we have $M(1) = S^2$.) Similarly for MSU. For MSp, the term $E_{4n+\epsilon}$ is $MSp(n) \wedge S^{\epsilon}$ for $\epsilon = 0, 1, 2, 3$.

These spectra arise in cobordism theory, as I said before.

We now define the homotopy groups of a spectrum. These are really stable homotopy groups. We have the following homomorphisms.

$$\pi_{n+r}(E_n) \longrightarrow \pi_{n+r+1}(SE_{n+1}) \xrightarrow{(\epsilon_n)_*} \pi_{n+r+1}(E_{n+1})$$

We define

$$\pi_r(E) = \lim_{n \to \infty} \pi_{n+r}(E_n);$$

here the homomorphisms of the direct system are those displayed above. If E is an Ω-spectrum or an Ω_0-spectrum, then the homomorphism

$$\pi_{n+r}(E_n) \longrightarrow \pi_{n+r+1}(E_{n+1})$$

is an isomorphism for $n+r \geq 1$; the direct limit is attained, and we have

$$\pi_r(E) = \pi_{n+r}(E_n) \qquad \text{for } n+r \geq 1.$$

Thus, in Example 2.1, The Eilenberg-Mac Lane spectrum, we have

$$\pi_r(E) = \begin{cases} \pi & (r = 0) \\ 0 & (r \neq 0) \end{cases} .$$

In Example 2.2, the BU-spectrum, we have

$$\pi_r(E) = \begin{cases} Z & (r \text{ even}) \\ 0 & (r \text{ odd}) \end{cases}$$

by the Bott periodicity theorem. For the BO-spectrum we have

$r =$ 0	1	2	3	4	5	6	7	8	mod 8
$\pi_r(E) = Z$	Z_2	Z_2	0	Z	0	0	0	Z	

by Bott periodicty again.

In Example 2.3 we have

$$E_n = \begin{cases} S^n X & (n \geq 0) \\ \text{pt.} & (n < 0) , \end{cases}$$

so that

$$\pi_r(E) = \lim_{n \to \infty} \pi_{n+r}(S^n X) .$$

The limit is attained for $n > r+1$. The homotopy groups of E are the stable homotopy groups of X.

In Example 2.4 the homotopy groups of the spectrum MO are precisely those which arise in Thom's work, namely

$$\pi_r(MO) = \lim_{n \to \infty} \pi_{n+r}(MO(n)).$$

The limit is attained for $n > r+1$. Similarly for the other Thom spectra.

In general, there is no reason why the limit $\lim_{n\to\infty} \pi_{n+r}(E_n)$ should be attained. Exercise: Construct a counterexample.

Similarly, of course, we can define relative homotopy groups. To do so we need subobjects. Let X be a spectrum; then a <u>subspectrum</u> A of X consists of subspaces $A_n \subset X_n$ such that the structure map $\xi_n: SX_n \longrightarrow X_{n+1}$ maps SA_n into A_{n+1}. Of course we take $\xi_n|SA_n$ as the structure map α_n for A. And if we think in terms of maps $\xi'_n: X_n \longrightarrow \Omega X_{n+1}$, we ask that ξ'_n maps A_n into ΩA_{n+1}.

In fact we want to define not only relative homotopy groups, but also boundary homomorphisms. For this purpose we want the exact homotopy sequences of the pairs (X_n, A_n) and (X_{n+1}, A_{n+1}) to fit into the following commutative diagram.

$$\begin{array}{ccccccccc}
\cdots \to & \pi_{n+r}(A_n) & \longrightarrow & \pi_{n+r}(X_n) & \longrightarrow & \pi_{n+r}(X_n, A_n) & \xrightarrow{\partial} & \pi_{n+r-1}(A_n) & \to \cdots \\
& \downarrow & & \downarrow & & \downarrow & & \downarrow & \\
\cdots \to & \pi_{n+r+1}(A_{n+1}) & \to & \pi_{n+r+1}(X_{n+1}) & \to & \pi_{n+r+1}(X_{n+1}, A_{n+1}) & \xrightarrow{\partial} & \pi_{n+r}(A_{n+1}) & \to \cdots
\end{array}$$

But here we must be careful of the signs. If $\partial E^m = S^{m-1}$, then with the usual conventions,

$$\partial(S^1 \wedge E^m) = -S^1 \wedge \partial E^m \quad \text{and} \quad \partial(E^m \wedge S^1) = S^{m-1} \wedge S^1.$$

So at this point we prefer to interpret SX_n as $X_n \wedge S^1$, as is done in Puppe's paper on stable homotopy theory. With this convention, the ladder diagram commutes; we can define

$$\pi_r(X, A) = \lim_{n\to\infty} \pi_{n+r}(X_n, A_n) .$$

and we obtain our exact homotopy sequence

$$\cdots \longrightarrow \pi_*(A) \longrightarrow \pi_*(X) \longrightarrow \pi_*(X, A) \longrightarrow \pi_*(A) \longrightarrow \cdots .$$

We have seen how to associate a spectrum to a generalised cohomology theory. The converse is also possible: with any spectrum E we can associate a generalized homology theory and a generalized cohomology theory. This is due to G. W. Whitehead, in a celebrated paper [17]. I'll get back to this later. If we have a spectrum E, it is very convenient to write E_* and E^* for this associated homology and cohomology theories. I will also reverse this. Ordinary homology and cohomology (with Z coefficients) are always written H_*, H^*; therefore, H will mean the Eilenberg-MacLane spectrum for the group Z. (For coefficients in a group G, we write HG.) This frees the letter K for other uses. Classical complex K-theory is always written K^*; therefore, K will mean the BU-spectrum. This is fine, because I would anyway need notation to distinguish the space BU from the BU-spectrum. Similarly, we write KO for the BO-spectrum.

The coefficient groups of the theories E_*, E^* will be given by

$$E_r(pt) = E^{-r}(pt) = \pi_r(E).$$

I take it that in Chicago I need not make propaganda for taking spectra as the objects of a category. For one thing only, I would like to define the E-cohomology of the spectrum X, in dimension 0, to be

$$E^0(X) = [X, E] ,$$

the set of morphisms from X to E in our category. (Morphisms will correspond to homotopy classes of maps.) In fact I would like to go further and construct a graded category, so that we can define

$$E^r(X) = [X, E]_{-r}$$

(morphisms which lower dimension by r).

Next I must explain why one would want to introduce smash products of spectra. First, we would like to define the E-homology of the spectrum X to be

$$E_r(X) = \pi_r(E \wedge X) = [S, E \wedge X]_r.$$

Secondly, we would like to introduce products, for example, cup-products in cohomology. In order to define cup-products in ordinary cohomology, say

$$H^n(X;A) \otimes H^m(X;B) \longrightarrow H^{n+m}(X;C)$$

we need a pairing $A \otimes B \longrightarrow C$. George Whitehead wanted to introduce cup-products in generalized cohomology

$$E^n(X) \otimes F^m(X) \longrightarrow G^{n+m}(X)$$

and he found he needed a pairing of spectra from E and F to G. Now it would be very nice if a pairing of spectra were just a morphism

$$\mu: E \wedge F \longrightarrow G$$

in our category. Thirdly, for example, we might want to restate a result of R. Wood in the form $KO \wedge CP^2 \simeq K$.

When we come to undertake a complicated piece of work, the convenience of having available smash products of spectra is so great that I, for one, would hate to do without it.

Now let me get on and define my category.

I say E is a CW-spectrum if

(i) the terms E_n are CW-complexes with base-point, and

(ii) each map $\epsilon_n: SE_n \longrightarrow E_{n+1}$ is an isomorphism from SE_n to a subcomplex of E_{n+1}.

Notes. (i) There is no essential loss of generality in restricting to CW-spectra. (See the exercise after 3.12 or the discussion of the telescope functor in §4.)

(ii) An isomorphism between CW-complexes is a homeomorphism h such that h and h^{-1} are cellular. The CW structure on SE_n is the obvious one on $E_n \wedge S^1$, where S^1 is regarded as a CW-complex with one 0-cell and one 1-cell. Thus SE_n has a base-point and one cell Sc_α for each cell c_α of E_n other than the base-point.

(iii) It would be possible to identify SE_n with its image under ϵ_n and so suppose $SE_n \subset E_{n+1}$. Sometimes it may be convenient to speak in this way. On the whole, it seems best to leave the definition as I've given it.

The ideas which come next are introduced to help in defining the morphisms of our category.

A <u>subspectrum</u> A of a CW-spectrum E will be a subspectrum as defined above, with the added condition that $A_n \subset X_n$ be a subcomplex for each n.

Let E be a CW-spectrum, E' a subspectrum of E. We say E' is <u>cofinal</u> in E (Boardman says <u>dense</u>) if for each n and each finite subcomplex $K \subset E_n$ there is an m (depending on n and K) such that $S^m K$ maps into E'_{m+n} under the obvious map

$$S^m E_n \xrightarrow{S^{m-1}\epsilon_n} S^{m-1}E_{n+1} \longrightarrow \cdots \longrightarrow SE_{m+n-1} \xrightarrow{\epsilon_{m+n-1}} E_{m+n}.$$

The essential point is that each cell in each E_n gets into E' after enough suspensions. I said that m depends on n and K, but there is no need to suppose that it does so in any particular way.

The construction of our category is in several steps. In particular, we will distinguish between "functions", "maps" and "morphisms".

A <u>function</u> f from one spectrum E to another F, and of degree r, is a sequence of maps $f_n : E_n \longrightarrow F_{n-r}$ such that the following diagrams are strictly commutative for each n.

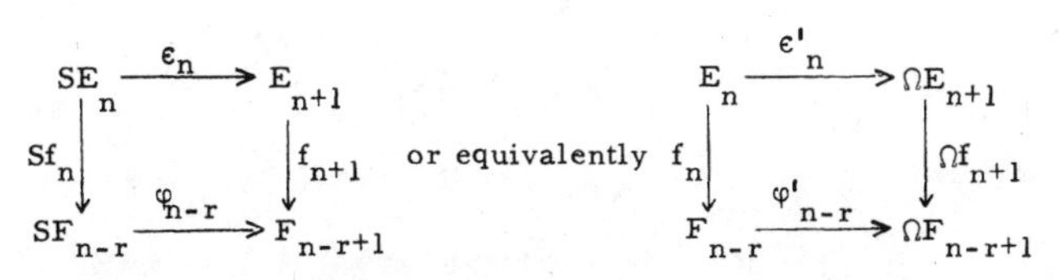

Notes. (i) The diagrams are to be strcitly commutative. If we allowed the diagrams to be commutative up to homotopy, then to make any further construction we would need to know what the homotopies were, so we would have to take the homotopies as part of the given structure of a function. It seems better to proceed as I said.

(ii) Composition of functions is done in the obvious way, and we have identity functions.

(iii) If E' is a subspectrum of E, the injection i of E' in E is a function in good standing. Restriction of functions from E to E' is the same as composition with i.

(iv) For graded functions, it is convenient if n runs over Z.

(v) The details of the grading are cooked up so that in the end we get $\pi_r(F) = [S, F]_r$.

If E is a CW-spectrum and F is an Ω-spectrum, then the functions from E to F are usable as they stand. But is is convenient to deal with spectra which are not Ω-spectra, and then there are examples to show that there are not enough functions to do what we want.

For one example, consider the Hopf map $S^3 \xrightarrow{\eta} S^2$. We would like to have a corresponding function $S \longrightarrow S$ of degree 1. But there are no candidates for the maps $S^1 \longrightarrow S^0$, or $S^2 \longrightarrow S^1$ required to make a function.

For another example, take two spectra with

$$E_n = S^{n+3} \vee S^{n+7} \vee S^{n+11} \vee \ldots$$

$$F_n = S^n .$$

We would like to have a function from E to F whose component from S^{n+4k-1} to S^n is a generator for the image of J in the stable (4k-1)-stem. But there is no single value of n for which all the requisite maps exist as maps into S^n; we have to concede that for the different cells of E the maps come into existence for different values of n.

So we need the following construction. Let E be a CW-spectrum and F a spectrum. Take all cofinal subspectra $E' \subset E$ and all functions $f': E' \longrightarrow F$. Say that two functions $f': E' \longrightarrow F$ and $f'': E'' \longrightarrow F$ are equivalent if there is a cofinal subspectrum E''' contained in E' and E'' such that the restrictions of f' and f'' to E''' coincide. (Check that this is an equivalence relation.)

Definition. A map from E to F is an equivalence class of such functions.

This amounts to saying that if you have a cell c in E_n, a map need not be defined on it at once; you can wait till E_{m+n} before defining the map on $S^m c$. The slogan is, "cells now--maps later."

Notes. (i) In order to prove that the relation is an equivalence relation, we use the following lemma.

Lemma 2.5. If E' and E'' are cofinal subspectra of E, then so is $E' \cap E''$.

The proof is trivial.

(ii) It would amount to the same to say that two functions $f': E' \longrightarrow F$, $f'': E'' \longrightarrow F$ are equivalent if their restrictions to $E' \cap E''$ coincide. This comes from the following fact: if $g,h: K \longrightarrow L$

are maps of CW-complexes with base-point, and $Sg = Sh$, then $g = h$.

Let E, F, G be spectra, of which E and F are CW-spectra. Then we define composition of maps by composition of representatives, choosing representatives for which composition is defined. For this purpose we need the following lemma.

Lemma 2.6. (i) Let $f: E \longrightarrow F$ be a function, and F' a cofinal subspectrum of F. Then there is a cofinal subspectrum E' of E such that f maps E' into F'.

(ii) If E' is a cofinal subspectrum of E, and E'' is a cofinal subspectrum of E', then E'' is a cofinal subspectrum of E.

The proof is trivial.

Restriction of maps is done by composition with the inclusion map, which is the class of the inclusion function.

We can piece maps together in the usual way. Let E be a CW-spectrum, and U, V subspectra of E.

Lemma 2.7. Let $u: U \longrightarrow F$, $v: V \longrightarrow F$ be maps whose restrictions to $U \cap V$ are equal. Then there exists one and only one map $w: U \cup V \longrightarrow F$ whose restrictions to U and V are u and v respectively.

The proof is easy.

A morphism in our category will be a homotopy class of maps, and a "homotopy" will be a map of a cylinder, just as in ordinary topology. So we begin by defining cylinders. Let I^+ be the union of the unit interval and a disjoint base-point. If E is a spectrum, we define the cylinder spectrum $Cyl(E)$ to have terms

$$(Cyl(E))_n = I^+ \wedge E_n$$

and maps

$$(I^+ \wedge E_n) \wedge S^1 \xrightarrow{1 \wedge \epsilon_n} I^+ \wedge E_{n+1} .$$

The cylinder spectrum is a functor: a map $f: E \longrightarrow F$ induces a map $\mathrm{Cyl}(f): \mathrm{Cyl}(E) \longrightarrow \mathrm{Cyl}(F)$ in the obvious way. We have obvious injection functions

$$i_0, i_1: E \longrightarrow \mathrm{Cyl}(E) ,$$

corresponding to the two ends of the cylinder. These are natural for maps of E. The other properties of the cylinder are as usual, and they are too obvious to list.

We say that two maps

$$f_0, f_1: E \longrightarrow F$$

are <u>homotopic</u> if there is a map

$$h: \mathrm{Cyl}(E) \longrightarrow F$$

such that $f_0 = hi_0$, $f_1 = hi_1$.

Homotopy is an equivalent relation. If E, F are spectra, with E a CW-spectrum, we write $[E, F]_r$ for the set of homotopy classes of maps of degree r from E to F. Composition passes to homotopy classes, as in the usual case.

The category in which we propose to work is as follows. The objects are the CW-spectra. The morphisms of degree r are homotopy classes of maps of degree r.

<u>Notes.</u> (i) Let X be a CW-spectrum consisting of X_n, $n \in Z$. Define X' by $X'_n = \begin{cases} X_n & (n \geq 0) \\ \mathrm{pt.} & (n < 0) \end{cases}$. Then X' is cofinal in X, and therefore equivalent to X in our category. For this reason it doesn't really make any difference whether we consider spectra indexed with

$n \in Z$ or with $n \in \{0, 1, 2, \ldots\}$.

(ii) Since we have our objects and maps open to direct inspection, we have no trouble elaborating these definitions. For example, suppose given a CW-spectrum X with a subspectrum A, and another spectrum Y with a subspectrum B. Then I have no trouble in defining

$$[X, A; Y, B].$$

To define maps $f: X, A \longrightarrow Y, B$ we consider functions $f': X', A' \longrightarrow Y, B$ where X' is cofinal in X, $A' \subset X'$ and A' is cofinal in A. Two such, $f': X', A' \longrightarrow Y, B$ and $f'': X'', A'' \longrightarrow Y, B$ are defined to be equivalent if there exist X''', A''' such that $f' | X''', A''' = f'' | X''', A'''$. A map $f: X, A \longrightarrow Y, B$ is an equivalence class of such functions. I can define homotopies

$$\mathrm{Cyl}(X), \mathrm{Cyl}(A) \longrightarrow Y, B$$

and the elements of $[X, A; Y, B]$ are homotopy classes of maps.

As long as we deal entirely with CW-spectra we can restrict attention to functions whose components $f_n: E_n \longrightarrow F_{n-r}$ are cellular maps. A construction in these terms leads to the same sets $[E, F]_r$. The proof is left as an exercise.

In order to validate our category we give one small result. Let K be a finite CW-complex, and let E be its suspension spectrum, so that $E_n = S^n K$ for $n \geq 0$. Let F be any spectrum.

<u>Proposition 2.8.</u> We have

$$[E, F]_r = \lim_{n \to \infty} [S^{n+r} K, F_n] .$$

In particular,

$$[S, F]_r = \pi_r(F).$$

Proof. For any map $f: S^{n+r}K \longrightarrow F_n$ we can define a corresponding map between spectra by taking its component on E_{n+r} to be $f: S^{n+r}K \longrightarrow F_n$; the higher components are then forced. In fact, they are

$$S^{m+n+r}K \xrightarrow{S^m f} S^m f_n \longrightarrow F_{m+n} .$$

Suppose two maps $f: S^{n+r}K \longrightarrow F_n$, $g: S^{m+r}K \longrightarrow F_m$ give the same element of the direct limit. Then for some p, the maps

$$S^{p+r}K \xrightarrow{S^{p-n}f} S^{p-n}F_n \longrightarrow F_p$$

$$S^{p+r}K \xrightarrow{S^{p-m}g} S^{p-m}F_m \longrightarrow F_p$$

are homotopic. This homotopy yields a homotopy between the corresponding maps of spectra. This shows we have a function

$$\lim_{n \to \infty} [S^{n+r}K, F_n] \xrightarrow{\theta} [E, F]_r .$$

Now every map from E to F arises in the way we have mentioned: this shows θ is onto. Also every homotopy arises in the way we have mentioned: this shows that θ is a 1-1 correspondence.

3. ELEMENTARY PROPERTIES OF THE CATEGORY OF CW-SPECTRA

We want to show that CW-spectra can be manipulated very much like CW-complexes. The standard way to make constructions for CW-complexes is by induction over the cells. Now we can define "stable cells" for CW-spectra. Let C_n be the set of cells in E_n other than the base-point. Then we get a function

$$C_n \longrightarrow C_{n+1} \quad \text{by} \quad c_\alpha \longrightarrow \epsilon_n(Sc_\alpha) .$$

This function is an injection. Let C be the direct limit

$$\lim_{n \to \infty} C_n ;$$

an element of C may be called a "stable cell." Unwrapping the definition, a stable cell is an equivalence class of cells; for each n such an equivalence class contains at most one cell in E_n. Take two cells, c_α in E_n and c_β in E_m, and suppose without loss of generality $n \leq m$; then c_α and c_β are equivalent if

$$C_n \longrightarrow C_{n+1} \longrightarrow \ldots \longrightarrow C_m$$

maps c_α into c_β.

Example. $E' \subset E$ is cofinal if and only if $C' \longrightarrow C$ is a bijection.

I said that the standard way to make constructions for CW-complexes is by induction over the cells. It is usual to order the cells of a CW-complex by dimension; first we take the cells of dimension 0, then the cells of dimension 1, and so on. For a CW-spectrum we can order the stable cells by "stable dimension," but this ordering is not inductive in general, because we can have stable cells of arbitrarily large negative stable dimension. Nevertheless we can perform inductive proofs, because each stable cell is attached to only a finite number of predecessors. More formally, we have:

Lemma 3.1. Let E be a CW-spectrum, and G a subspectrum of E which is not cofinal. Then E has a subspectrum F such that $E \supset F \supset G$ and F contains just one more stable cell than G.

Proof. G is not cofinal, so there exists a stable cell c in E not in G. It has a representative c_α, which is contained in a finite subcomplex $K \subset E_n$. So there exist finite subcomplexes K containing representatives for stable cells in E not in G. Among such K, choose one with fewest cells. Let $K = L \cup e$, where e is a top-dimensional cell of K. Then L fails to satisfy the conditions, for it has fewer cells than K. So all the stable cells in L represent stable cells in G. Then there exists m such that $S^m L$ gives a finite subcomplex of G_{m+n}. Form F by adjoining $S^r e$ to G_{n+r} for $r \geq m$.

We illustrate the use of this lemma by proving the homotopy extension theorem. Actually we prove something slightly more general.

Lemma 3.2. Let X, A be a pair of CW-spectra, and Y, B a pair of spectra such that $\pi_*(Y, B) = 0$. Suppose given a map $f: X \longrightarrow Y$ and a homotopy $h: \mathrm{Cyl}(A) \longrightarrow Y$ from $f|A$ to a map $g: A \longrightarrow B$. Then the homotopy can be extended over $\mathrm{Cyl}(X)$ so as to deform f to a map $X \longrightarrow B$.

The homotopy extension theorem is the special case $B = Y$.

Proof. Work at the level of functions. Suppose f is represented by a function $f': X' \longrightarrow Y$, and h by a function $h': \mathrm{Cyl}(A') \longrightarrow Y$, where $X' \supset A'$, X' is cofinal in X and A' is cofinal in A. We make our induction using Zorn's Lemma. The objects to be ordered are pairs (U, k') where $A' \subset U \subset X'$ and $k': \mathrm{Cyl}(U) \longrightarrow Y$ is a function which deforms $f'|U$ to a function into B. The set of such pairs is non-empty since (A', h') qualifies; and it is clearly inductive. So we can choose a maximal element (U, k'). I claim the maximal element has U cofinal in X'. If not, then by 3.1 we can find $U \subset V \subset X'$ where V contains just one more stable cell than U, say $V_n = U_n \cup e^m$. Then the maps

$$f'_n | 0 \wedge e: 0 \wedge e \longrightarrow Y_{n-r}$$

$$k'_n | I^+ \wedge \partial e: I^+ \wedge \partial e, 1 \wedge \partial e \longrightarrow Y_{n-r}, B_{n-r}$$

define an element of $\pi_m(Y_{n-r}, B_{n-r})$. Now $\pi_*(Y, B) = 0$, so that this element vanishes after sufficiently many suspensions. So on passing to $V_{n+p} = U_{n+p} \cup e^{m+p}$, we can extend k'_{n+p} to a map

$$k''_{n+p}: I^+ \wedge e, 1 \wedge e \longrightarrow Y_{p+n-r}, B_{p+n-r}.$$

Then define k''_{n+q} for $q > p$ by suspension. This extension of k' shows that (U, k') is not maximal, a contradiction. This contradiction shows that U is cofinal in X', i.e., U is cofinal in X. This gives the required map of $\mathrm{Cyl}(X)$.

A generalized version of 3.2 works when the inclusion $B \longrightarrow Y$ is replaced by a general function.

Lemma 3.2'. Let X, A be a pair of CW-spectra and $\emptyset: B \longrightarrow Y$ a function of spectra such that $\emptyset_*: \pi_*(B) \longrightarrow \pi_*(Y)$ is an isomorphism. Suppose given maps $f: X \longrightarrow Y$ and $g: A \longrightarrow B$ and a homotopy $h: \mathrm{Cyl}(A) \longrightarrow Y$ from $f|A$ to $\emptyset g$. Then we can extend g over X and h over $\mathrm{Cyl}(X)$ so that h becomes a homotopy from f to $\emptyset g: X \longrightarrow Y$.

The proof is similar to that of 3.2, except that we order triples (U, k', g') where $g': U \longrightarrow B$ and $k' i_1 = \emptyset g'$. The element

$$k'_n | I^+ \wedge \partial e: \ I^+ \wedge \partial e \longrightarrow Y_{n-r}$$

can be patched together with a contracting homotopy for $f|\partial e$ to define an element of $\pi_m(\emptyset_{n-r})$, say k''_n, which under the hypotheses must vanish on passing to $\emptyset_{p+n-r}$ for some p.

For later use I also record:

Lemma 3.3. Suppose that $\pi_*(Y) = 0$, and X, A is a pair of CW-spectra. Then any map $f: A \longrightarrow Y$ can be extended over X.

Proof. Exercise. Either copy the proof of 3.2 or else quote the result of 3.2.

Theorem 3.4. Let $f: E \longrightarrow F$ be a function between spectra such that $f_*: \pi_*(E) \longrightarrow \pi_*(F)$ is an isomorphism. Then for any CW-spectrum X,

$$f_*: [X, E]_* \longrightarrow [X, F]_*$$

is a (1-1) correspondence.

I emphasize that E and F are not assumed to be CW-spectra.

By analogy with the case of CW-complexes, a function $f\colon E \longrightarrow F$ between spectra such that $[X, E]_* \xrightarrow{f_*} [X, F]_*$ is a 1-1 correspondence for all CW-spectra X would be called a weak equivalence.

Proof of 3.4. (First argument). Without loss of generality we can suppose that f is an inclusion; for if not, replace F by the spectrum M in which M_n is the mapping cylinder of f_n. Then $\pi_*(F, E) = 0$ by the exact sequence. Now we see that f_* is an epimorphism by applying 3.2, taking the pair X mod A to be X mod pt. Similarly, we see that f_* is a monomorphism by applying 3.2, taking the pair X mod A to be Cyl(X) mod its ends.

(Second argument). Instead of using the mapping cylinder spectrum, use Lemma 3.2' in the above argument.

Corollary 3.5. (Compare the theorem of J. H. C. Whitehead.) Let $f\colon E \longrightarrow F$ be a morphism between CW-spectra such that

$$f_*\colon \pi_*(E) \longrightarrow \pi_*(F)$$

is an isomorphism. Then f is an equivalence in our category.

The deduction of 3.5 from 3.4 is a triviality, valid in any category.

Example. Let $f\colon E \longrightarrow F$ be a function such that $f_n\colon E_n \longrightarrow F_n$ is a homotopy equivalence for each n. Then f is an equivalence in our category.

Exercise. Use (3.5) to show that any CW-spectrum Y is equivalent in our category to an Ω_0-spectrum.

Hint: Construct a functor T_n from CW-complexes to spectra by

$$(T_n X)_r = \begin{cases} S^{r-n}X & \text{for } r \geq n \\ \text{pt.} & \text{for } r < n \end{cases}$$

Form the set of morphisms in our category

$$[T_n X, Y]_0 ,$$

and check that it is a representable functor, represented say by Z_n. Observe that the Z_n give the components of an Ω_0-spectrum Z; construct a function $Y \longrightarrow Z$ and apply 3.5.

Now I must reveal that we would really like a relative form of the theorem of J.H.C. Whitehead. If X is a spectrum, let Cone(X) be the spectrum whose n^{th} term is $I \wedge X_n$, with maps $(I \wedge X_n) \wedge S^1 \xrightarrow{1 \wedge \epsilon_n} I \wedge X_{n+1}$. (We take the base-point in I to be 0.) We have an obvious inclusion function $i: X \longrightarrow \text{Cone}(X)$ (use the end of the cone).

<u>Theorem 3.6.</u> Let $f: E, A \longrightarrow F, B$ be a function between pairs of spectra such that

$$f_*: \pi_*(E, A) \longrightarrow \pi_*(F, B)$$

is an isomorphism. Then for any CW-spectrum X,

$$f_*: [\text{Cone}(X), X; E, A]_* \longrightarrow [\text{Cone}(X), X; F, B]_*$$

is a 1-1 correspondence.

<u>Sketch proof.</u> Construct a new spectrum R (for relative) with $R_n = L(E_n, A_n)$ (the space of paths in E_n starting at the base-point and finishing in A_n) and structure maps ρ_n given by

$$L(E_n, A_n) \xrightarrow{L\epsilon'_n} L(E_{n+1}\Omega, A_{n+1}\Omega) \cong (L(E_{n+1}, A_{n+1}))\Omega ,$$

where the Ω is written on the right to keep the "loops" coordinate out of the way of the path coordinate. Similarly, construct S (not, for the moment, the sphere spectrum) with $S_n = L(F_n, B_n)$. Then f induces a function of spectra $R \longrightarrow S$, inducing an isomorphism of absolute homotopy groups. By 3.4,

$$[X, R]_* \longrightarrow [X, S]_*$$

is a 1-1 correspondence. Unwrapping this, it says

$$f_*: [\text{Cone}(X), X; E, A]_* \longrightarrow [\text{Cone}(X), X; F, B]_*$$

is a 1-1 correspondence.

This application shows why I specified that E and F in 3.4 need not be CW-spectra.

Now for any spectrum X, we will define $\text{Susp}(X)$ so that its nth term is $S^1 \wedge X_n$ and its structure maps are $(S^1 \wedge X_n) \wedge S^1 \xrightarrow{1 \wedge \xi_n} S^1 \wedge X_{n+1}$. Susp is obviously a functor.

Theorem 3.7. Susp: $[X, Y]_* \longrightarrow [\text{Susp}(X), \text{Susp}(Y)]_*$ is a 1-1 correspondence.

This theorem assures us that in some sense we did succeed in getting into a stable situation.

Proof. We have the following commutative diagram.

$$\begin{array}{ccc} [X, Y]_* & \xrightarrow{\text{Cone}} & [\text{Cone}(X), X; \text{Cone}(Y), Y]_* \\ \downarrow \text{Susp} & & \downarrow j_* \\ [\text{Susp}(X), \text{Susp}(Y)]_* & \xrightarrow{j^*} & [\text{Cone}(X), X; \text{Susp}(Y), \text{pt.}]_* \end{array}$$

Now the map "Cone" is clearly injective (since restriction gives an inverse for it) and surjective (by 3.3). Also j^* is clearly a 1-1 correspondence. The proof will be complete as soon as we show that j_* is a 1-1 correspondence, by quoting 3.6 and proving:

Lemma 3.8. $j_*: \pi_*(\text{Cone}(Y), Y) \longrightarrow \pi_*(\text{Susp}(Y), \text{pt.})$ is a 1-1 correspondence.

Consider the following commutative diagram.

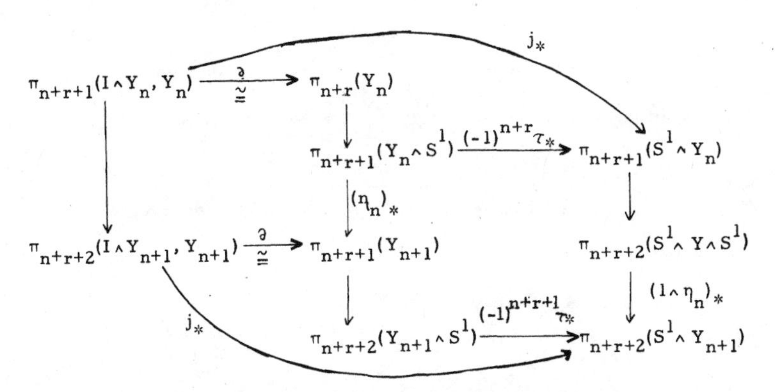

$\pi_*(\text{Cone}(Y), Y)$ is the direct limit of the left-hand column, and the diagram shows it is isomorphic to $\lim_{n\to\infty} \pi_{n+r}(Y_n)$. $\pi_*(\text{Susp}(Y), \text{pt.})$ is the direct limit of the right-hand column, and the diagram shows that it is isomorphic to the direct limit of the system in which the groups are $\pi_{n+r+1}(Y_n \wedge S^1)$ and the maps are the vertical arrows in the center column. But the center column shows that these two direct limits are the same. This proves Lemma 3.8, which proves Theorem 3.7.

Now we can remark that $[\text{Susp}(X), Z]$ is obviously a group, because in Susp(X) we have a spare suspension coordinate out in front to manipulate. And for the same reason, $[\text{Susp}^2(X), Z]$ is an abelian group. But now we can give $[X, Y]$ the structure of an abelian group, because $[X, Y]$ is in 1-1 correspondence with $[\text{Susp}^2(X), \text{Susp}^2(Y)]$, and we pull back the group structure on that. So now our sets of morphisms $[X, Y]$ are abelian groups, and it's easy to see that composition is bilinear.

Actually there is a unique way to give each set of morphisms $[X, Y]$ the structure of an abelian group so that composition is bilinear; this is standard once I've said the usual categorical things about sums and products.

Well, now I would like to say that I have an additive category. The existence of a trivial object is easy: we take the spectrum $E_n = \text{pt.}$ for all n. Then $[X, \text{pt.}] = 0$, and $[\text{pt.}, X] = 0$.

I claim this category has arbitrary sums (= coproducts). In fact, given spectra X_α for $\alpha \in A$, we form $X = \bigvee_\alpha X_\alpha$ by $X_n = \bigvee_\alpha (X_\alpha)_n$ with the obvious structure maps

$$X_n \wedge S^1 = (\bigvee_\alpha (X_\alpha)_n) \wedge S^1 = \bigvee_\alpha (X_\alpha) \wedge S^1 \xrightarrow{\bigvee_\alpha \xi_{\alpha n}} \bigvee_\alpha (X_\alpha)_{n+1} .$$

This obviously has the required property:

$$[\bigvee_\alpha X_\alpha, Y] \xrightarrow{\cong} \prod_\alpha [X_\alpha, Y] .$$

Now I must talk about cofiberings. Suppose given a map $f: X \longrightarrow Y$ between CW-spectra. It is represented by a function $f': X' \longrightarrow Y$, where X' is a cofinal subspectrum. Without loss of generality I can suppose f' is cellular, i.e., f'_n is a cellular map of CW-complexes for each n. We form the mapping cone $Y \cup_{f'} CX$ as follows: its nth term is $Y_n \cup_{f'_n} (I \wedge X'_n)$ and the structure maps are the obvious ones. If we replace X' by a smaller cofinal subspectrum X'', we get $Y \cup_{f''} CX''$ which is smaller than $Y \cup_{f'} CX'$, but cofinal in it, and so equivalent. So the construct depends essentially only on the map f, and we can write it $Y \cup_f CX$. If we vary f by a homotopy, $Y \cup_{f_0} CX$ and $Y \cup_{f_1} CX$ are equivalent, but the equivalence depends on the choice of homotopy.

Let X be a CW-spectrum, A a subspectrum. I will say A is <u>closed</u> if for every finite subcomplex $K \subset X_n$, $S^m K \subset A_{m+n}$ implies $K \subset A_n$. That is, if a cell gets into A later, I put it into A to start with. It is equivalent to saying that $A \subset B \subset X$, A cofinal in B implies that $A = B$.

Suppose that $i: X \longrightarrow Y$ is the inclusion of a closed subspectrum. Then we can form Y/X, with n^{th} term Y_n/X_n. In this case there is a map

$$r: Y \cup_i CX \longrightarrow Y/X$$

with components

$$Y_n \cup_{i_n} CX_n \longrightarrow Y_n/X_n.$$

The map r is an equivalence, by 3.5.

Let's return to the general case. We have morphisms

$$X \xrightarrow{f} Y \xrightarrow{i} Y \cup_f CX.$$

Proposition 3.9. For each Z the sequence

$$[X,Z] \xleftarrow{f^*} [Y,Z] \xleftarrow{i^*} [Y \cup_f CX, Z]$$

is exact.

The proof is the same as for CW-complexes, and is trivial, because homotopies were defined in terms of maps of cylinders.

The sequence $X \xrightarrow{f} Y \xrightarrow{i} Y \cup_f CX$, or anything equivalent to it, is called a cofibre sequence or Puppe sequence. We can extend co-fiberings to the right, by taking

$$X \xrightarrow{f} Y \xrightarrow{i} Y \cup_f CX \longrightarrow (Y \cup_f CX) \cup_i CY.$$

The last spectrum is equivalent to $(Y \cup_f CX)/Y = \mathrm{Susp}(X)$. If we continue the sequence further, we get

$$X \xrightarrow{f} Y \xrightarrow{i} Y \cup_f CX \xrightarrow{j} \mathrm{Susp}(X) \xrightarrow{-\mathrm{Susp}\, f} \mathrm{Susp}(Y),$$

as for CW-complexes. It follows that the exact sequence of Proposition 3.9 can also be extended to the right.

Proposition 3.10. The sequence

$$[W,X] \xrightarrow{f_*} [W,Y] \xrightarrow{i_*} [W, Y \cup_f CX]$$

is exact.

In other words, in our category cofiberings are the same as fiberings.

<u>Proof.</u> Since $if \sim 0$, $i_* f_* = 0$. Suppose given $g: W \longrightarrow Y$ such that $ig \sim 0$. Then we can construct the following diagram of cofiberings.

$$\begin{array}{ccccccccc}
X & \xrightarrow{f} & Y & \xrightarrow{i} & Y \cup_f CX & \xrightarrow{j} & \mathrm{Susp}(X) & \xrightarrow{-\mathrm{Susp}\, f} & \mathrm{Susp}(Y) \\
 & & \uparrow{\scriptstyle g} & & \uparrow{\scriptstyle h} & & \uparrow{\scriptstyle k} & & \uparrow{\scriptstyle \mathrm{Susp}\, g} \\
W & \xrightarrow{1} & W & \xrightarrow{i} & CW & \xrightarrow{j} & \mathrm{Susp}(W) & \xrightarrow{-1} & \mathrm{Susp}(W)
\end{array}$$

(The homotopy $ig \sim 0$ gives us h, and the rest follows automatically.)

Now by Theorem 3.7 we have $k = \mathrm{Susp}\, \ell$ for some $\ell \in [W, X]$, and

$$(-\mathrm{Susp}\, f)(\mathrm{Susp}\, \ell) \simeq (\mathrm{Susp}\, g)(-1)$$

i.e.,

$$(\mathrm{Susp}(f\ell) \simeq \mathrm{Susp}\, g$$

so using Theorem 3.7 again, we have $f\ell \simeq g$. This proves Proposition 3.10.

<u>Proposition 3.11.</u> Finite sums are products.

In fact,

$$X \longrightarrow X \vee Y \longrightarrow Y$$

is clearly a cofibering, because $(X \vee Y) \cup CX \simeq Y$. So by 3.10,

$$[W, X] \longrightarrow [W, X \vee Y] \longrightarrow [W, Y]$$

is exact; but it is clearly split short exact, so $[W, X \vee Y] \cong [W, X] \oplus [W, Y]$ and $X \vee Y$ is also the product of X and Y.

Now I know that my category is an additive category.

<u>Theorem 3.12.</u> The Representability Theorem of E.H. Brown is valid in the category of CW-spectra and morphisms of degree 0.

The proof is as usual, but arrange the induction right.

Exercise. Use 3.12 to show that any spectrum Y is weakly equivalent to a CW-spectrum. (Consider the functor $[X, Y]_0$.)

Proposiition 3.13. The stable category has arbitrary products.

Proof. The functor of X given by

$$\prod_\alpha [X, Y_\alpha]_0$$

satisfies the data of Brown's theorem, so it is representable. Now we see that this representing object works for maps of degree r as well.

Note next that for any collection of X_α we have a morphism

$$\bigvee_\alpha X_\alpha \longrightarrow \prod_\alpha X_\alpha$$

Namely, for each α and β I have to give a component which is a map $X_\alpha \longrightarrow X_\beta$; I take it to be 1 if $\alpha = \beta$, 0 if $\alpha \neq \beta$.

Proposition 3.14. (This form is due to Boardman). Suppose that for each n, $\pi_n(X_\alpha) = 0$ for all but a finite number of α. Then the map

$$\bigvee_\alpha X_\alpha \longrightarrow \prod_\alpha X_\alpha$$

is an equivalence.

Proof. First note that

$$\pi_n(X_1 \vee X_2) \cong \pi_n(X_1) \oplus \pi_n(X_2)$$

under the obvious maps. (See 3.11.) (Exercise. Prove this directly from the definitions of π_* and $X_1 \vee X_2$.) By induction, we have

$$\pi_n(X_1 \vee \dots \vee X_m) \cong \sum_{i=1}^{m} \pi_n(X_i)$$

for finite wedges. Now we have

$$\pi_n(\bigvee_\alpha X_\alpha) = \sum_\alpha \pi_n(X_\alpha)$$

by passing to direct limits. Also

$$\pi_n(\prod_\alpha X_\alpha) = \prod_\alpha \pi_n(X_\alpha), \text{ by definition.}$$

Now the data was chosen precisely so that $\sum_\alpha \pi_n(X_\alpha) \longrightarrow \prod_\alpha \pi_n(X_\alpha)$ is an isomorphism. Therefore $\bigvee_{\alpha} X_{\alpha} \longrightarrow \coprod_{\alpha} X_{\alpha}$ is an equivalence, by 3.5.

Remark. If we use the direct proof that

$$\pi_n(X_1 \vee X_2) \cong \pi_n(X_1) \oplus \pi_n(X_2)$$

this gives a proof that finite sums are products, independently of 3.7, but depending on Brown's theorem. This can be used, in a way which is familiar to categorists, to define an addition in the sets $[X, Y]$; this way of introducing the addition is independent of 3.7. Of course you have to show that the addition makes the sets $[X, Y]$ into abelian groups; the main point is to establish the existence of inverses. I recommend making use of an argument which is standard for H-spaces, as follows. Since $X \vee X$ is both a sum and product, you can make a map

$$X \vee X \longrightarrow X \vee X$$

with components $\begin{bmatrix} 1 & 1 \\ 0 & 1 \end{bmatrix}$. Check that it satisfies the hypotheses of 3.5, so it has an inverse. The inverse has the form $\begin{bmatrix} 1 & \nu \\ 0 & 1 \end{bmatrix}$. But you know the inverse of $\begin{bmatrix} 1 & 1 \\ 0 & 1 \end{bmatrix}$ is $\begin{bmatrix} 1 & -1 \\ 0 & 1 \end{bmatrix}$; so you use ν for inversion and it works.

4. SMASH PRODUCTS

In this section we will construct smash products of spectra. More precisely, we will construct from any two CW-spectra X and Y a CW-spectrum $X \wedge Y$, so as to have the properties stated in the following theorem, among other properties.

Theorem 4.1. (a) $X \wedge Y$ is a functor of two variables, with arguments and values in the (graded) stable homotopy category.

(b) The smash-product is associative, commutative, and has the sphere-spectrum S as a unit, up to coherent natural equivalences.

We explain that statement (a) is to be taken in the graded sense. That is, if

$$f \in [X, X']_r, \quad g \in [Y, Y']_s$$

then

$$f \wedge g \in [X \wedge Y, X' \wedge Y']_{r+s},$$

and besides $1 \wedge 1 = 1$, we have

$$(f \wedge g)(h \wedge k) = (-1)^{bc}(fh) \wedge (gk)$$

if $f \in [X', X'']_a$, $h \in [X, X']_b$, $g \in [Y', Y'']_c$, $k \in [Y, Y']_d$.

We explain statement (b). It claims that there are the following equivalences in our category.

$$a = a(X, Y, Z) : (X \wedge Y) \wedge Z \longrightarrow X \wedge (Y \wedge Z),$$

$$c = c(X, Y) : X \wedge Y \longrightarrow Y \wedge X,$$

$$\ell = \ell(Y) : S \wedge Y \longrightarrow Y,$$

$$r = r(X) : X \wedge S \longrightarrow X.$$

They are all of degree 0. They are all natural as X, Y, and Z vary over the stable category; in the case of c this means that the diagram

$$\begin{array}{ccc} X \wedge Y & \xrightarrow{c} & X \wedge Y \\ \downarrow{\scriptstyle f \wedge g} & & \downarrow{\scriptstyle g \wedge f} \\ X' \wedge Y' & \xrightarrow{c} & Y' \wedge X' \end{array}$$

is commutative up to a sign $(-1)^{pq}$, if $f \in [X, X']_p$, $g \in [Y, Y']_q$. The other naturality conditions are the obvious ones and don't involve signs. The equivalences make the following diagrams commute in our category. (If one thinks in terms of representative maps, one says that these diagrams are homotopy-commutative.)

(i)

$$
\begin{array}{c}
((W \wedge X) \wedge Y) \wedge Z \xrightarrow{a_1} (W \wedge X) \wedge (Y \wedge Z) \xrightarrow{a_2} W \wedge (X \wedge (Y \wedge Z)) \\
((W \wedge X) \wedge Y) \wedge Z \xrightarrow{a_3 \wedge 1} (W \wedge (X \wedge Y)) \wedge Z \xrightarrow{a_4} W \wedge ((X \wedge Y) \wedge Z) \xleftarrow{1 \wedge a_5} W \wedge (X \wedge (Y \wedge Z))
\end{array}
$$

Here $a_1 = a(W \wedge X, Y, Z)$ $\quad a_4 = a(W, X \wedge Y, Z)$

$a_2 = a(W, X, Y \wedge Z)$ $\quad a_5 = a(X, Y, Z)$

$a_3 = a(W, X, Y)$.

(ii)

$$
X \wedge Y \xrightarrow{c_1} Y \wedge X \xrightarrow{c_2} X \wedge Y, \qquad X \wedge Y \xrightarrow{1} X \wedge Y
$$

Here $c_1 = c(X, Y)$

$c_2 = c(Y, X)$.

(iii)

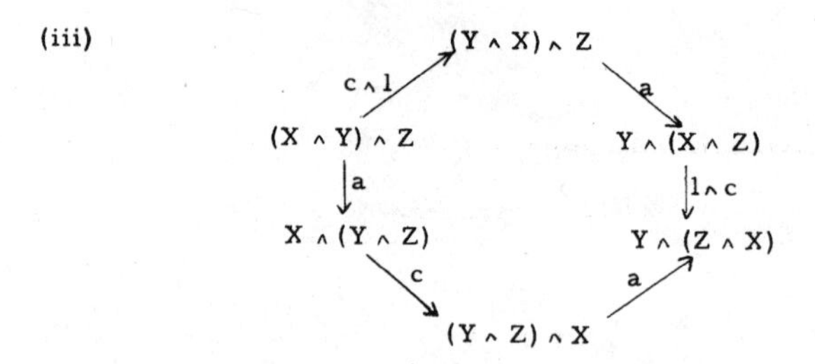

Here the morphisms can be made precise, as in (i) and (ii).

(iv)

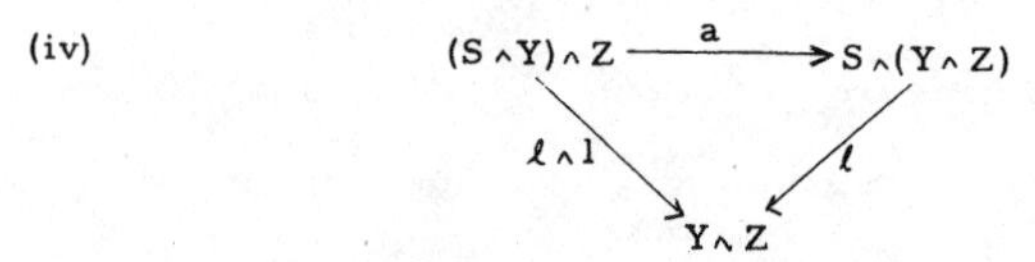

(v)

$$
(X \wedge S) \wedge Z \xrightarrow{a} X \wedge (S \wedge Z), \qquad (X \wedge S) \wedge Z \xrightarrow{r \wedge 1} X \wedge Z \xleftarrow{1 \wedge \ell} X \wedge (S \wedge Z)
$$

(vi)

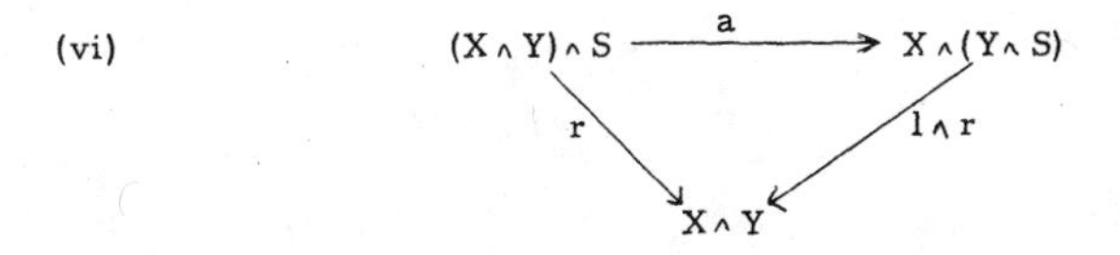

(vii)

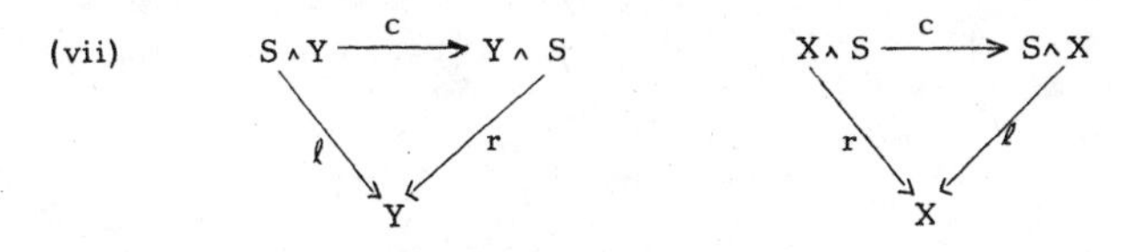

(These are equivalent, in view of (iii).)

(viii) $S \wedge S \overset{c}{\underset{1}{\rightrightarrows}} S \wedge S$

It follows from these properties that every other diagram constructed from a, c, ℓ, and r which you might conceivably wish to prove commutative, is commutative; see MacLane [8].

The properties stated in this theorem are not intended to be a complete list. We also want our smash-products to be compatible with those which we already have for CW-complexes. We can take it as a guiding idea that if X is a CW-spectrum with terms X_n, and Y is a CW-spectrum with terms Y_m, then we want $X \wedge Y$ to be the thing to which $X_n \wedge Y_m$ tends as n and m tend to infinity. It is therefore tempting to define a product spectrum P so that

$$P_p = X_{n(p)} \wedge Y_{m(p)} ,$$

where $n(p)$ and $m(p)$ are fixed functions such that $n(p) + m(p) = p$, while $n(p) \longrightarrow \infty$ and $m(p) \longrightarrow \infty$ as $p \longrightarrow \infty$. This approach gives the "handicrafted smash products" (in later versions, "naive smash products") of Boardman. Of course, there are many different ways of choosing the functions $n(p)$ and $m(p)$, and these give rise to different "handicrafted smash products"; it is obviously desirable to prove that these different products are related by natural equivalences. For later work it is also desirable to have a notation more convenient than that of functions $n(p)$ and $m(p)$; it is for this purpose that we introduce the details which follow next.

Let A be an ordered set, isomorphic to the ordered set $\{0, 1, 2, 3, \ldots\}$. (The reason that we do not take A to be the ordered set $\{0, 1, 2, 3, \ldots\}$ is that we will later want to take A to be a subset of $\{0, 1, 2, 3, \ldots\}$.) Let B be a subset of A; then we define a corresponding function

$$\beta: A \longrightarrow \{0, 1, 2, 3, \ldots\}$$

as follows: $\beta(a)$ is the number of elements $b \in B$ such that $b < a$. Then β is monotonic, and $\beta | B$ is an order-preserving isomorphism between B and some initial segment of $\{0, 1, 2, 3, \ldots\}$. The notation β emphasizes the dependence of β on B rather than on A; this is legitimate, for if we have $B \subset A \subset A'$, then the function β_A defined on A is the restriction to A of the function $\beta_{A'}$ defined in A'.

Next suppose given a partition of A into two subsets B and C, so that $A = B \cup C$, $B \cap C = \emptyset$. A suitable illustration is obtained by taking

$$A = \{0, 1, 2, 3, \ldots\}$$

$$B = \{0, 2, 4, 6, \ldots\}$$

$$C = \{1, 3, 5, 7, \ldots\}$$

but there are many other equally suitable choices. Then we define a smash-product functor which assigns to any two CW-spectra X and Y a CW-spectrum $X \wedge_{BC} Y$. It is convenient to display only B and C in the notation, but of course the product depends on the ordering of $B \cup C$.

The terms of the product spectrum

$$P = X \wedge_{BC} Y$$

are given by

$$P_{\alpha(a)} = X_{\beta(a)} \wedge Y_{\gamma(a)} .$$

Note that α is an isomorphism from the ordered set $A = B \cup C$ to $\{0, 1, 2, 3, \ldots\}$ and β, γ are monotonic functions from $A = B \cup C$ to the set $\{0, 1, 2, 3, \ldots\}$ such that $\beta(a) + \gamma(a) = \alpha(a)$.

The maps of the product spectrum are defined as follows. We have

$$P_{\alpha(a)} \wedge S^1 \longrightarrow X_{\beta(a)} \wedge Y_{\gamma(a)} \wedge S^1 .$$

Here it is convenient to regard S^1 as R^1 compactified by adding a point at infinity, which becomes the base-point. This allows us to define a map of degree -1 from S^1 to S^1 by $t \longmapsto -t$.

If $a \in B$, then

$$P_{\alpha(a)+1} = X_{\beta(a)+1} \wedge Y_{\gamma(a)}$$

and we define the map

$$\pi_{\alpha(a)} : SP_{\alpha(a)} \longrightarrow P_{\alpha(a)+1}$$

by

$$\pi_{\alpha(a)}(x \wedge y \wedge t) = \xi_{\beta(a)}(x \wedge (-1)^{\gamma(a)} t) \wedge y .$$

If $a \in C$, then

$$P_{\alpha(a)+1} = X_{\beta(a)} \wedge Y_{\gamma(a)+1}$$

and we define the map

$$\pi_{\alpha(a)} : SP_{\alpha(a)} \longrightarrow P_{\alpha(a)+1}$$

by

$$\pi_{\alpha(a)}(x \wedge y \wedge t) = x \wedge \eta_{\gamma(a)}(y \wedge t) .$$

Here

$$x \in X_{\beta(a)}, \quad y \in Y_{\gamma(a)}, \quad t \in S^1 ,$$

and

$$\xi_{\beta(a)}: X_{\beta(a)} \wedge S^1 \longrightarrow X_{\beta(a)}, \qquad \eta_{\gamma(a)}: Y_{\gamma(a)} \wedge S^1 \longrightarrow Y_{\alpha(a)+1}$$

are the appropriate maps from the spectra X, Y. The sign $(-1)^{\gamma(a)}$ is introduced, of course, because we have moved S^1 across $Y_{\gamma(a)}$.

It is clear that $P = X \wedge_{BC} Y$ is functorial for functions of X and Y of degree 0. Next we point out that we have not assumed that the sets B and C are infinite. In the obvious applications they are infinite, so that $\beta(a) \longrightarrow \infty$ and $\gamma(a) \longrightarrow \infty$; but it is convenient to allow B and C to be finite. For example, let $\underline{S}^1$ be the suspension spectrum of S^1; then $\underline{S}^1 \wedge_{\emptyset, A} Y = \mathrm{Susp}(Y)$. If B is infinite, and X' is a cofinal subspectrum of X, then $X' \wedge_{BC} Y$ is a cofinal subspectrum $X \wedge_{BC} Y$. So in this case $X \wedge_{BC} Y$ is natural for maps of Y of degree 0. Next we observe that $(\mathrm{Cyl}(X) \wedge_{BC} Y$ and $X \wedge_{BC} (\mathrm{Cyl}(Y))$ can be identified with $\mathrm{Cyl}(X \wedge_{BC} Y)$. It follows that the homotopy class of $f \wedge_{BC} g$ depends only on the homotopy class of f (if B is infinite) or g (if C is infinite).

We propose to construct $X \wedge Y$ to have the properties stated in the following theorem.

THEOREM 4.2. For each choice of B, C there is a morphism

$$eq_{BC}: X \wedge_{BC} Y \longrightarrow X \wedge Y \qquad \text{(of degree 0)}$$

with the following properties.

(i) If B is infinite and $f: X \longrightarrow X'$ is a morphism of degree 0, then the following diagram is commutative.

$$\begin{array}{ccc} X \wedge_{BC} Y & \xrightarrow{eq_{BC}} & X \wedge Y \\ \Big\downarrow {\scriptstyle f \wedge_{BC} 1} & & \Big\downarrow {\scriptstyle f \wedge 1} \\ X' \wedge_{BC} Y & \xrightarrow{eq_{BC}} & X' \wedge Y \end{array}$$

(ii) If C is infinite and $g: Y \longrightarrow Y'$ is a morphism of degree 0, then the following diagram is commutative.

$$\begin{array}{ccc} X \wedge_{BC} Y & \xrightarrow{eq_{BC}} & X \wedge Y \\ {\scriptstyle 1 \wedge_{BC} g}\downarrow & & \downarrow{\scriptstyle 1 \wedge g} \\ X \wedge_{BC} Y' & \xrightarrow{eq_{BC}} & X \wedge Y' \end{array}$$

(iii) The morphism $eq_{BC}: X \wedge_{BC} Y \longrightarrow X \wedge Y$ is an equivalence if any one of the following conditions is satisfied.

(a) B and C are infinite.

(b) B is finite, say with d elements, and $\xi_r: SX_r \longrightarrow X_{r+1}$ is an isomorphism for $r \geq d$.

(c) C is finite, say with d elements, and $\eta_r: SY_r \longrightarrow Y_{r+1}$ is an isomorphism for $r \geq d$.

Let me show how Theorem 4.2 will help to prove Theorem 4.1(b). Consider first the associativity. The point is that the "handicrafted smash products" are actually associative if you pick the right product at each point. More precisely, take a set A and partition it into three disjoint subsets B, C, and D, such that $B \cup C$ and $C \cup D$ are infinite. Let X, Y, and Z be CW-spectra. Then we can form the spectra

$$(X \wedge_{BC} Y) \wedge_{B \cup C, D} Z \quad \text{and} \quad X \wedge_{B, C \cup D} (Y \wedge_{CD} Z).$$

(Now one begins to see the purpose for which the notation was designed.) These two spectra are actually the same spectrum. For the terms of each are given by

$$P_{\alpha(a)} = X_{\beta(a)} \wedge Y_{\gamma(a)} \wedge Z_{\delta(a)} .$$

The maps of each are described in the same way as before. We have

$$SP_{\alpha(a)} = X_{\beta(a)} \wedge Y_{\gamma(a)} \wedge Z_{\delta(a)} \wedge S^1 .$$

If $a \in B$, then

$$P_{\alpha(a)+1} = Z_{\beta(a)+1} \wedge Y_{\gamma(a)} \wedge Z_{\delta(a)},$$

and we have

$$\pi_{\alpha(a)}(x \wedge y \wedge z \wedge t) = \xi_{\beta(a)}(x \wedge (-1)^{\gamma(a)+\delta(a)} t) \wedge y \wedge z .$$

If $a \in C$ then

$$P_{\alpha(a)+1} = X_{\beta(a)} \wedge Y_{\gamma(a)+1} \wedge Z_{\delta(a)}$$

and we have

$$\pi_{\alpha(a)}(x \wedge y \wedge z \wedge t) = x \wedge \eta_{\gamma(a)}(y \wedge (-1)^{\delta(a)} t) \wedge z .$$

If $a \in D$ then

$$P_{\alpha(a)+1} = X_{\beta(a)} \wedge Y_{\gamma(a)} \wedge Z_{\delta(a)+1}$$

and we have

$$\pi_{\alpha(a)}(x \wedge y \wedge z \wedge t) = x \wedge y \wedge \zeta_{\delta(a)}(z \wedge t).$$

Here, of course, we have $x \in X_{\beta(a)}$, $y \in Y_{\gamma(a)}$, $z \in Z_{\delta(a)}$, $t \in S^1$ and $\xi_{\beta(a)}$, $\eta_{\gamma(a)}$, $\zeta_{\delta(a)}$ are the appropriate maps of the spectra X, Y, Z.

We will arrange our construction to have the following property.

THEOREM 4.3. The equivalence

$$a = a(X, Y, Z): (X \wedge Y) \wedge Z \longrightarrow X \wedge (Y \wedge Z)$$

makes the following diagram commutative for each choice of B, C, D.

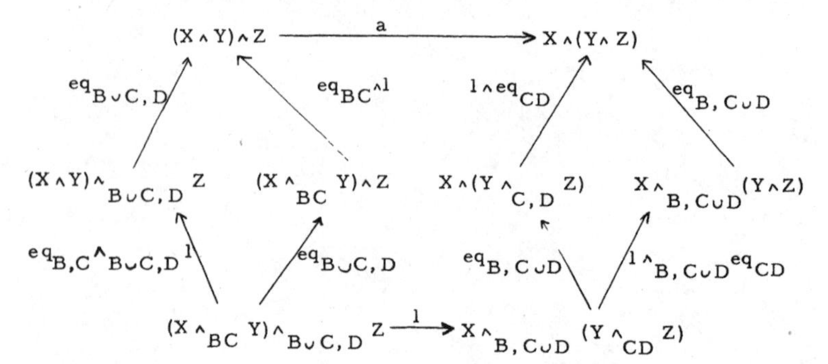

Note that the squares are commutative by the naturality of eq; we can apply 4.2 (i) and (ii) since $B \cup C$ and $C \cup D$ are infinite.

Let us now show how to check the commutativity of diagram (i) in Theorem 4.1(b) (the pentagon diagram). Take a set A and partition it into four infinite subsets B, C, D and E. Then by Theorems 4.2 and 4.3, all we have to do is check that the following diagram is commutative.

$$\begin{array}{ccc}
& (W \wedge_{BC} X) \wedge_{B\cup C,\, D\cup E} (Y \wedge_{DE} Z) & \\
\overset{1}{\nearrow} & & \overset{1}{\searrow} \\
((W \wedge_{BC} X) \wedge_{B\cup C,\, D} Y) \wedge_{B\cup C\cup D,\, E} Z & & W \wedge_{B,\, C\cup D\cup E} (X \wedge_{C,\, D\cup E} (Y \wedge_{DE} Z)) \\
\overset{1}{\searrow} & & \overset{1}{\nearrow} \\
(W \wedge_{B,\, C\cup D} (X \wedge_{CD} Y)) \wedge_{B\cup C\cup D,\, E} Z & \xrightarrow{\ 1\ } & W \wedge_{B,\, C\cup D\cup E} ((X \wedge_{CD} Y) \wedge_{C\cup D,\, E} Z)
\end{array}$$

This diagram is commutative as a diagram of functions before we pass to homotopy classes.

Similarly, the "handicrafted smash products" are commutative if you pick the right product at each point. It is tempting to partition A as $B \cup C$, and consider $X \wedge_{BC} Y$ and $Y \wedge_{CB} X$. Corresponding terms of these spectra are isomorphic; it is tempting to define

$$c_{\alpha(a)} : X_{\beta(a)} \wedge Y_{\gamma(a)} \longrightarrow Y_{\gamma(a)} \wedge X_{\beta(a)}$$

by

$$c_{\alpha(a)}(x \wedge y) = y \wedge x \ .$$

However, these components do not give a function between spectra, because the relevant diagrams do not commute. We should have inserted a sign $(-1)^{\beta(a)\gamma(a)}$, and we do not have a spare suspension coordinate to reverse. The answer is easy; we need only consider partitions $A = B \cup C$ such that $\beta(a)\gamma(a)$ is always even. This amounts to the following condition. Elements number 0 and 1 in A must be either two elements of B, or else two elements of C. Similarly for elements number 2 and 3, and similarly for elements number $2r$ and $2r+1$ for each r.

Now that we realize we can restrict the choice of partition in this way, we see that it is easy and useful to go further. In fact, we now introduce the following restriction on the partition $A = B \cup C$.

Condition (4.4). Elements number $0, 1, 2,$ and 3 in A are either four elements of B, or else four elements of C; similarly for elements number $4, 5, 6$ and 7 in A, and similarly for elements number $4r$, $4r+1$, $4r+2$ and $4r+3$ for each r.

With this restriction, we define an isomorphism

$$c = c_{BC}: X \wedge_{BC} Y \longrightarrow Y \wedge_{CB} X$$

in the manner suggested:

$$c_n(x \wedge y) = y \wedge x.$$

This is clearly natural for functions of X and Y. It is also natural for maps of X if B is infinite; similarly for Y if C is infinite.

We will arrange our constructions to have the following property.

THEOREM 4.5. The equivalence $c = c(X, Y): X \wedge Y \longrightarrow Y \wedge X$ makes the following diagram commutative for each choice of B, C satisfying (4.4).

$$\begin{array}{ccc} X \wedge Y & \xrightarrow{c} & Y \wedge X \\ \uparrow{\scriptstyle eq_{BC}} & & \uparrow{\scriptstyle eq_{CB}} \\ X \wedge_{BC} Y & \xrightarrow{c_{BC}} & Y \wedge_{CB} X \end{array}$$

Let us now show how to check the commutativity of diagram (iii) in Theorem 4.1(b) (the hexagon diagram). Take a set A and partition it into three infinite subsets B, C and D satisfying the obvious analogue of condition 4.4. Then by Theorems 4.2, 4.3 and 4.5, all we have to do is to check that the following diagram is commutative.

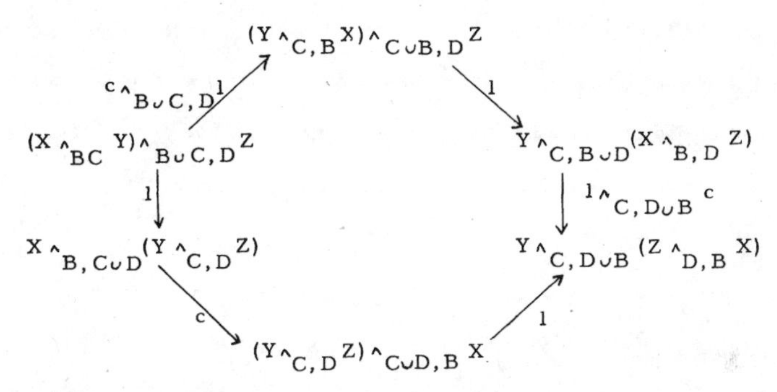

This diagram is commutative as a diagram of functions.

Similarly, suppose we wish to check the commutativity of diagram (ii) in Theorem 4.1(b). By Theorems 4.2 and 4.5, all we have to do is check that the following diagram is commutative.

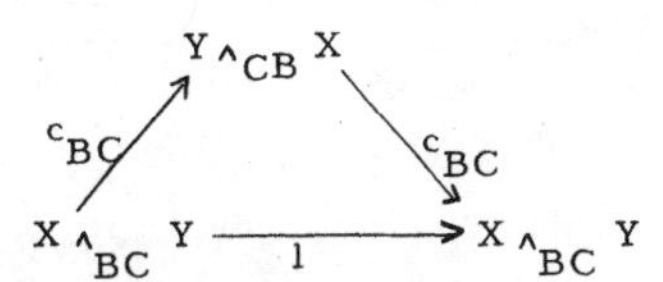

This diagram, too, is commutative as a diagram of functions.

Similarly, the "handicrafted smash products" have S as a unit if you pick the right product at each point. More precisely, suppose we partition A as $\emptyset \cup A$. This is a legitimate partition satisfying the condition (4.4); this was the reason that we allowed the set B to be finite. We can form the spectrum $S \wedge_{\emptyset A} Y$, and it is isomorphic to Y; the obvious isomorphism has as its components the isomorphisms $S^o \wedge Y_n \cong Y_n$. This isomorphism is natural for morphisms of degree 0. We can now define

$$\ell : S \wedge Y \longrightarrow Y$$

to be the composite

$$S\wedge Y \xleftarrow{eq_{\emptyset,A}} S\wedge_{\emptyset A} Y \cong Y\ .$$

Here $eq_{\emptyset A}$ is an equivalence by 4.2(iii)(b). Similarly, we can form the spectrum $X \wedge_{A\emptyset} S$, and it is isomorphic to X; the obvious isomorphism has as its components the isomorphisms $X_n \wedge S^0 \cong X_n$. As before, this isomorphism is natural for morphisms of degree 0. We now define

$$r: X\wedge S \longrightarrow X$$

to be the composite

$$X\wedge S \xleftarrow{eq_{A\emptyset}} X \wedge_{A\emptyset} S \cong X.$$

Here $eq_{A\emptyset}$ is an equivalence by 4.2(iii)(c).

To check the commutativity of diagrams (iv), (v), (vi) and (vii) in Theorem 4.1(b), we have only to check that the following diagrams are commutative.

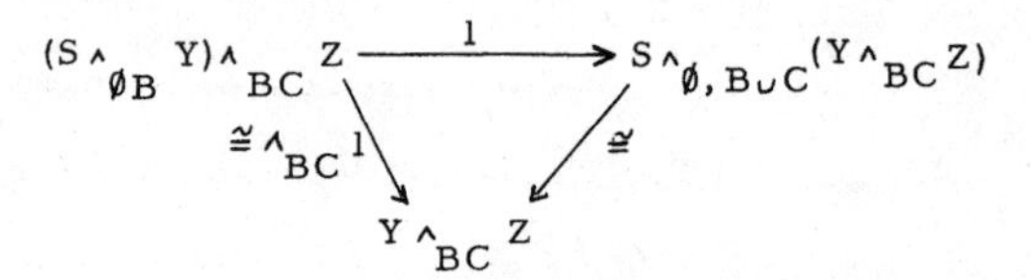

$$\begin{array}{ccc} (X\wedge_{B\emptyset} S)\wedge_{BC} Z & \xrightarrow{\quad 1 \quad} & X\wedge_{BC}(S\wedge_{\emptyset C} Z) \\ & {\scriptstyle \cong\wedge_{BC}1}\searrow \quad \swarrow{\scriptstyle 1\wedge_{BC}\cong} & \\ & X\wedge_{BC} Z & \end{array}$$

$$\begin{array}{ccc} (X\wedge_{BC} Y)\wedge_{B\cup C,\emptyset} S & \longrightarrow & X\wedge_{BC}(Y\wedge_{C\emptyset} S) \\ & {\scriptstyle \cong}\searrow \quad \swarrow{\scriptstyle 1\wedge_{BC}\cong} & \\ & X\wedge_{BC} Y & \end{array}$$

These diagrams are all commutative as diagrams of functions.

Finally, we comment on part (viii) of Theorem 4.1(b). If you believe any of these results you must believe that $S \wedge S$ is equivalent to S. So $[S \wedge S, S \wedge S]_0 \cong [S, S]_0 = Z$. So all we have to do is check that $c: S \wedge S \longrightarrow S \wedge S$ has degree 1; but we shall make all our constructions to have the obvious effect on orientations.

We now turn to the constructions necessary to prove Theorems 4.1, 4.2, 4.3 and 4.5. First we give a simple construction which is used in proving Theorem 4.2; this is the telescope functor. If $f_n : X_n \longrightarrow Y_n$ is a sequence of maps of CW-complexes, we can form the iterated mapping cylinder, or telescope. If the f_n are taken to be cellular, the telescope is a CW-complex. We apply this construction to the terms of a spectrum of certain form. Let X be a spectrum consisting of CW-complexes X_n with base-point and cellular maps $\xi_n : X_n \wedge S^1 \longrightarrow X_{n+1}$; we need not even assume that ξ_n is an isomorphism from $X_n \wedge S^1$ to a subcomplex of X_{n+1}; the telescope functor Tel will convert a spectrum X which does not have this property into one which does.

We take the half-line $i \geq 0$ and divide it into 0-cells $[i]$ and 1-cells $[i, i+1]$ for $i = 0, 1, 2, \ldots$. We define the nth term of Tel(X) as a quotient space of the following wedge-sum:

$$\left(\bigvee_{i=0}^{n-1} [i, i+1]^+ \wedge X_i \wedge S^{n-i} \right) \vee \left(\bigvee_{i=0}^{n} [i]^+ \wedge X_i \wedge S^{n-i} \right).$$

Here it is convenient to regard S^m as R^m compactified by adding a point at infinity, which becomes the base-point. In this way the isomorphism $R^m \times R^n \longrightarrow R^{m+n}$ gives an isomorphism $S^m \wedge S^n \longrightarrow S^{m+n}$ which is convenient for later use. The following identifications are to be made. Identify the point

$$i \wedge x \wedge t \in [i, i+1]^+ \wedge X_i \wedge S^{n-i}$$

with the point

$$i \wedge x \wedge t \in [i]^+ \wedge X_i \wedge S^{n-i}.$$

Identify the point

$$(i+1) \wedge x \wedge t \wedge u \in [i, i+1]^+ \wedge X_i \wedge S^1 \wedge S^{n-i-1}$$

with the point

$$(i+1) \wedge \xi_i(x \wedge t) \wedge u \in [i+1]^+ \wedge X_{i+1} \wedge S^{n-i-1} .$$

We give $Tel(X)_n$ the obvious structure as a CW-complex.

The nth map of the spectrum $Tel(X)$ is obtained by passing to quotients from the obvious isomorphism of

$$\{(\bigvee_{i=0}^{n} [i]^+ \wedge X_i \wedge S^{n-i}) \vee (\bigvee_{i=0}^{n-1} [i, i+1]^+ \wedge X_i \wedge S^{n-i})\} \wedge S^1$$

with

$$(\bigvee_{i=0}^{n} [i]^+ \wedge X_i \wedge S^{n-i+1}) \vee (\bigvee_{i=0}^{n-1} [i, i+1]^+ \wedge X_i \wedge S^{n-i+1}) .$$

There is an obvious homotopy equivalence $r_n \colon Tel(X)_n \longrightarrow X_n$ (collapse the telescope to its right-hand end $[n]^+ \wedge X_n \wedge S^0$). These equivalences give the components of a function $r \colon Tel(X) \longrightarrow X$. This function is a weak equivalence, by 3.4.

We pause to observe that this construction is functorial. It is clear that a function $f \colon X \longrightarrow Y$ induces $Tel(f) \colon Tel(X) \longrightarrow Tel(Y)$. If X' is a subspectrum of X, then $Tel(X')$ is a subspectrum of $Tel(X)$. Unfortunately, if X is a CW-spectrum and X' is cofinal in X, it does not follow that $Tel(X')$ is cofinal in $Tel(X)$. So we avoid saying that a map of X induces a map of $\mathrm{Tel}(X)$. However, the injection $\mathrm{Tel}(X') \longrightarrow \mathrm{Tel}(X)$ is a homotopy equivalence, as we see using Theorem 3.5. Moreover, we can identify $\mathrm{Tel}(\mathrm{Cyl}(X))$ with $\mathrm{Cyl}(\mathrm{Tel}(X))$. It follows that a homotopy class of maps of X induces a homotopy class of maps of $\mathrm{Tel}(X)$. We can now remark that r is a natural transformation. These facts are, of course, fairly trivial, but we need to cite this passage later; it is for this reason that I have avoided a short-cut--one could define Tel on morphisms by requiring that r be natural.

We propose to arrange for Theorem 4.2 to be true by constructing $X \wedge Y$ so that it contains a copy of $\mathrm{Tel}(X \wedge_{BC} Y)$ for each choice of B and C. The morphism

$$eq_{BC}: X \wedge_{BC} Y \longrightarrow X \wedge Y$$

will be defined as the following composite:

$$X \wedge_{BC} Y \xleftarrow{\ r\ } \mathrm{Tel}(X \wedge_{BC} Y) \longrightarrow X \wedge Y.$$

The construction of $X \wedge Y$ (call it P) is as a "double telescope." That is, just as the parts of $\mathrm{Tel}(X)$ corresponded to the cells of a cell-decomposition of the half-line $i \geq 0$, so here we make a similar use of the quarter-plane $i \geq 0$, $j \geq 0$. We divide the half-line $i \geq 0$ with 0-cells $[i]$ and 1-cells $[i, i+1]$, $i = 0, 1, 2, 3 \ldots$. We divide the half-line $j \geq 0$ similarly, and we divide the quarter-plane $i \geq 0$, $j \geq 0$ into the products of these cells. Thus we have four cells e_{ij} with bottom left-hand corner at (i, j):

the 0-cell $[i] \times [j]$

the 1-cells $[i, i+1] \times [j]$ and $[i] \times [j, j+1]$

and the 2-cell $[i, i+1] \times [j, j+1]$.

To construct P_n we use those cells e_{ij} which lie entirely in the part of the quarter-plane given by $x+y \leq n$. The condition for this is

$$i+j + \dim(e_{ij}) \leq n.$$

Let us start from

$$\bigvee_{i+j\leq n} ([i]\times[j])^+ \wedge X_i \wedge Y_j \wedge S^{n-i-j}$$

and attach

$$\bigvee e_{ij}^+ \wedge X_i \wedge Y_j \wedge S^1 \wedge S^{n-i-j-1}$$

where e_{ij} runs over the 1-cells $[i,i+1]\times[j]$ and $[i]\times[j,j+1]$ such that $i+j+1 \leq n$. The identifications are obvious. The point

$$(i,j)\wedge x\wedge y\wedge s\wedge t \quad \text{in} \quad e_{ij}^+ \wedge X_i \wedge Y_j \wedge S^1 \wedge S^{n-i-j-1}$$

is to be identified with

$$(i,j)\wedge x\wedge y\wedge (s\wedge t) \quad \text{in} \quad ([i]\times[j])^+ \wedge X_i \wedge Y_j \wedge S^{n-i-j}.$$

The point

$$(i+1,j)\wedge x\wedge y\wedge s\wedge t \quad \text{in} \quad ([i,i+1]\times[j])^+ \wedge X_i \wedge Y_j \wedge S^1 \wedge S^{n-i-j-1}$$

is to be identified with

$$(i+1,j)\wedge \xi_i(x\wedge(-1)^j s)\wedge y\wedge t \quad \text{in} \quad ([i+1]\times[j])^+ \wedge X_{i+1} \wedge Y_j \wedge S^{n-i-j-1}$$

The point

$$(i,j+1)\wedge x\wedge y\wedge s\wedge t \qquad \text{in} \quad ([i]\times[j,j+1])^+ \wedge X_i \wedge Y_j \wedge S^1 \wedge S^{n-i-j-1}$$

is to be identified with

$$(i,j+1)\wedge x\wedge \eta_j(y\wedge s)\wedge t \quad \text{in} \quad ([i]\times[j+1])^+ \wedge X_i \wedge Y_{j+1} \wedge S^{n-i-j-1}.$$

Consider now a cell $e = [i,i+1]\times[j,j+1]$ such that $i+j+2 \leq n$. We have just described the subcomplex of P_n corresponding to ∂e. Moreover, it contains a family of subspaces $X_i \wedge Y_j \wedge S^2 \wedge S^{n-i-j-2}$, parametrized by the points of ∂e. Unfortunately, this family is not a product family, at least, not in a completely trivial way. Let us start from the point

$$(i,j)\wedge x\wedge y\wedge s\wedge t\wedge u \quad \text{in} \quad ([i]\times[j])^+ \wedge X_i \wedge Y_j \wedge S^1 \wedge S^1 \wedge S^{n-i-j-2}.$$

If we first increase i and then increase j, we get first to $(i+1,j)\wedge\xi_i(x\wedge(-1)^j s)\wedge y\wedge t\wedge u$ and then to $(i+1,j+1)\wedge\xi_i(x\wedge(-1)^j s)\wedge\eta_j(y\wedge t)\wedge u$. If we first increase j and then increase i, we get first to $(i,j+1)\wedge x\wedge\eta_j(y\wedge s)\wedge t\wedge u$ and then to $(i+1,j+1)\wedge\xi_i(x\wedge(-1)^{j+1}t)\wedge\eta_j(y\wedge s)\wedge u$. If I wanted to turn the first formula into the second I would have to substitute s for t, $-t$ for s.

We conclude, then, that the family of subspaces we have considered is best described as

$$X_i\wedge Y_j\wedge M(\tau)\wedge S^{n-i-j-2},$$

Here $M(\tau)$ is the Thom complex of a certain 2-plane bundle τ over ∂e; more precisely, τ is obtained from $I\times R^2$ by identifying the two ends under the homeomorphism $\begin{pmatrix}0 & -1\\ 1 & 0\end{pmatrix}$. So τ is an $SO(2)$-bundle over $\partial e = S^1$; it can be extended to a bundle over e. Of course there are different ways of extending τ to a bundle over e, since $\pi_1(SO(2)) = Z$. But τ is essentially independent of n, i, j, X and Y; this follows from the description given above; or else one can use coordinates to write down explicit isomorphisms which increase i by 1 or j by 1. (The isomorphisms start from the identity map of R^2 over $[i]\times[j]$, and each suspension coordinate is either preserved or reversed according to the demands of the signs.) All that is essential is that we choose an extension of τ which is similarly independent of n, i, j, X and Y. For example, with the description of τ given above, we can trivialize τ by using a geodesic path of length $\pi/2$ in $SO(2)$.

We take the part of P_n corresponding to $e = e_{ij}$ to be

$$X_i\wedge Y_j\wedge M(\tau)\wedge S^{n-i-j-2},$$

where τ now refers to the bundle as extended over e_{ij}. The identification with the part of P_n already constructed is automatic.

This completes the construction of $P_n = (X \wedge Y)_n$. The structure maps are obvious.

To summarize, we have constructed $(X \wedge Y)_n$ as a quotient space of

$$\bigvee X_i \wedge Y_j \wedge M(\tau_d) \wedge S^{n-i-j-d} .$$

Here the sum runs over cells e_{ij} such that $i+j + \dim(e_{ij}) \leq n$, and $d = \dim(e_{ij})$, and τ_d is a suitable d-plane bundle over e_{ij}. (For $d = 0$ and $d = 1$, τ_d was introduced as an explicitly trivialized bundle.) The identifications are obvious: we regard $X_i \wedge S^1$ as embedded in X_{i+1},

$$X_i \wedge S^1 \wedge Y_j \wedge M(\tau) \wedge S^{n-i-j-d}$$

as

$$X_i \wedge Y_j \wedge M(1 \oplus \tau) \wedge S^{n-i-j-d} ;$$

and similarly for Y_j. Also we regard

$$X_i \wedge Y_j \wedge M(\tau) \wedge S^1 \wedge S^{n-i-j-d-1}$$

as

$$X_i \wedge Y_j \wedge M(\tau \oplus 1) \wedge S^{n-i-j-d-1} .$$

The discussion of the functoriality of $X \wedge Y$ goes exactly as for the telescope functor. More precisely, suppose X' is cofinal in X, and we are given a function $f: X' \longrightarrow Z$. Then $X' \wedge Y$ is not cofinal in $X \wedge Y$, but we have the following functions.

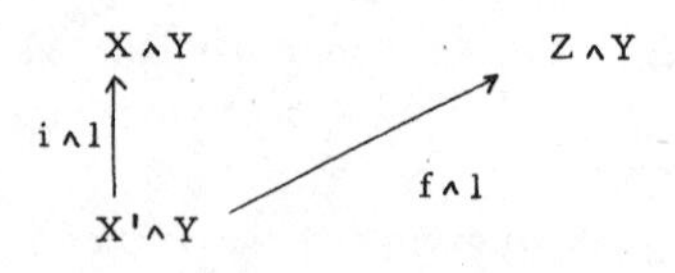

When we pass to morphisms, $i' \wedge 1$ is an equivalence, by 3.5, so we obtain a morphism from $X \wedge Y$ to $Z \wedge Y$. Since cylinders work right, we conclude that this morphism depends only on the homotopy class of f.

It is clear how one embeds $\mathrm{Tel}(X \wedge_{BC} Y)$ in $X \wedge Y$. The functions $\beta\alpha^{-1}$, $\gamma\alpha^{-1}$ give a function

$\{0, 1, 2, 3, \ldots\} \longrightarrow \{0, 1, 2, 3, \ldots\} \times \{0, 1, 2, 3, \ldots\}$. In other words, they give the corners of a stepwise path in the quarter-plane $i \geq 0$, $j \geq 0$. We extend it to a function $\theta: \{k \geq 0\} \longrightarrow \{i \geq 0\} \times \{j \geq 0\}$ so that if $a \in B$,

$$\theta[\alpha(a), \alpha(a) + 1] \subset [\beta(a), \beta(a)+1] \times [\gamma(a)]$$

and if $a \in C$,

$$\theta[\alpha(a), \alpha(a)+1] \subset [\beta(a)] \times [\gamma(a), \gamma(a)+1].$$

The choice of θ is immaterial; two choices are homotopic through maps θ satisfying the same restrictions.

A typical part of $\mathrm{Tel}(X \wedge_{BC} Y)$ is

$$[k, k+1]^{+} \wedge X_i \wedge Y_j \wedge S^1 \wedge S^{n-k-1}$$

where $i = \beta\alpha^{-1}k$, $j = \gamma\alpha^{-1}k$. We take the point

$$t \wedge x \wedge y \wedge u \wedge v$$

and map it to

$$\theta(t) \wedge x \wedge y \wedge u \wedge v \quad \text{in} \quad e_{ij}^{+} \wedge X_i \wedge Y_j \wedge S^1 \wedge S^{n-i-j-1}$$

where e_{ij} is the appropriate 1-cell. Similarly for $[k]^{+} \wedge X_i \wedge Y_j \wedge S^{n-k}$.

It is clear that changing the choice of θ only changes the resulting function

$$\mathrm{Tel}(X \wedge_{BC} Y) \longrightarrow X \wedge Y$$

by a homotopy. For any choice of θ, the function $\mathrm{Tel}(X \wedge_{BC} Y) \longrightarrow X \wedge Y$ is natural for functions of X and Y of degree 0. From this one has no difficulty in obtaining the naturality properties of eq_{BC} in Theorem 4.2.

We now prove Theorem 4.2(iii). First we consider case (a). So we suppose that B and C are infinite. We define a subspectrum Q of P as follows. Let $Q_{\alpha(a)}$ be the subcomplex of $P_{\alpha(a)}$ corresponding to the cells e_{ij} in the part of the quarter-plane given by $i' \leq \beta(a)$, $j' \leq \gamma(a)$. $Q_{\alpha(a)}$ admits a deformation retraction on $X_{\beta(a)} \wedge Y_{\gamma(a)}$, and $\mathrm{Tel}(X \wedge_{BC} Y)_{\alpha(a)}$ admits a deformation retraction on $X_{\beta(a)} \wedge Y_{\gamma(a)}$. Hence, in the diagram

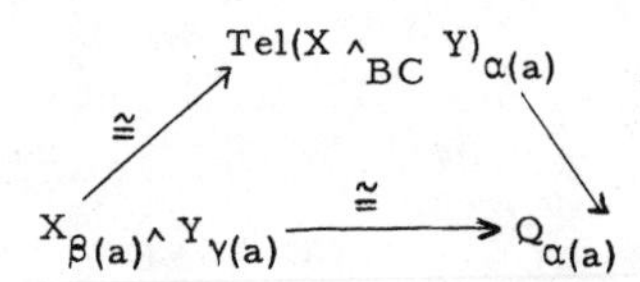

the two inclusions marked induce isomorphisms of homotopy groups, so the third one does also; passing to direct limits and applying 3.5, the inclusion

$$\mathrm{Tel}(X \wedge_{BC} Y) \longrightarrow Q$$

is an equivalence.

It remains to consider cases (b) and (c), which are similar. Let us consider case (b), so that B is finite with d members, and

$$\xi_r : SX_r \longrightarrow X_{r+1}$$

is an isomorphism for $r \geq d$. We now make a small change in the definition of $Q_{\alpha(a)}$ for "a" such that $\beta(a) \geq d$. For such a, we define $Q_{\alpha(a)}$ to be the subcomplex of $P_{\alpha(a)}$ corresponding to the cells e_{ij} in the part of the quarter-plane given by $i' + j' \leq \alpha(a)$, $j' \leq \gamma(a)$. Then $Q_{\alpha(a)}$ still admits a deformation retraction onto $X_{\beta(a)} \wedge Y_{\gamma(a)}$, since the relevant map

$$X_d \wedge Y_j \wedge S^{n-d} \longrightarrow X_{d+e} \wedge Y_j \wedge S^{n-d-e} \quad (e \geq 0)$$

is an isomorphism. Also Q is cofinal in P, so the proof carries over.

We now turn to the proof of Theorem 4.5.

LEMMA 4.6. There is a spectrum Q with homotopy equivalences $i_o: X \wedge Y \longrightarrow Q$, $i_1: Y \wedge X \longrightarrow Q$ so that the following diagram is commutative for each choice of B and C satisfying condition 4.4.

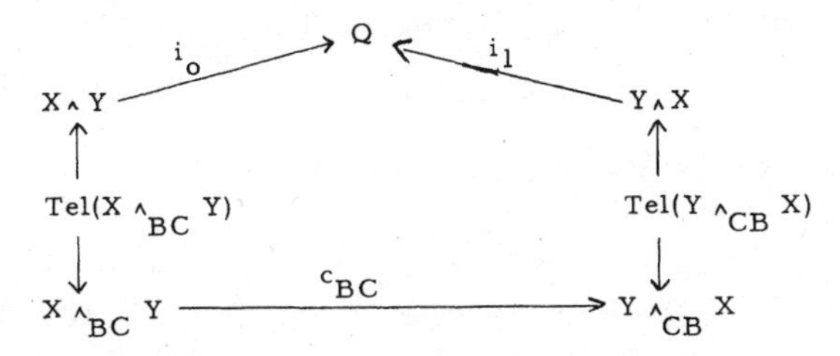

This will certainly prove Theorem 4.5; we have only to define c to be $i_1^{-1} i_o$. Note that we do not have to discuss the naturality of $i_1^{-1} i_o$; it follows from that of the other morphisms in 4.6.

To construct Q, we begin by taking a copy of $X \wedge Y$ and a copy of $Y \wedge X$. The remainder of the construction will be indexed over the product of the quarter-plane $i \geq 0$, $j \geq 0$ and the interval I. The endpoint 0 of I will correspond to $X \wedge Y$ and the endpoint 1 of I will correspond to $Y \wedge X$.

First we observe that we can make a construction over the following cells:

$$([i] \times [j] \times I)^+ = ([i] \times [j])^+ \wedge I^+ \qquad (i \text{ or } j \text{ even})$$

$$([i, i+1] \times [j] \times I)^+ \qquad (j \text{ even})$$

$$([i] \times [j, j+1] \times I)^+ \qquad (i \text{ even}) \; .$$

The nth term of the construction consists in taking the appropriate part of $(X \wedge Y)_n \wedge I^+$, identifying the end 0 of the cylinder with the appropriate part of $(X \wedge Y)_n$, attaching the end 1 of the cylinder to the appropriate part of $(Y \wedge X)$, by the following map: the point

$$(s,t)\wedge x\wedge y\wedge u\wedge v \quad \text{in} \quad e_{ij}^{+}\wedge X_i\wedge Y_j\wedge S^d\wedge S^{n-i-j-d}$$

is to be identified with

$$(t,s)\wedge y\wedge x\wedge u\wedge v \quad \text{in} \quad e_{ji}^{+}\wedge Y_j\wedge X_i\wedge S^d\wedge S^{n-i-j-d}.$$

These identifications are consistent.

Consider now a cell $e = [2i, 2i+2] \times [2j, 2j+2] \times I$. We have just described the part of Q_n corresponding to the boundary ∂e of e. Moreover, it contains a subcomplex of the following form:

$$X_{2i}\wedge Y_{2j}\wedge M(\tau')\wedge S^{n-2i-2j-4}.$$

Here τ' is a certain 4-plane bundle over ∂e. This 4-plane bundle depends only on the permutations and signs in our construction and on the extension τ chosen in the construction of $X\wedge Y$; it does not depend on n, i, j, X or Y. It is classified by an element

$$\alpha \in \pi_1(SO) = Z_2.$$

Suppose now that we consider the four cells, like the cell e just considered, which make up the cell

$$e' = [4i, 4i+4] \times [4j, 4j+4] \times I.$$

Call them e_1, e_2, e_3 and e_4. The part of Q so far constructed, corresponding to these cells, has a subcomplex of the form

$$S^{n-4i-4j-8}\wedge M(\tau'')\wedge X_{4i}\wedge Y_{4j}.$$

Here $M(\tau'')$ is the Thom complex of a certain 8-plane bundle over $\partial e_1 \cup \partial e_2 \cup \partial e_3 \cup \partial e_4$. Over each ∂e_i it restricts to the Whitney sum of the previous bundle τ' and a trivial 4-plane bundle. Therefore the restriction of τ'' to $\partial e'$ is classified by $4\alpha = 0$. Therefore τ'' can be extended over e'.

From the previous construction, we now retain only $X\wedge Y$, $Y\wedge X$, and the parts of the cylinder $(X\wedge Y)_n\wedge I^+$ with i divisible by 4 or j divisible

divisible by 4. We now add

$$X_{4i} \wedge Y_{4j} \wedge M(\tau'') \wedge S^{n-4i-4j-8}$$

for each i, j, and n such that $n \geq 4i + 4j + 8$. This completes the construction of Q.

The injections of $X \wedge Y$ and $Y \wedge X$ into Q are clearly homotopy equivalences, by 3.5. It is also clear that the diagram of Lemma 4.6 is commutative, because the relevant part of the cylinder $\mathrm{Cyl}(X \wedge Y)$ was put in for that purpose.

This completes the proof of Lemma 4.6 and, therefore, of Theorem 4.5.

We now turn to the proof of Theorem 4.3. The constructions $(X \wedge Y) \wedge Z$ and $X \wedge (Y \wedge Z)$ are "quadruple telescopes," indexed by a cell-decomposition of the positive cone in 4-space. We arrange to replace $(X \wedge Y) \wedge Z$ by an equivalent construction P', and $X \wedge (Y \wedge Z)$ by an equivalent construction P'', so that both P' and P'' are "triple telescopes, indexed by a cell-decomposition of the positive cone in 3-space. It will then be apparent that P' and P'' are equivalent. More formally, we have the following lemma.

LEMMA 4.7. There is a spectrum P' and a homotopy equivalence $i': P' \longrightarrow (X \wedge Y) \wedge Z$ (both independent of B, C and D) such that the following diagram is commutative for each choice of B, C and D.

$$\begin{array}{ccccc}
(X \wedge_{BC} Y) \wedge Z & \xleftarrow{r \wedge 1} & (\mathrm{Tel}(X \wedge_{BC} Y)) \wedge Z & \xrightarrow{i \wedge 1} & (X \wedge Y) \wedge Z \\
\uparrow i & \nearrow j' & & & \uparrow i' \\
\mathrm{Tel}((X \wedge_{BC} Y) \wedge_{B \cup C, D} Z) & & \xrightarrow{k'} & & P' \\
\downarrow r & & & & \\
(X \wedge_{BC} Y) \wedge_{B \cup C, D} Z & & & &
\end{array}$$

Similarly for $X \wedge (Y \wedge Z)$ with i', j', k' and P' replaced by i'', j'', k'', and P''. Moreover, there is a homotopy equivalence $P' \xrightarrow{e} P''$ such that the following diagram is commutative.

$$\begin{array}{ccc} P' & \xrightarrow{\quad e \quad} & P'' \\ \uparrow{\scriptstyle k'} & & \uparrow{\scriptstyle k''} \\ \mathrm{Tel}((X \wedge_{BC} Y) \wedge_{B \cup C, D} Z) & = & \mathrm{Tel}(X \wedge_{B, C \cup D} (Y \wedge_{CD} Z)) \end{array}$$

<u>Proof.</u> By definition, the nth term of $(X \wedge Y) \wedge Z$ is a union

$$\bigcup_{e_{hk}} (X \wedge Y)_h \wedge Z_k \wedge M(\tau_\delta) \wedge S^{n-h-k-\delta}$$

where the union extends over cells e_{hk} such that

$$h + k + \dim e_{hk} \leq n \quad ,$$

$\delta = \dim e_{hk}$ and τ_δ is a δ-plane bundle. That is, it is a union

$$\bigcup_{e_{ij}, e_{hk}} X_i \wedge Y_j \wedge M(\tau_d) \wedge S^{h-i-j-d} \wedge Z_k \wedge M(\tau_\delta) \wedge S^{n-h-k-\delta}$$

where e_{ij} runs over cells with

$$i + j + \dim e_{ij} \leq h \quad ,$$

$d = \dim (e_{ij})$, and τ_d is a d-plane bundle. We rearrange this as

$$\bigcup_{e_{ij}, e_{hk}} X_i \wedge Y_j \wedge Z_k \wedge M(\tau_d \oplus \tau_\delta) \wedge S^{h-i-j-d} \wedge S^{n-h-k-\delta} \quad .$$

Thus the construction is indexed over a cell-decomposition of the positive cone $i \geq 0$, $j \geq 0$, $h \geq 0$, $k \geq 0$ in 4-space. Call this cone C^4. Let C^3 be the positive cone $i \geq 0$, $j \geq 0$, $k \geq 0$ in 3-space, and divide C^3 into cells in the obvious way, so that the cells are r-cubes of side 1 for $r = 0, 1, 2, 3$.

We construct P' by giving a suitable cellular map θ from C^3 to C^4 and "pulling back" the bundles and complexes we have associated with the parts of C^4. Actually we construct θ to preserve the k-coordinate, so it is only necessary to construct a map φ from the positive cone $i \geq 0$, $j \geq 0$ to the positive cone $i \geq 0$, $j \geq 0$, $h \geq 0$.

Our idea in defining θ and φ is to use only cells $e_{ij} \times e_{hk}$ such that

$$i + j + \dim e_{ij} = h;$$

firstly because the other parts of $(X \wedge Y) \wedge Z$ are redundant, and secondly because by keeping $S^{h-i-j-d} = S^0$ we avoid suspension coordinates in the wrong place.

We first indicate into which subcomplexes the cells are to be mapped.

$$\varphi([i] \times [j]) = [i] \times [j] \times [i+j]$$

$$\varphi([i,i+1] \times [j]) \subset ([i] \times [j] \times [i+j,i+j+1]) \cup ([i,i+1] \times [j] \times [i+j+1])$$

$$\varphi([i] \times [j,j+1]) \subset ([i] \times [j] \times [i+j,i+j+1]) \cup ([i] \times [j,j+1] \times [i+j+1])$$

$$\begin{aligned}\varphi([i,i+1] \times [j,j+1]) \subset & ([i] \times [j] \times [i+j,i+j+1]) \\ & \cup ([i,i+1] \times [j] \times [i+j+1,i+j+2]) \\ & \cup ([i] \times [j,j+1] \times [i+j+1,i+j+2]) \\ & \cup ([i,i+1] \times [j,j+1] \times [i+j+2]).\end{aligned}$$

In each case the proposed subcomplex is contractible, so the construction of φ is possible and unique up to homotopy. In each case, the image of φ must be the whole subcomplex given, so we can refer to the subcomplex as $\varphi(e_{ij})$. Similarly for $\theta(e_{ijk})$.

We next note that for each cell e_{ijk} such that $i + j + k + \dim(e_{ijk}) \leq n$, the part of $((X \wedge Y) \wedge Z)_n$ associated with $\theta(e_{ijk})$ contains a subcomplex of the form

$$X_i \wedge Y_j \wedge Z_k \wedge M(\tau_d) \wedge S^{n-i-j-k-d} \quad ,$$

where $d = \dim(e_{ijk})$ and τ_d is a d-plane bundle over $\theta(e_{ijk})$. We define the corresponding part of P' to be

$$X_i \wedge Y_j \wedge Z_k \wedge M(\theta^* \tau_d) \wedge S^{n-i-j-k-d},$$

where $\theta^* \tau_d$ is the induced bundle over e_{ijk}. The map i' on this part of P' is induced by the map of bundles $\theta^* \tau_d \longrightarrow \tau_d$ over the map θ of spaces. The identifications to be made in assembling P' are automatic; one just pulls back the identifications in $(X \wedge Y) \wedge Z$.

We make the structure of P' more explicit. Corresponding to the 0-cells e_{ijk} we have

$$\bigvee_{i+j+k \leq n} X_i \wedge Y_j \wedge Z_k \wedge ([i] \times [j] \times [k])^+ \wedge S^{n-i-j-k}.$$

Corresponding to the 1-cells we have

$$\bigvee_{i+j+k+1 \leq n} e^+_{ijk} \wedge X_i \wedge Y_j \wedge Z_k \wedge S^1 \wedge S^{n-i-j-k-1}.$$

Here the attaching maps are the obvious ones, involving the obvious signs.

For each 2-cell $e = e_{ijk}$, the bundle $\theta^* \tau$ over e is exactly as described in the construction of $X \wedge Y$.

For each 3-cell $e = e_{ijk}$, there is only one bundle over e extending the given bundle $\theta^* \tau$ over ∂e, since $\pi_3(BSO(3)) = \pi_2(SO(3)) = 0$. So we need not worry which bundle arises.

On the other hand, the description of P'' is exactly the same as the description we have just given for P'. This provides the map $e: P' \longrightarrow P''$.

The map

$$k': \mathrm{Tel}((X \wedge_{BC} Y) \wedge_{B \cup C, D} Z) \longrightarrow P'$$

is basically obvious. The functions $\beta\alpha^{-1}$, $\gamma\alpha^{-1}$, $\delta\alpha^{-1}$ give a function $\theta': \{0, 1, 2, 3, \ldots\} \longrightarrow \{0, 1, 2, 3, \ldots\}^3$. We extend it to a function θ'' mapping each cell of C^1 (the positive half-line with our usual cell structure) into the obvious cell of C^3. Now we construct the map k' as we constructed the map $\mathrm{Tel}(X \wedge_{BC} Y) \longrightarrow X \wedge Y$.

We now observe that the function

$i'k'$: $\mathrm{Tel}((X \wedge_{BC} Y) \wedge_{B\cup C, D} Z) \longrightarrow (X \wedge Y) \wedge Z$ actually maps into $(\mathrm{Tel}(X \wedge_{BC} Y)) \wedge Z$; this defines the function

$$j' \colon \mathrm{Tel}(X \wedge_{BC} Y)) \wedge_{B\cup C, D} Z \longrightarrow (\mathrm{Tel}(X \wedge_{BC} Y)) \wedge Z.$$

The function $(r \wedge 1)j'$ satisfies the definition for i. (Some of the cylinders spend some of their time stationary and the rest hurrying to make up for it, but this is allowed.)

This completes the proof of Lemma 4.7, which completes the proof of 4.3, which completes the proof of Theorem 4.1 so far as it refers to maps of degree 0.

We now propose to go back and recover the properties of our constructions with respect to maps of non-zero degree.

First we introduce sphere-spectra of different stable dimensions. Let us define the spectrum $\underline{S}^i$ by

$$(\underline{S}^i)_n = \begin{cases} S^{n+1} & (n+i \geq 0) \\ \text{pt.} & (n+i < 0) \end{cases}.$$

PROPOSITION 4.8. We have an equivalence $\underline{S}^i \wedge \underline{S}^j \xrightarrow{e} \underline{S}^{i+j}$ such that the following diagrams are commutative.

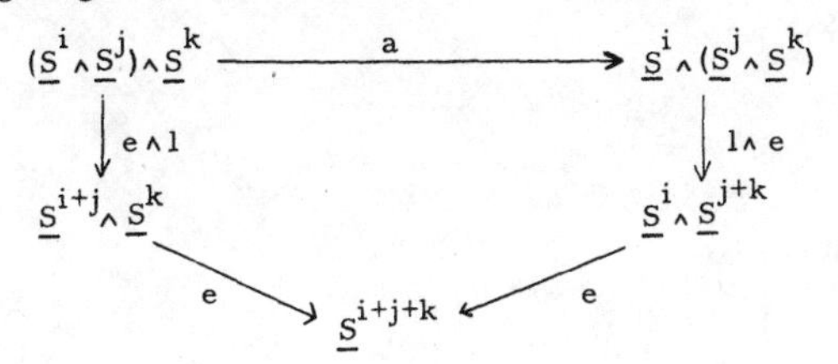

$$\begin{array}{ccc} (\underline{S}^i \wedge \underline{S}^j) \wedge \underline{S}^k & \xrightarrow{a} & \underline{S}^i \wedge (\underline{S}^j \wedge \underline{S}^k) \\ \downarrow e \wedge 1 & & \downarrow 1 \wedge e \\ \underline{S}^{i+j} \wedge \underline{S}^k & & \underline{S}^i \wedge \underline{S}^{j+k} \\ & \searrow e \quad \underline{S}^{i+j+k} \quad e \swarrow & \end{array}$$

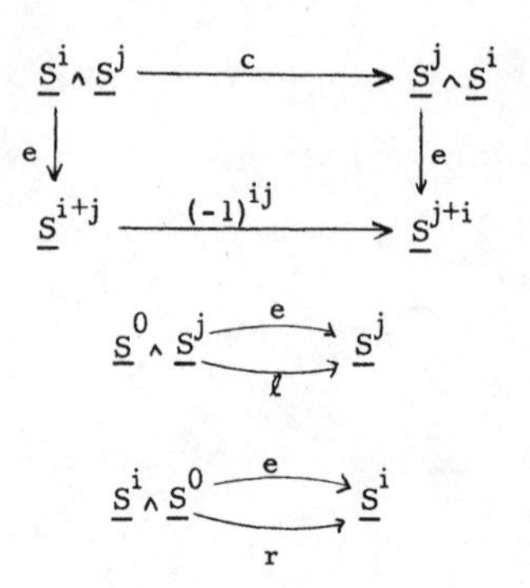

Proofs. (i) Any handicrafted smash-product of $\underline{S}^i$ and $\underline{S}^j$ gives a spectrum which has the same terms as $\underline{S}^{i+j}$ from some point onwards. We just take care to pick an equivalence which is orientation-preserving.

(ii) $[\underline{S}^i, \underline{S}^i] = \lim_{n \to \infty} [S^{n+i}, S^{n+i}] = Z$; so to check the commutativity of any such diagram, we have only to check the degree of a map. We have been careful to make all our constructions so as to do the right thing on orientations.

Proposition 4.9. We have equivalences

$$\gamma_r : X \longrightarrow \underline{S}^r \wedge X \qquad \text{(of degree } r\text{)}$$

with the following properties.

(i) γ_r is natural for maps of X of degree 0. (This is all we can ask, because we have not yet made $\underline{S}^r \wedge X$ functorial for maps of non-zero degree.)

(ii) $\gamma_0 = \ell^{-1}$.

(iii) The following diagram is commutative for each r and s.

$$\begin{array}{ccc}
\underline{S}^{r+s} \wedge X & \xleftarrow{e \wedge 1} & (\underline{S}^r \wedge \underline{S}^s) \wedge X \\
\uparrow \gamma_{r+s} & & \downarrow a \\
& & \underline{S}^r \wedge (\underline{S}^s \wedge X) \\
& & \uparrow \gamma_r \\
X & \xrightarrow{\gamma_s} & \underline{S}^s \wedge X
\end{array}$$

Proof. Clearly if we take $\gamma_o = \ell^{-1}$, it is natural for maps of X of degree 0. Consider now

$$\underline{S}^0 \wedge_{1, \{2,3, \ldots\}} X \quad \text{and} \quad \underline{S}^1 \wedge_{\emptyset, \{1,2,3, \ldots\}} X .$$

On the left, the nth term is $S^1 \wedge X_{n-1}$; on the right, the $(n-1)$-st term is $S^1 \wedge X_{n-1}$. The structure maps are the same in both cases. So the identity maps $S^1 \wedge X_{n-1} \longrightarrow S^1 \wedge X_{n-1}$ give the components of an equivalence of degree $+1$

$$\underline{S}^0 \wedge X \longrightarrow \underline{S}^1 \wedge X .$$

It is clearly natural for maps of X of degree 0. Composing with ℓ^{-1}, we obtain an equivalence γ_1.

Note that at this point I have essentially picked up the Puppe Desuspension Theorem, without restrictive hypotheses.

Now I define γ_s for all other values of s by induction upwards and downwards over s, making the following diagram commutative.

$$\begin{array}{ccc}
\underline{S}^{1+s} \wedge X & \xleftarrow{e \wedge 1} & (\underline{S}^1 \wedge \underline{S}^s) \wedge X \\
\uparrow \gamma_{1+s} & & \downarrow a \\
& & \underline{S}^1 \wedge (\underline{S}^s \wedge X) \\
& & \uparrow \gamma_1 \\
X & \xrightarrow{\gamma_s} & \underline{S}^s \wedge X
\end{array}$$

One has to check that this is consistent for $s = 0$. Note also that γ_{1+s} or γ_s, whichever is being defined, is natural for maps of X of degree

0, because all the ingredients of its definition are so.

We now prove the commutativity of the diagram

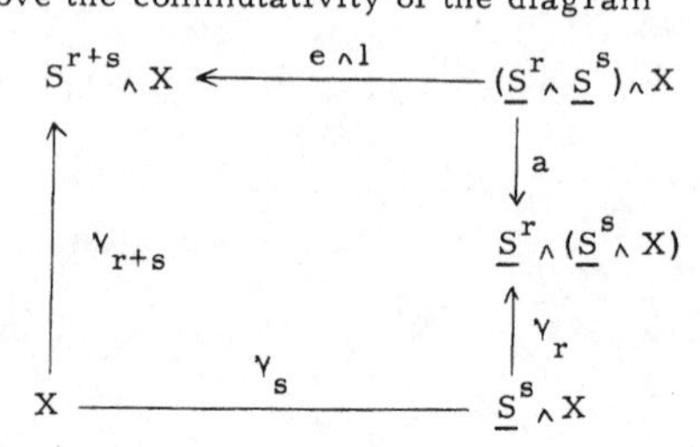

by induction upwards and downwards over r. Here we start from the cases $r = 0$ (which is a trivial verification) and $r = 1$ (which holds by the construction of γ_s). The inductive step is diagram-chasing.

We are now ready to replace our original graded category by one which appears slightly different. In the new category, the objects are the CW-spectra just as before; but the morphisms of degree r are given by

$$[\underline{S}^r \wedge X, Y]_0$$

in the old category. Composition is done as follows. Suppose given

$$\underline{S}^r \wedge X \xrightarrow{f} Y, \qquad \underline{S}^s \wedge Y \xrightarrow{g} Z$$

of degree 0; take their composite to be

$$S^{s+r} \wedge X \xleftarrow{e \wedge 1} (\underline{S}^s \wedge \underline{S}^r) \wedge X \xrightarrow{a} \underline{S}^s \wedge (\underline{S}^r \wedge X) \xrightarrow{1 \wedge f} \underline{S}^s \wedge Y \xrightarrow{g} Z.$$

One has to check that composition is associative, and that $\ell : \underline{S}^0 \wedge X \longrightarrow X$ is an identity map. This is easy.

FROPOSITION 4.10. The new graded category is isomorphic to the old, under the isomorphism sending

$$\underline{S}^r \wedge X \xrightarrow{f} Y \qquad \text{(in the new category)}$$

to

$$X \xrightarrow{\gamma_r} \underline{S}^r \wedge X \xrightarrow{f} Y \qquad \text{(in the old category)} .$$

Proof. Since γ_r is an equivalence in the old category, it is clear that this gives a 1-1 correspondence between $[\underline{S}^r \wedge X, Y]_0$ (that is, the set of morphisms of degree r in the new category) and $[X, Y]_r$ (morphisms of degree r in the old category). It remains only to check that this 1-1 correspondence preserves composition and identity maps. But this is immediate from the properties of γ_r in Proposition 4.9.

If you want to see what you are doing with maps of degree r, I really recommend considering them as maps $\underline{S}^r \wedge X \longrightarrow Y$ of degree 0. In particular, it is easy to see how to make $X \wedge Y$ functorial on the new category. More precisely, suppose given morphisms in the new category

$$S^r \wedge X \xrightarrow{f} X', \qquad \underline{S}^s \wedge Y \xrightarrow{g} Y'.$$

Then we define their smash-product to be

$$\underline{S}^{r+s} \wedge X \wedge Y \xleftarrow{e \wedge 1 \wedge 1} \underline{S}^r \wedge S^s \wedge X \wedge Y \xrightarrow{1 \wedge c \wedge 1} \underline{S}^r \wedge X \wedge \underline{S}^s \wedge Y \xrightarrow{f \wedge g} X' \wedge Y'.$$

To prove that this has all the properties mentioned in Theorem 4.1 is now a routine exercise in diagram-chasing. At the same time, we check that we have not altered the definition of $f \wedge g$ if f and g happen to be of degree 0.

This completes the proof of Theorem 4.1.

Exercise. Show that the naturality of γ_r with respect to maps of degree s is as follows: the diagram

$$\begin{array}{ccc} X & \xrightarrow{\gamma_r} & \underline{S}^r \wedge X \\ f \downarrow & (-1)^{rs} & \downarrow 1 \wedge f \\ Y & \xrightarrow[\gamma_r]{} & \underline{S}^r \wedge Y \end{array}$$

is commutative up to a sign of $(-1)^{rs}$ if $f \in [X, Y]_s$.

PROPOSITION 4.11. The smash-product is distributive over the wedge-sum. Let $X = \bigvee_\alpha X_\alpha$; let $i_\alpha : X_\alpha \longrightarrow X$ be a typical inclusion. Then the morphism

$$\bigvee_\alpha (X_\alpha \wedge Y) \xrightarrow{\{i_\alpha \wedge 1\}} (\bigvee_\alpha X_\alpha) \wedge Y$$

is an equivalence.

Proof. Use a suitable handicrafted smash-product.

PROPOSITION 4.12. Let $X \xrightarrow{f} Y \xrightarrow{i} Z$ be a cofibering (it is sufficient to consider morphisms of degree zero). Then

$$W \wedge X \xrightarrow{1 \wedge f} W \wedge Y \xrightarrow{1 \wedge i} W \wedge Z$$

is also a cofibering.

Proof. It suffices to check for the case in which $f: X \longrightarrow Y$ is the inclusion of a closed subspectrum, $i: Y \longrightarrow Z$ is the projection $Y \longrightarrow Y/X$ and $\wedge = \wedge_{BC}$.

5. SPANIER-WHITEHEAD DUALITY

Suppose I have a compact subset $X \subset S^n$, say $X \neq \emptyset$, $X \neq S^n$. Then I know that the homology of the complement $\mathcal{C}X$ of X is determined by the cohomology of X. This is given by the Alexander duality theorem:

$$\tilde{\check{H}}^r(X) \cong \check{H}^r(X, pt.) \cong H_{n-r}(\mathcal{C} pt., \mathcal{C}X) \cong H_{n-r-1}(\mathcal{C}X, pt.) \cong \tilde{H}_{n-r-1}(\mathcal{C}X).$$

However, the homotopy type of $\mathcal{C}X$ is clearly not determined by X; it depends on the embedding. For example, take $X = S^1$, $n = 3$; we can embed S^1 in S^3 as a knotted circle or an unknotted circle, and make $\pi_1(\mathcal{C}X)$ different in the two cases. It would be reasonable to ask the following question. Suppose X is a good subset, e.g., a finite simplicial complex linearly embedded in $\partial\sigma^{n+1}$. (We make this assumption to avoid pathologies.) How far does X determine anything about $\mathcal{C}X$ beyond its bare homology groups?

It was proved by Spanier and Whitehead that X does determine the stable homotopy type of $\mathcal{C}X$; even the stable homotopy type of X suffices to do this. This may easily be seen as follows. First, suppose that I take $X \subset S^n$. Now embed S^n as an equatorial sphere in S^{n+1}, and embed the suspension SX of X in S^{n+1} by joining to the two poles. Then the complement of SX in S^{n+1} is homotopy-equivalent to the complement of X in S^n. So if somebody gives me $X \subset S^n$, $Y \subset S^m$ and a homotopy equivalence $f: S^pX \longrightarrow S^qY$, I may as well embed S^pX in S^{n+p} and S^qY in S^{m+q}, because I can do so without changing the complements. So without loss of generality I can suppose I have $X' \subset S^{n'}$, $Y' \subset S^{m'}$ and a homotopy equivalence $f: X' \longrightarrow Y'$. I can even suppose that f is PL.

Now suppose we take $X' \subset S^{n'}$ and embed $S^{n'}$ as an equatorial sphere in $S^{n'+1}$ without changing X'. Then the complement of X' suspends; more precisely, the complement of X' in $S^{n'+1}$ is the suspension of that in $S^{n'}$. So now consider $S^{n'} * S^{m'}$. In this sphere we can embed the mapping-cylinder M of f'. In this sphere we have

$$S^{m'+n'+1} - X = S^{m'+1}(S^{n'} - X)$$

$$S^{m'+n'+1} - Y = S^{n'+1}(S^{m'} - Y)$$

and two maps

$$S^{m'+n'+1} - X \xleftarrow{\ f\ } S^{m'+n'+1} - M \xrightarrow{\ g\ } S^{m'+n'+1} - Y .$$

But the injections

$$X \longrightarrow M \longleftarrow Y$$

induce isomorphisms of cohomology. The Alexander duality isomorphism is natural for inclusion maps, and therefore f and g induce isomorphisms of homology. But now I can suspend further if necessary to make everything simply-connected. So f and g are stable homotopy equivalences, and we have proved the result.

With a little more attention to detail we can show that the passage from X to the stable homotopy type of its complement in a sphere is essentially functorial; a map $f: X \longrightarrow Y$ induces a stable class of maps $f^*: \mathcal{C}Y \longrightarrow \mathcal{C}X$. The functor is contravariant, as we would expect.

The next step was taken by Spanier, and it was to eliminate the embedding in S^n. More precisely, suppose I have two finite simplicial complexes K and L embedded in S^n so as to be disjoint. I am really interested in the case when the inclusions $L \longrightarrow \mathcal{C}K$, $K \longrightarrow \mathcal{C}L$ are homotopy equivalences, but this is not necessary for the construction. Run a PL path from some point in K to some point in L; without loss of generality we can suppose the first point is the only point where it meets K, and the last point is the only point where it meets L. Without loss of generality we can suppose these points are vertices and take them as the base-points in K and L, writing bpt. for either. Take some point in the middle of the path as the point at ∞. Then we have an embedding of K and L in R^n. Define a map

$$\mu: K \times L \longrightarrow S^{n-1}$$

by

$$\mu(k, \ell) = \frac{k - \ell}{||k - \ell||}.$$

The maps $u|K \times \text{bpt.}$ and $\mu|\text{bpt.} \times L$ are null homotopic, so we get a map

$$\mu: K \wedge L \longrightarrow S^{n-1}.$$

Spanier's essential step was to realize that everything could be said in terms of this map μ. To begin with he considered maps $\mu: K \wedge L \longrightarrow S^{n-1}$

whose homological behavior was such as you would expect. In order to explain what you would expect, I need slant products, which I have not done yet.

So we use the framework we already have. Let X be a CW-spectrum. Then we can form

$$[W \wedge X, S]_0 \ .$$

With X fixed this is a contravariant functor of W, and it satisfies the axioms of E.H. Brown. So it is representable; there is a spectrum X^* and a natural isomorphism

$$[W \wedge X, S]_0 \xleftarrow{T} [W, X^*]_0 \ .$$

Taking $W = X^*$ and $1: X^* \longrightarrow X^*$ on the right, we see that there is a map

$$e: X^* \wedge X \longrightarrow S \ .$$

Using the fact that T is natural, we see that T carries

$$f: W \longrightarrow X^*$$

into

$$W \wedge X \xrightarrow{f \wedge 1} X^* \wedge X \xrightarrow{e} S \ .$$

Of course, this prescription defines

$$T: [W, X^*]_r \longrightarrow [W \wedge X, S]_r \ ,$$

and by applying the canonical isomorphism to a different choice of W we see that

$$T: [W, X^*]_r \longrightarrow [W \wedge X, S]_r$$

is an isomorphism also.

We think of this as being like duality for vector-spaces over a field K. In that case we have

$$V^* = \mathrm{Hom}_K(V, K);$$

there is a canonical evaluation map

$$e: V^* \otimes V \longrightarrow K ;$$

and there is a 1-1 correspondence

$$\mathrm{Hom}_K(U \otimes V, K) \xleftarrow{T} \mathrm{Hom}_K(U, \mathrm{Hom}_K(V, K)) .$$

The dual X^* is a contravariant functor of X. For if we take a map $g: X \longrightarrow Y$, it induces a natural transformation

$$\begin{array}{ccc} [W \wedge X, S]_0 & \xleftarrow{(1 \wedge g)^*} & [W \wedge Y, S]_0 \\ \| & & \| \\ [W, X^*]_0 & & [W, Y^*]_0 \end{array}$$

and this natural transformation must be induced by a unique map

$$g^*: Y^* \longrightarrow X^* .$$

(We go through the usual argument of substituting $W = Y^*$ and $1: Y^* \longrightarrow Y^*$ on the right.) In terms of maps e, the relation between g and g^* is that the following diagram commutes.

$$\begin{array}{ccc} Y^* \wedge X & \xrightarrow{1 \wedge g} & Y^* \wedge Y \\ {\scriptstyle g^* \wedge 1}\downarrow & & \downarrow{\scriptstyle e_Y} \\ X^* \wedge X & \xrightarrow{e_X} & S \end{array}$$

Let Z be a third spectrum; we can make a map

$$[W, Z \wedge X^*]_r \xrightarrow{T} [W \wedge X, Z]_r$$

as follows. Given

$$W \xrightarrow{f} Z \wedge X^*$$

we take

$$W \wedge X \xrightarrow{f \wedge 1} Z \wedge X^* \wedge X \xrightarrow{1 \wedge e} Z .$$

T is clearly a natural transformation if we vary Z.

<u>Remark 5.1.</u> T is an isomorphism if Z is the spectrum S^n. (The case $n = 0$ has already been considered, and changing n just changes the degrees.)

Remark 5.2. Suppose given a cofibering

$Z_1 \to Z_2 \to Z_3 \to Z_4 \to Z_5$. If T is an isomorphism for Z_1, Z_2, Z_4 and Z_5, then it is an isomorphism for Z_3.

Proof. Use the five lemma.

$$\begin{array}{ccccccccc} [W, Z_1 \wedge X^*]_r & \to & [W, Z_2 \wedge X^*]_r & \to & [W, Z_3 \wedge X^*]_r & \to & [W, Z_4 \wedge X^*]_r & \to & [W, Z_5 \wedge X^*]_r \\ \downarrow & & \downarrow & & \downarrow & & \downarrow & & \downarrow \\ [W \wedge X, Z_1]_r & \to & [W \wedge X, Z_2]_r & \to & [W \wedge X, Z_3]_r & \to & [W \wedge X, Z_4]_r & \to & [W \wedge X, Z_5]_r \end{array}$$

Remark 5.3. T is an isomorphism if Z is any finite spectrum.

This is immediate by induction, using 5.1 and 5.2.

PROPOSITION 5.4. If W and X are finite spectra, then

$$T: [W, Z \wedge X^*]_r \longrightarrow [W \wedge X, Z]_r$$

is an isomorphism for any spectrum Z.

Proof. Pass to direct limits from the case of finite spectra.

LEMMA 5.5. If X is a finite spectrum, then X^* is equivalent to a finite spectrum.

The proof is postponed until section 6, for a reason which will appear.

PROPOSITION 5.6. Let X be a finite spectrum, Y any spectrum. Then we have an equivalence $(X \wedge Y)^* \xrightarrow{h} X^* \wedge Y^*$ which makes the following diagram commute.

$$\begin{array}{ccc} (X \wedge Y)^* \wedge X \wedge Y & \xrightarrow{e_{X \wedge Y}} & S \\ {\scriptstyle h \wedge 1} \downarrow & & \uparrow {\scriptstyle e_X \wedge e_Y} \\ X^* \wedge Y^* \wedge X \wedge Y & \xrightarrow{1 \wedge c \wedge 1} & X^* \wedge X \wedge Y^* \wedge Y \end{array} .$$

Proof. By 5.5 we can assume that X^* is a finite spectrum. By 5.3,

$$[W, X^* \wedge Y^*]_r \xrightarrow{T_Y} [W \wedge Y, X^*]_r$$

is an isomorphism for any spectrum W, and so is

$$[W \wedge Y, X^*]_r \xrightarrow{T_X} [W \wedge Y \wedge X, S]_r$$

by the original property of X^*, applied to the spectrum $W \wedge Y$. This state of affairs reveals $X^* \wedge Y^*$ as the dual of $Y \wedge X$, with $T_{Y \wedge X} = T_X T_Y$. Writing this equation in terms of maps e, we obtain the diagram given by a little diagram-chasing.

I should perhaps emphasize that I have only done what I need later. In particular, I have not proved that S-duality converts a cofibering of finite spectra into another cofibering. This is true, but it needs a slightly more precise argument, given in Spanier's exercises. Also, I have only talked about maps <u>into</u> X^* or $Z \wedge X^*$. Once we have the result on cofiberings we can talk about maps <u>from</u> X^*, at least when X is a finite spectrum, and so prove $X^{**} \simeq X$.

6. HOMOLOGY AND COHOMOLOGY

Suppose given a spectrum E. Then we define the E-homology and E-cohomology of other spectra X as follows.

(i) $E_n(X) = [S, E \wedge X]_n$.

(ii) $E^n(X) = [X, E]_{-n}$.

These definitions are due to G. W. Whitehead [17].

In order to convince ourselves these functors do deserve the name of generalized homology and cohomology, let's list their trivial properties.

<u>Proposition 6.1.</u> (i) $E_*(X)$ is a covariant functor of two variables E, X in our category, and with values in the category of graded abelian groups.

(Note: A morphism $f\colon X \longrightarrow Y$ of degree r induces $f_*\colon E_n(X) \longrightarrow E_{n+r}(Y)$, etc.)

The same is true for $E^*(X)$, except that it is covariant in E and contravariant in X.

(ii) If we vary E or X along a cofibering, we obtain an exact sequence. That is, if

$$X \xrightarrow{f} Y \xrightarrow{g} Z$$

is a cofiber sequence, then

$$E_n(X) \xrightarrow{f_*} E_n(Y) \xrightarrow{g_*} E_n(Z)$$

and

$$E^n(X) \xleftarrow{f^*} E^n(Y) \xleftarrow{g^*} E^n(Z)$$

are exact; if $E \xrightarrow{i} F \xrightarrow{j} G$ is a cofiber sequence, then

$$E_n(X) \xrightarrow{i_*} F_n(X) \xrightarrow{j_*} G_n(X)$$

and

$$E^n(X) \xrightarrow{i_*} F^n(X) \xrightarrow{j_*} G^n(X)$$

are exact.

(iii) There are natural isomorphisms

$$E_n(X) \cong E_{n+1}(S^1 \wedge X) ,$$

$$E_n(X) \cong E^{n+1}(S^1 \wedge X).$$

(iv) $$E_n(S) = E^{-n}(S) = \pi_n(E) .$$

The proofs are mostly easy. Part (ii) uses 4.12, 3.9 and 3.10. Part (iii) uses 4.9--the fact that we have an equivalence $X \longrightarrow S^1 \wedge X$ of degree 1.

These statements give the analogues for a theory defined on spectra of the Eilenberg-Steenrod axioms.

Once we have defined homology and cohomology of spectra, of course we can define homology and cohomology of CW-complexes. That is, if L is a CW-complex, we define $\tilde{E}_n(L)$ to be E_n applied to the suspension spectrum of the complex L, and similarly for $\tilde{E}^n$. The theory on complexes satisfies the same axioms.

For example, let $H\pi$ be an Eilenberg-MacLane spectrum with a single non-vanishing homotopy group π in dimension 0; then $(H\pi)_*$ is a homology theory defined on spectra with a single non-vanishing coefficient group, π in dimension 0. Apply $(H\pi)_*$ to the suspension spectrum of a complex L; it must coincide with the ordinary homology theory of L. If one happens to have seen the ordinary homology groups of a spectrum defined before, then $(H\pi)_*$ is the same thing, as we see by passing to limits.

THEOREM 6.2. (G. W. Whitehead). $E_n(X) \cong X_n(E)$.

Proof. $E \wedge X \xrightarrow{c} X \wedge E$ is an equivalence, so

$$[S, E \wedge X]_n \cong [S, X \wedge E]_n.$$

COROLLARY 6.3. $(H\pi)_n(HG) \cong (HG)_n(H\pi)$.

This was found empirically by Cartan, but it is non-trivial to prove directly. G. W. Whitehead's discovery of the proof just given was probably an important step in his thinking about the connection between spectra and homology theories.

PROPOSITION 6.4. If X is a finite spectrum, $E_n(X^*) \cong E^{-n}(X)$.

Proof. $[S, E \wedge X^*]_n \xrightarrow{T} [X, E]_n$ is an isomorphism by 5.4.

This shows that generalized homology and cohomology behave correctly under S-duality.

Proof of 5.5: that is, if X is a finite spectrum, then X^* is equivalent to a finite spectrum.

Let X be a finite spectrum. Then $[S, X^*] \cong [X, S]_n$, and the right-hand side is zero if n is negative with sufficiently large absolute value. But $H_n(X^*) = H^{-n}(X)$, which is finitely generated in each dimension and zero outside a finite range of dimensions. Therefore X^* is equivalent to a finite spectrum.

Remark 6.5. Every generalized homology or cohomology theory defined on the category of CW-complexes arises by G. W. Whitehead's construction from some spectrum E.

In order to have a proper statement, it is necessary to spell out the assumptions we make on the homology or cohomology of infinite complexes. In the case of homology we assume that

$$\varinjlim_{\alpha} \widetilde{E}_n(L_\alpha) \longrightarrow \widetilde{E}_n(L)$$

is an isomorphism, where L_α runs over the finite subcomplexes of L. In the case of cohomology we assume the Wedge Axiom of Milnor and Brown, that is,

$$\widetilde{E}^n(\bigvee_\alpha L_\alpha) \longrightarrow \prod_\alpha \widetilde{E}^n(L_\alpha)$$

is an isomorphism.

I propose to omit the proof of Remark 6.5. In the case of cohomology the results is fairly easily deduced from E.H. Brown's theorem in the category of CW-complexes, and this was done in G. W. Whitehead's original paper [17]. The argument is essentially that given in section 2. In the case of homology we first obtain a homology theory on spectra in an obvious way. One then converts one's homology theory into a cohomology theory defined only on finite spectra, by the definition

$$E^{-n}(X) = E_n(X^*) .$$

(So one only needs the homology theory on finite spectra, in which case it is trivial to define it.) One then has a contravariant functor defined on finite spectra or finite complexes, and we have the task of representing it. I have proved the required result (Topology 10 (1971) pp. 185-198).

We now consider generalized homology and cohomology groups with coefficients. Let G be an abelian group. We can take a resolution $0 \longrightarrow R \xrightarrow{i} F \longrightarrow G \longrightarrow 0$ by free Z-modules (a subgroup of a free abelian group is free). Take $\bigvee_{\alpha\in A} S$, $\bigvee_{\beta\in B} S$, such that

$$\pi_0(\bigvee_{\alpha\in A} S) = R$$

$$\pi_0(\bigvee_{\beta\in B} S) = F .$$

Take a map $f: \bigvee_{\alpha\in A} S \longrightarrow \bigvee_{\beta\in B} S$ inducing i. Form $M = (\bigvee_{\beta\in B} S) \cup_f C(\bigvee_{\alpha\in A} S)$; this is a Moore spectrum of type G. That is, we have

$$\pi_r(M) = 0 \qquad \text{for } r < 0$$

$$\pi_0(M) = H_0(M) = G ,$$

$$H_r(M) = 0 \qquad \text{for } r > 0 .$$

Now for any spectrum E, we define the corresponding spectrum with coefficients in G by

$$EG = E \wedge M .$$

<u>Example.</u> SG means $S \wedge M = M$, so a Moore spectrum of type G may be written SG.

PROPOSITION 6.6. (i) There exists an exact sequence

$$0 \longrightarrow \pi_n(E) \otimes G \longrightarrow \pi_n(EG) \longrightarrow \mathrm{Tor}_1^Z(\pi_{n-1}(E), G) \longrightarrow 0.$$

(This need not split, e.g., take $E = KO$, $G = Z_2$.)

(ii) More generally, there exist exact sequences

$$0 \longrightarrow E_n(X) \otimes G \longrightarrow (EG)_n(X) \longrightarrow \mathrm{Tor}_1^Z(E_{n-1}(X), G) \longrightarrow 0$$

and (if X is a finite spectrum or G is finitely generated)

$$0 \longrightarrow E^n(X) \otimes G \longrightarrow (EG)^n(X) \longrightarrow \mathrm{Tor}_1^Z(E^{n+1}(X), G) \longrightarrow 0.$$

Proof. $\bigvee_\alpha S \longrightarrow \bigvee_\beta S \longrightarrow M$ is a cofibering, hence the top row of

$$\begin{array}{ccccc} E \wedge (\bigvee_\alpha S) & \longrightarrow & E \wedge (\bigvee_\beta S) & \longrightarrow & E \wedge M \\ \downarrow \simeq & & \downarrow \simeq & & \\ \bigvee_\alpha E & & \bigvee_\beta E & & \end{array}$$

is a cofibering. Similarly

$$\begin{array}{ccccc} E \wedge (\bigvee_\alpha S) \wedge X & \longrightarrow & E \wedge (\bigvee_\beta S) \wedge X & \longrightarrow & E \wedge M \wedge X \\ \downarrow \simeq & & \downarrow \simeq & & \\ \bigvee_\alpha E \wedge X & & \bigvee_\beta E \wedge X & & \end{array}$$

is a cofibering. Therefore we get exact sequences

$$\begin{array}{ccccccc} \longrightarrow & \pi_n(\bigvee_\alpha E) & \longrightarrow & \pi_n(\bigvee_\beta E) & \longrightarrow & \pi_n(E \wedge M) & \longrightarrow \\ & \updownarrow \cong & & \updownarrow \cong & & & \\ & R \otimes \pi_n(E) & \xrightarrow{i \otimes 1} & F \otimes \pi_n(E) & & & \end{array}$$

and more generally,

$$\begin{array}{ccccccc} \longrightarrow & [S, \bigvee_\alpha E \wedge X]_n & \longrightarrow & [S, \bigvee_\beta E \wedge X]_n & \longrightarrow & [S, E \wedge M \wedge X]_n & \longrightarrow \\ & \updownarrow \cong & & \updownarrow \cong & & & \\ & R \otimes [S, E \wedge X]_n & \xrightarrow{i \otimes 1} & F \otimes [S, E \wedge X]_n & & & \end{array}$$

$$\begin{array}{ccccccc} \longrightarrow & [X, \bigvee_\alpha E]_{-n} & \longrightarrow & [X, \bigvee_\beta E]_{-n} & \longrightarrow & [X, E \wedge M]_{-n} & \longrightarrow \\ & \updownarrow \cong & & \updownarrow \cong & & & \\ & R \otimes [X, E]_{-n} & \xrightarrow{i \otimes 1} & F \otimes [X, E]_{-n} & & & \end{array}$$

To get the isomorphisms in the last case we assume either that X is a finite spectrum or that α and β run over finite sets, which we can arrange if G is finitely generated. Now the cokernel and kernel of $i \otimes 1$ are, according to the case,

$$G \otimes \pi_n(E) \quad \text{and} \quad \mathrm{Tor}_1^Z(G, \pi_n(E))$$
$$G \otimes E_n(X) \quad \text{and} \quad \mathrm{Tor}_1^Z(G, E_n(X))$$
$$G \otimes E^n(X) \quad \text{and} \quad \mathrm{Tor}_1^Z(G, E^n(X)) .$$

Example. If H means an Eilenberg-Mac Lane spectrum of type Z, then HZ does indeed mean the Eilenberg-Mac Lane spectrum of type G.

Proof. The Tor term is zero in

$$0 \longrightarrow Z \otimes G \longrightarrow \pi_*(HG) \longrightarrow \mathrm{Tor}_Z^1(Z, G) \longrightarrow 0 .$$

PROPOSITION 6.7. If G is torsion-free, then

$$\pi_*(E) \otimes G \longrightarrow \pi_*(EG)$$

and

$$E_*(X) \otimes G \longrightarrow (EG)_*(X)$$

are isomorphisms, and if X is finite or G finitely generated,

$$E^*(X) \otimes G \longrightarrow (EG)^*(X)$$

is an isomorphism.

Proof.
$$\mathrm{Tor}_1^Z(\pi_*(E), G) = 0$$
$$\mathrm{Tor}_1^Z(E_*(X), G) = 0$$
and
$$\mathrm{Tor}_1^Z(E^*(X), G) = 0 .$$

Example. Take $G = Q$, and take a map $i: S \longrightarrow H$ representing a generator of $\pi_0(H) = Z$. Then i induces an equivalence $SQ \xrightarrow{\simeq} HQ$, i.e., the Moore spectrum for Q is the same as the Eilenberg-Mac Lane spectrum.

Proof. In the diagram

$$\begin{array}{ccc} \pi_n(S) \otimes Q & \longrightarrow & \pi_n(SQ) \\ {\scriptstyle i_* \otimes 1} \Big| & & \Big| \\ \pi_n(H) \otimes Q & \longrightarrow & \pi_n(HQ) \end{array}$$

the top and bottom rows are isomorphisms by 6.7. But by a theorem of Serre, $\pi_n(S) \otimes Q = 0$ for $n \neq 0$; and for $n = 0$, $i_*: \pi_0(S) \longrightarrow \pi_0(H)$ is an isomorphism.

Example. The map $i: S \longrightarrow H$ induces

$$\pi_*(X) \otimes Q \longrightarrow H_*(X) \otimes Q,$$

that is, rational stable homotopy is the same as rational homology.

Proof. $\pi_*(X) = S_*(X)$. Again by 6.7 the top and bottom rows of the following diagram are isomorphisms.

$$\begin{array}{ccc} S_*(X) \otimes Q & \longrightarrow & SQ_*(X) \\ \downarrow & & \downarrow \\ H_*(X) \otimes Q & \longrightarrow & HQ_*(X) \end{array}$$

By the previous example $SQ \longrightarrow HQ$ is an equivalence, so the right-hand arrow is an isomorphism.

Now we give a checklist of the standard spectra corresponding to the usual generalized homology and cohomology theories.

(i) HG, the Eilenberg-MacLane spectrum for the group G, so that

$$\pi_n(HG) = \begin{cases} G & (n = 0) \\ 0 & (n \neq 0) \end{cases}.$$

The theories $(HG)_*$, $(HG)^*$ are ordinary homology and cohomology with coefficients in G.

For greater interest, let G_* be a graded group, and define $H(G_*) = \bigvee_n H(G_n, n) \cong \prod_n H(G_n, n)$; the second map is an equivalence by

3.14. Then by the first form

$$H(G_*)_r(X) = \sum_n H_{r-n}(X;G_n)$$

and by the second

$$H(G_*)^r(X) = \prod_n H^{r+n}(X;G_n) .$$

(ii) S, the sphere spectrum. The corresponding homology and cohomology theories are stable homotopy and stable cohomotopy. With all due respect to anyone who is interested in them, the coefficient groups $\pi_n(S)$ are a mess. There is a lot of detailed information known about them, but I won't try to summarize it.

(iii) K, the classical BU-spectrum. This is an Ω- or Ω_0-spectrum; each even term is the space BU or $Z \times BU$; each odd term is the space U.

The corresponding homology and cohomology theories are complex K-homology and K-cohomology. In fact it is rather easy to see that for a finite-dimensional CW-complex X, $[X, Z \times BU]$ agrees with the Atiyah-Hirzebruch definition of $K(X)$ or $\tilde{K}(X)$ in terms of complex vector-bundles over X. (Here we have to take $\tilde{K}(X)$ if $[X, Z \times BU]$ means homotopy classes of maps preserving the base-point, or $K(X)$ if we work without base-points.) This shows that our definition of $K^*(X)$ agrees with the Atiyah-Hirzebruch definition if X is a finite-dimensional CW-complex. For infinite-dimensional complexes our $K^*(X)$ is the variant called "representable K-theory", i.e., we take $[X, Z \times BU]$ as the definition.

The coefficient groups are given by the Bott periodicity theorem:

$$\pi_n(K) = \begin{cases} Z & (n \text{ even}) \\ 0 & (n \text{ odd}) . \end{cases}$$

We have a map $K \simeq S \wedge K \xrightarrow{i \wedge 1} H \wedge K \longrightarrow H(\pi_*(K) \otimes Q)$. This map is the universal Chern character.

(iv) K-theory with coefficients. Suppose we are willing to localize Z at the prime p; i.e., let Q_p be the ring of fractions a/b with b prime to p. Then we can form KQ_p. It splits as the sum or product of $(p-1)$ similar spectra E. The typical one has

$$\pi_n(E) = \begin{cases} Q_p & (n \equiv 0 \bmod 2(p-1)) \\ 0 & \text{otherwise} . \end{cases}$$

Of course you may just want to split K into the sum or product of d similar spectra, so that a typical one has

$$\pi_n(E) = \begin{cases} R & (n \equiv 0 \bmod 2d) \\ 0 & \text{otherwise} , \end{cases}$$

where R is a subring of Q. In this case one need only invert those primes p such that $p \not\equiv 1 \bmod d$. For example, for $d = 2$ take $R = Z[1/2]$. See [1].

(v) Connective K-theory. bu is a spectrum having a map $bu \longrightarrow K$ such that

$$\pi_r(bu) \longrightarrow \pi_r(K) \text{ is an isomorphism for } r \geq 0, \text{ and}$$
$$\pi_r(bu) = 0 \quad \text{for } r < 0.$$

We may take the 0-th term of bu to be $Z \times BU$ and the second term to be BU. If X is a complex, we have

$$bu^0(X) = K^0(X),$$

but the groups $bu^n(X)$ and $K^n(X)$ are different in general for $n > 0$.

(vi) Similarly, one can consider connective K-theory with coefficients.

(vii) KO, the classical BO-spectrum. This is an Ω- or Ω_0-spectrum; every term E_{8r} is the space BO or $Z \times BO$; every term

E_{8r+4} is the space BSp or $Z \times BSp$. The other terms are the ones which come in Bott's periodicity theorem for the real case:

$$O, O/U, U/Sp, BSp, Sp, Sp/U, U/O, BO.$$

The corresponding homology and cohomology theories are real K-homology and real K-cohomology. In fact (as for the complex case) for a finite-dimensional CW-complex X, $[X, Z \times BO]$ agrees with the Atiyah-Hirzebruch definition of $KO(X)$ or $\widetilde{KO}(X)$ in terms of real vector-bundles over X. So our definition of $KO^*(X)$ agrees with Atiyah and Hirzebruch if X is a finite-dimensional CW-complex.

The coefficient groups are given by the Bott periodicity theorem:

$n \equiv 0$	1	2	3	4	5	6	7	8	mod 8
$\pi_n(KO) = Z$	Z_2	Z_2	0	Z	0	0	0	Z .	

(viii) KO-theory with coefficients. The quickest thing to say is that by a theorem of Reg Wood, $KO \wedge (S^0 \cup_\eta e^2) \simeq K$. Here $S^0 \cup_\eta e^2$ means the suspension spectrum whose second term is CP^2. The attaching map η is stably of order 2. So $SZ[1/2] \xrightarrow{\eta \wedge 1} SZ[1/2]$ factors through $(S^0 \cup_2 e^1)Z[1/2]$, which is contractible. So

$$\begin{aligned} KZ[1/2] &\simeq KO \wedge (S^0 \cup_\eta e^2)Z[1/2] \\ &\simeq KO \wedge (S^0 \vee S^2)Z[1/2] \\ &\simeq KOZ[1/2](S^0 \vee S^2) . \end{aligned}$$

So the two summands into which $KZ[1/2]$ splits are actually copies of $KOZ[1/2]$. It follows that if you introduce a ring of coefficients containing $1/2$, K cannot be distinguished from two copies of KO. Of course this is classical, by a more direct proof.

(ix) Connective real K-theory. bo is a spectrum having a map $bo \longrightarrow KO$ with properties like those of $bu \longrightarrow K$.

(x) KSC, the self-conjugate K-theory of Anderson and Green. The quickest way to say it is this. To each bundle ξ we have its complex conjugate $\bar{\xi}$ which has the same underlying space but a new C-module structure on each fiber; the new action of z is the old action of $\bar{z}$. Stably, this is induced by a map $Z \times BU \xrightarrow{t} Z \times BU$. We can define a map of spectra $T: K \longrightarrow K$ which has components t in dimensions divisible by 4, and $-t$ in dimensions of the form $4r+2$. Now take KSC to be the fiber of

$$K \xrightarrow{1-\tau} K .$$

You can read its homotopy groups off from the exact sequence of this fibering: we have

$n \equiv$	0	1	2	3	4 mod 4
$\pi_n(KSC) =$	Z	Z_2	0	Z	Z .

(xi) MO, the Thom spectrum for the group O. The corresponding theories are unoriented bordism and cobordism. To connect our definition of $MO_*(X)$ with a geometrical definition in terms of manifolds one has to make use of a transversality theorem at some point; see e.g., [5].

We have

$$MO \simeq H(\pi_*(MO)) .$$

$\pi_*(MO)$ is a polynomial algebra over Z_2 with one generator in every dimension $d > 0$ such that $d + 1$ is not a power of 2. The decomposition of MO as a wedge of copies of HZ_2 shows that the theories MO_* and MO^* are not very powerful, but they are good for studying unoriented manifolds.

(xii) MSO. The corresponding theories are oriented bordism and cobordism. We have

$$MSO \simeq H(\pi_*(MSO)) .$$

$\pi_*(MSO)$ is a direct sum of copies of Z and Z_2. It is known, but somewhat complicated to describe.

(xiii) MU. The corresponding theories are complex bordism and cobordism. $\pi_*(MU)$ is a polynomial algebra over Z with generators of dimension $2, 4, 6, 8 \ldots$. There is a very good map $MU \longrightarrow K$ due to Atiyah-Hirzebruch, Conner-Floyd [3], [5]. The theories MU_*, MU^* are powerful.

(xiv) MU with coefficients. If one takes MUQ_p, it splits as a sum of suspensions of similar spectra. A typical one is BP, the Brown-Peterson spectrum. $\pi_*(BP)$ is a polynomial algebra over Q_p on generators of dimension $2(p^f-1)$ for $f = 1, 2, \ldots$.

(xv) MSpin, MSU, MSp. $\pi_*(MSpin)$ and $\pi_*(MSU)$ are known, but $\pi_*(MSp)$ is not yet known.

For a general reference on bordism and cobordism, I suggest Stong [16].

We now consider the elementary additive properties of generalized homology and cohomology theories.

Recall that I had my theories E_*, E^* defined on spectra, and then I defined them on CW-complexes with base-point by saying

$$\tilde{E}_*(L) = E_*(\bar{L})$$
$$\tilde{E}^*(L) = E^*(\bar{L}) ,$$

where $\bar{L}$ is the suspension spectrum of L. I should say how one defines relative groups $E_*(X,A)$, $E^*(X,A)$. This is well enough known. One defines X/A to be the quotient complex in which A is identified to a new point, which becomes the base-point. In particular, $X/\emptyset = X \cup pt.$, also written X^+. Alternatively, one constructs the unreduced cone CA and forms $X \cup CA$, taking the base-point at the vertex. This happens to be the same as the reduced cone $X^+ \cup CA^+$. Then one has a map $X \cup CA \xrightarrow{r} X/A$, which is a homotopy equivalence. Then one defines

$$E_*(X, A) = \tilde{E}_*(X \cup CA) = \tilde{E}_*(X/A) ,$$

using the isomorphism r_* to identify the last two groups. Similarly,

$$E^*(X, A) = \tilde{E}^*(X \cup CA) = \tilde{E}^*(X/A) .$$

Note that $E_*(X, pt.) = \tilde{E}_*(X)$, as it should be, and similarly for $E^*(X, pt.) = \tilde{E}^*(X)$.

The induced homomorphisms are obvious: a map $f: X, A \longrightarrow Y, B$ induces

$$\begin{array}{ccc} X \cup CA & \longrightarrow & Y \cup CB \\ \downarrow r & & \downarrow r \\ X/A & \longrightarrow & Y/B \end{array}$$

and we take the induced homomorphisms of $\tilde{E}_*$ or $\tilde{E}^*$.

Excision is now obvious. Suppose a CW-complex is the union of two subcomplexes U, V. Then

$$U/U \cap V \longrightarrow U \cup V/V$$

is actually a homeomorphism, so it surely induces an isomorphism of $\tilde{E}_*$ and $\tilde{E}^*$. Homotopy is equally obvious. Now we would like to have boundary maps and exactness. Given an inclusion $X \longrightarrow Y$, we have a cofibering

$$X^+ \xrightarrow{i} Y^+ \xrightarrow{j} Y^+/X^+ \simeq Y^+ \cup CX^+ \longrightarrow SX^+ \xrightarrow{Si} SY^+ \longrightarrow \dots .$$

So applying $\tilde{E}_n$ we have the following exact sequence.

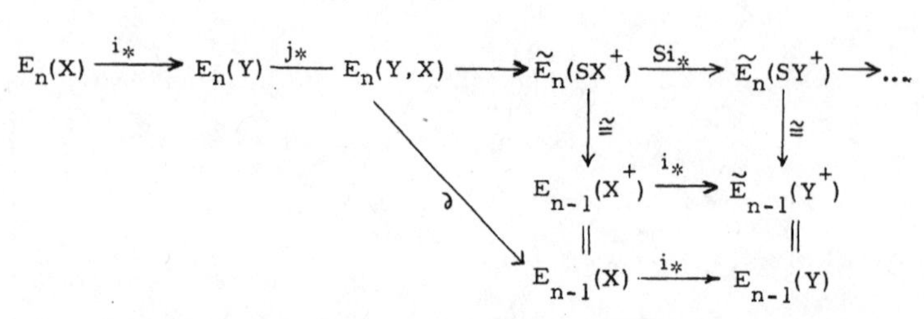

If I define ∂ to be the composite indicated, the sequence will be exact. So in order to fix the boundary map and have it natural I simply want to make some quite explicit choice of isomorphism $\tilde{E}_n(X^+) \cong \tilde{E}_{n+1}(S^1 \wedge X^+)$.

Let's recall that almost the last thing I did in section 4 was to make the smash-product a functor of maps of degrees other than zero. So I look at the sphere-spectrum

$$S = (S^0, S^1, S^2, \ldots)$$

and the S^1-spectrum

$$S^1 = (S^1, S^2, S^3, \ldots)$$

and I make a map from one to the other by taking the identity map from S^n, the nth component of S, to S^n, the (n-1)-st component of S^1. This gives me a morphism of degree 1, say $\sigma: \underline{S} \longrightarrow \underline{S}$. (This is actually γ_1 for the spectrum S, but you are allowed to have forgotten about γ_1 by now.) σ is clearly an equivalence. Since I have smash-products of morphisms of nonzero degree, I am entitled to form

$$X \simeq S \wedge X \xrightarrow{\sigma \wedge 1} S^1 \wedge X.$$

This is an equivalence too. (Of course, the smash-product of morphisms of nonzero degree was defined in terms of the maps γ_r, and if you go back to the definition and unwrap it, you find that this is just the map γ_1 for the spectrum X.) I now say that this map

$$X \xrightarrow{\sigma \wedge 1} S^1 \wedge X$$

is the one to be used in inducing

$$E_n(X) \xrightarrow{\cong} E_{n+1}(S^1 \wedge X)$$

$$E^n(X) \longleftarrow E^{n+1}(S^1 \wedge X).$$

This gets my suspension isomorphism in a form convenient for later work, and makes the boundary and coboundary quite precise.

Now we would like to assure ourselves that all the contents of Eilenberg-Steenrod, Chapter I, go through. But we can also put the question in this form: is there anything in Eilenberg-Steenrod Chapter I which can't be derived from our constructions? The grand conclusion should be that the homology groups of spheres are the right thing, and we already know that

$$\begin{aligned} \tilde{E}_r(S^n) &= [S^n, E]_r \\ &\cong [S^0, E]_{r-n} \\ &\cong \pi_{r-n}(E) . \end{aligned}$$

The only problem is to compute $\pi_*(E)$ for a given E. So what about the other things in Eilenberg-Steenrod Chapter I? One very useful thing is the exact sequence of a triple. Suppose we have CW-complexes $X \supset Y \supset Z$. We would like to know that the following sequence is exact.

$$E_n(Y,Z) \xrightarrow{i_*} E_n(X,Z) \xrightarrow{j_*} E_n(X,Y) \xrightarrow{\Delta} E_{n-1}(Y,Z) \longrightarrow \cdots$$

Here Δ is the composite

$$E_n(X,Y) \xrightarrow{\partial} E_{n-1}(Y) \xrightarrow{j_*} E_{n-1}(Y,Z) .$$

No special proof is needed. We know that the following is a cofibering:

$$Y^+/Z^+ \longrightarrow X^+/Z^+ \longrightarrow X^+/Y^+ \longrightarrow S(Y^+/Z^+) \longrightarrow S(X^+/Z^+) .$$

Therefore I know that I have an exact sequence

$$E_n(Y,Z) \xrightarrow{i_*} E_n(X,Z) \xrightarrow{j_*} E_n(X,Y) \xrightarrow{\partial} E_{n-1}(Y,Z) \xrightarrow{i_*} E_{n-1}(X,Z)$$

provided that ∂ is induced by the top line of the following commutative diagram.

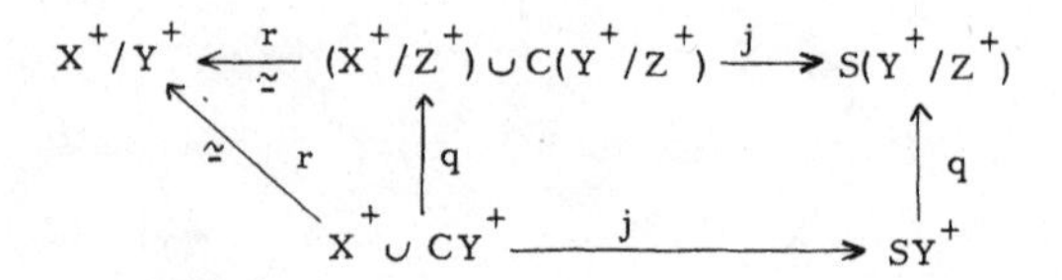

The rest of the diagram shows that ∂ is the same as Δ .

There is however a moral to be drawn. We know how to display the various groups and homomorphisms involved here in a sine wave diagram.

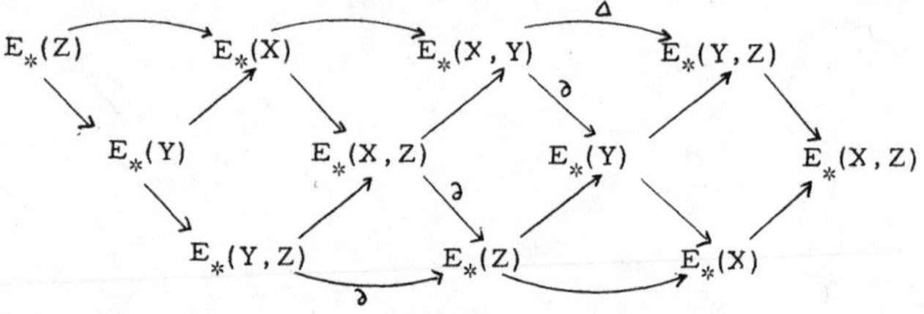

It is useful to know that we can obtain this whole diagram from a diagram of cofiberings.

LEMMA 6.8. Suppose given a commutative diagram

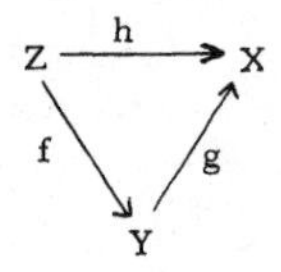

of CW-complexes with base-point. Then there exists the following commutative diagram of cofiber sequences.

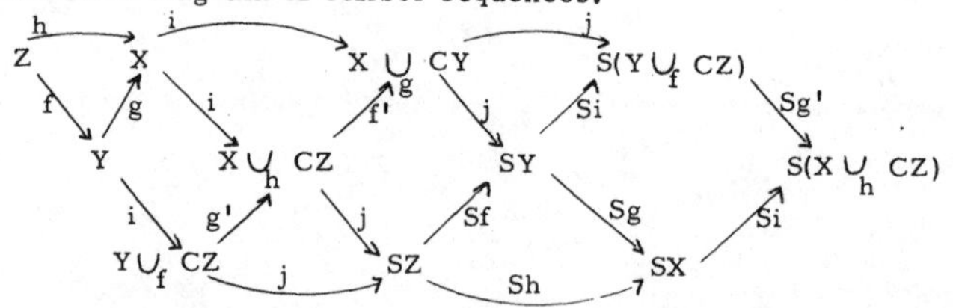

Here g' is induced from g, etc. If the original diagram is only homotopy-commutative, then by choosing a homotopy you can reduce to the case in which it is commutative.

This is sometimes known as Verdier's axiom. The proof is elementary. One way to say it is this: you can assume without loss of generality that f and g are inclusions, and then I have told you everything necessary already. Since the constructions are elementary, they commute with suspensions on the right and carry over to spectra. So the corresponding lemma is true for spectra. In a fully Bourbakized treatment this lemma would go in section 3.

The next thing we would like to know is the Mayer-Vietoris sequence. This needs no special proof either. Suppose that we have a CW-complex which is the union of two subcomplexes U and V. We wish to know the relationship of $E_*(U\cup V), E_*(U), E_*(V), E_*(U\cap V)$. We may replace these by $E_*(S(U\cup V)^+)$, etc. So we take $S(U\cap V)^+$ and $S(U^+\vee V^+)$ and make a map from one to the other by taking $i_1 - i_2$, where $i_1 : (U\cap V)^+ \longrightarrow U^+$ and $i_2 : (U\cap V)^+ \longrightarrow V^+$ are the inclusions. Now let me form the cofiber sequence

$$S(U\cap V)^+ \longrightarrow S(U^+\vee V^+) \longrightarrow S(U^+\vee V^+)\cup_{i_1 - i_2} CS(U\cap V)^+ \longrightarrow S^2(U\cap V)^+ \longrightarrow \ldots$$

The third term is the same as

$$SU^+\vee SV^+ \cup \mathrm{Cyl}(S(U\cap V)^+) ,$$

where the (reduced) cylinder is attached by i_1 to SU^+ and i_2 to SV^+. But this clearly has the same homotopy type as $S(U\cup V)^+$. So we get a cofibering

$$S(U\cap V)^+ \xrightarrow{i_1 - i_2} S(U^+\vee V^+) \longrightarrow S(U\cup V)^+ \longrightarrow S^2(U\cap V)^+ \longrightarrow \ldots .$$

Here the third map can be written either as

$$S(U\cup V)^+ \longrightarrow S(U\cup V/V) \xleftarrow{\cong} S(U/U\cap V) \longrightarrow S^2(U\cap V)^+$$

or as minus

$$S(U\cup V)^+ \longrightarrow S(U\cup V/U) \xleftarrow{\cong} S(V/U\cap V) \longrightarrow S^2(U\cap V)^+ .$$

So we get the following long exact sequence.

$$E_n(U\cap V) \xrightarrow{(i_{1*}, -i_{2*})} E_n(U) \oplus E_n(V) \xrightarrow{(j_{1*}, j_{2*})} E_n(U\cup V) \xrightarrow{\Delta} E_{n-1}(U\cap V) \to ..$$

Here the boundary map is given by

$$E_n(U\cup V) \to E_n(U\cup V, V) \xleftarrow{\cong} E_n(U, U\cap V) \xrightarrow{\partial} E_{n-1}(U\cap V)$$

or minus

$$E_n(U\cup V) \to E_n(U\cup V, U) \xleftarrow{\cong} E_n(V, U\cap V) \xrightarrow{\partial} E_{n-1}(U\cap V) .$$

We proceed similarly in cohomology.

Of course this construction also carries over to spectra. In fact for spectra we need not bother about writing the suspension, because up to equivalence everything is a suspension. We obtain:

LEMMA 6.9. Suppose a CW-spectrum is the union of two closed subspectra U, V. Then there is a cofibering

$$U\cap V \xrightarrow{(i_1, -i_2)} U \vee V \xrightarrow{(j_1, j_2)} U\cup V \to \mathrm{Susp}(U\cap V) \to \dots$$

in which the third morphism is

$$U\cup V \to U\cup V/V \xleftarrow{\cong} U/U\cap V \to \mathrm{Susp}(U\cap V)$$

or minus

$$U\cup V \to U\cup V/U \xleftarrow{\cong} V/U\cap V \to \mathrm{Susp}(U\cap V).$$

We may call this the Mayer-Vietoris cofibering.

7. THE ATIYAH-HIRZEBRUCH SPECTRAL SEQUENCE

In this section we study the machine which plays the same role in the study of generalized homology theories as the Eilenberg-Steenrod uniqueness theorem plays for ordinary homology theories. Let us suppose for convenience that X is a finite-dimensional CW-complex.

THEOREM. For each CW-spectrum F there exist spectral sequences

$$H_p(X;\pi_q(F)) \underset{p}{\Longrightarrow} F_{p+q}(X)$$
$$H^p(X;\pi_q(F)) \underset{p}{\Longrightarrow} F^{p+q}(X) .$$

These spectral sequences were probably first invented by G. W. Whitehead, but he got them just after he wrote the paper [18] in which they ought to have appeared. They then became a folk-theorem and were eventually published by Atiyah and Hirzebruch, who needed them for the case $F = K$.

It is probably desirable to give the first part of the construction in greater generality. Suppose I have a CW-complex X with a finite filtration by subcomplexes,

$$\emptyset = X_{-1} \subset X_0 \subset X_1 \subset X_2 \subset \ldots \subset X_n = X .$$

To get the Atiyah-Hirzebruch spectral sequence you take $X_r = X^r$, the r-skeleton; but other choices of filtration are possible, and sometimes useful. If we then apply a functor F_* or F^* to all the available pairs and triples, we get a maze of interlocking exact sequences. The spectral sequence helps us to find our way through this maze and to distill out the essential information.

There are two ways to present the distillation. The first is due to Massey, and it is the method of exact couples. We observe that we have an exact sequence, which we write in a triangle like this.

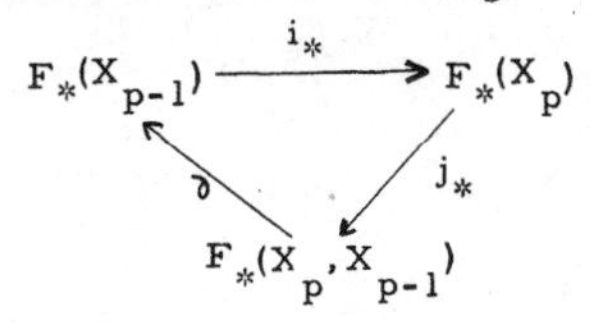

If we add over p, we obtain

$$\begin{array}{ccc} \sum_p F_*(X_{p-1}) & \xrightarrow{\;i_*\;} & \sum_p F_*(X_p) \\ & {\scriptstyle \partial}\nwarrow \quad \swarrow{\scriptstyle j_*} & \\ & \sum_p F_*(X_p, X_{p-1}) & \end{array} .$$

Here we interpret $F_*(X_p)$ as 0 for $p < 0$ and as $F_*(X)$ for $p \geq n$.

Now we have a triangle of the following form.

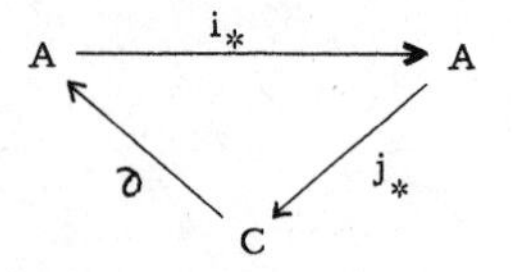

Massey called such a triangle an <u>exact couple</u>, and he showed that from such an exact couple you could obtain a derived exact couple

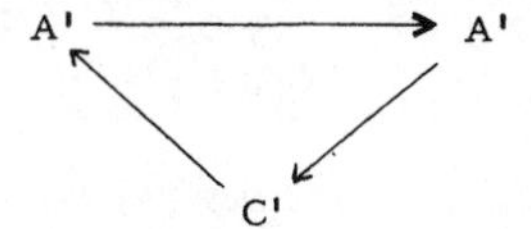

For example you define $d_1 = j_* \partial : C \longrightarrow C$ and define $C' = \operatorname{Ker} d_1 / \operatorname{Im} d_1$.

Iterating this procedure, you obtain at C', C'', C''' etc. all the terms of the spectral sequence. A suitable reference is Massey [9].

The second method probably goes back to Eilenberg, and it is essentially equivalent; it consists simply of writing down explicit definitions of the desired groups and homomorphisms. For example, we define

$$\begin{aligned} Z^r_{p,q} &= \operatorname{Ker} \{F_{p+q}(X_p, X_{p-1}) \xrightarrow{\;\partial\;} F_{p+q-1}(X_{p-1}, X_{p-r})\} \\ &= \operatorname{Im} \{F_{p+q}(X_p, X_{p-r}) \xrightarrow{\;j_*\;} F_{p+q}(X_p, X_{p-1})\} , \\ B^r_{p,q} &= \operatorname{Im} \{F_{p+q+1}(X_{p+r-1}, X_p) \xrightarrow{\;\partial\;} F_{p+q}(X_p, X_{p-1})\} \\ &= \operatorname{Ker} \{F_{p+q}(X_p, X_{p-1}) \xrightarrow{\;i_*\;} F_{p+q}(X_{p+r-1}, X_{p-1})\} , \end{aligned}$$

check $B^r_{p,q} \subset Z^r_{p,q}$, and define

$$E^r_{p,q} = Z^r_{p,q}/B^r_{p,q} .$$

We define the boundary maps d_r by passing to the quotient from boundary maps ∂ in an appropriate way. We prove $\mathrm{Ker}\, d_r/\mathrm{Im}\, d_r \cong E^{r+1}_{p,q}$ by diagram-chasing. For r sufficiently large the groups $Z^r_{p,q}$, $B^r_{p,q}$ and $E^r_{p,q}$ become independent of r, and may be written $Z^\infty_{p,q}$, $B^\infty_{p,q}$ and $E^\infty_{p,q}$.

We filter the groups $F_m(X)$ by taking the images of the maps

$$F_m(X_p) \longrightarrow F_m(X) ;$$

the image of $F_m(X_n)$ is the whole of $F_m(X)$, the image of $F_m(X_{-1})$ is zero, and the quotients of the successive filtration subgroups are isomorphic to the groups $E^\infty_{p,q}$ for $p+q = m$, as one sees with a little diagram-chasing.

So one gets a spectral sequence with

$$E^1_{p,q} = F_{p+q}(X_p, X_{p-1}) \underset{p}{\Longrightarrow} F_{p+q}(X) .$$

A similar construction works in cohomology.

Now we revert to the case in which we take the skeleton filtration on X, so that $X_r = X^r$ and $X = X^n$. Then we have

$$\begin{aligned} E^1_{p,q} &= F_{p+q}(X^p, X^{p-1}) \\ &= \widetilde{F}_{p+q}(X^p/X^{p-1}) \\ &= \widetilde{F}_{p+q}(\bigvee_\alpha S^p) \\ &= \sum_\alpha \pi_q(F) \\ &= C_p(X;\pi_q(F)) \end{aligned}$$

, the cellular chains of X with coefficients in $\pi_q(F)$.

Now we need to know that we have the following commutative diagram.

$$\begin{array}{ccc} E^1_{p,q} & \cong & C_p(X;\pi_q(F)) \\ d_1\downarrow & & \downarrow\partial \\ E^1_{p-1,q} & \cong & C_{p-1}(X;\pi_q(F)) \end{array}$$

If so, then we have $E^2_{p,q} \cong H_p(X;\pi_q(F))$. For this purpose there are two alternative methods of proceeding.

(i) Suppose we know that $\pi_p(S^p) = Z$. Then we argue that we simply have to find one component of our map ∂, say

$$\begin{array}{ccc} \sum_\alpha \pi_q(F) & \longrightarrow & \sum_\beta \pi_q(F) \\ i_\alpha \uparrow & & \downarrow p_\beta \\ \pi_q(F) & & \pi_q(F) \end{array}$$

One sees by diagram-chasing that this is the homomorphism of $\tilde{F}_{p+q-1}(S^{p-1})$ induced by the following map.

$$S^{p-1} \longrightarrow X^{p-1} \longrightarrow X^{p-1}/S^{p-2} = \bigvee_\beta S^{p-1} \longrightarrow S^{p-1}$$

Here the first map is the attaching map for the cell indexed by α, and the last is the projection to that indexed by β. This composite map has to have a degree ν, and the homomorphism of $\widetilde{F_{p+q-1}}(S^{p-1})$ which it induces is multiplication by ν. But then ν is also the incidence number between the cells e^p_α and e^{p-1}_β which figures in the definition of

$$\partial: C_p(X;G) \longrightarrow C_{p-1}(X;G) .$$

(ii) If you deny me the knowledge that $\pi_p(S^p) = Z$, then I have to begin by assuming that X is a finite simplicial complex. In this case

$$\partial: C_p(X;G) \longrightarrow C_{p-1}(X;G)$$

is given by a combinatorial formula. I arrange the proof that

$$F_{p+q}(X^p, X^{p-1}) \cong C_p(X;\pi_q(F))$$

with slightly more care and diagram-chasing, so as to incorporate a proof that the isomorphism takes d_1 onto the boundary ∂ given by the combinatorial formula. This is essentially as in Eilenberg-Steenrod, where they prove the uniqueness theorem. It issues in the result that when X is a finite simplicial complex, you can take the H in

$$H_p(X;\pi_q(F)) \underset{p}{\Longrightarrow} F_{p+q}(X)$$

to mean finite simplicial homology. Of course this form of the result is the one which includes the Eilenberg-Steenrod uniqueness theorem: for a finite simplicial complex, any ordinary homology theory agrees with finite simplicial homology with the same coefficients.

Example. Take $F = K$, the classical BU-spectrum, and $X = CP^n$.

We have

$$H^p(X;\pi_{-q}(K)) = \begin{cases} Z & p \text{ even}, \ 0 \leq p \leq 2n, \ q \text{ even} \\ 0 & \text{otherwise} . \end{cases}$$

The E_2-term is illustrated as follows.

$$\begin{array}{c|cccccccc}
q & & & & & & & & \\
 & & & & \cdots & & & & \\
 & Z & Z & Z & Z & & Z & 0 & \\
 & Z & Z & Z & Z & \cdots & Z & & \\
\hline
 & Z & Z & Z & Z & & Z & 0 & \to p \\
 & Z & Z & Z & Z & & Z & & \\
 & & & & \cdots & & & &
\end{array}$$

Since the terms with either grading odd are zero, the spectral sequence collapses, and

$$K^{2m}(CP^n) = \sum_{0}^{n} Z .$$

The Atiyah-Hirzebruch spectral sequence works for infinite complexes, but we need the discussion of limits in the following section.

The spectral sequence also works for spectra X, provided they are bounded below, i.e., there exists ν such that $\pi_r(X) = 0$ for $r < \nu$. For spectra which are not bounded below you can still formally set up the spectral sequence, but the convergence is so bad that the spectral sequence is unusable in practice.

8. THE INVERSE LIMIT AND ITS DERIVED FUNCTORS

Let I be a partially ordered set of indices α. We assume I is directed, that is, for any α, β there is a γ with $\alpha < \gamma$ and $\beta < \gamma$. An inverse system $\underline{G}$ of abelian groups indexed over I consists of abelian groups G_α (one for each $\alpha \in I$) and homomorphisms $g_{\alpha\beta}: G_\alpha \longleftarrow G_\beta$ (one for each pair of indices with $\alpha < \beta$ in I). Such inverse systems form the objects of a category; a morphism $\underline{\theta}: \underline{G} \longrightarrow \underline{H}$ in this category is a list $\{\theta_\alpha\}$ of homomorphisms $\theta_\alpha: G_\alpha \longrightarrow H_\alpha$ such that $\theta_\alpha g_{\alpha\beta} = h_{\alpha\beta}\theta_\alpha$ whenever $\alpha < \beta$. We define $\underleftarrow{\mathrm{Lim}}\, \underline{G}$ to be the subgroup of $\prod_\alpha G_\alpha$ consisting of lists $\{x_\alpha\}$, $x_\alpha \in G_\alpha$, which satisfy $x_\alpha = g_{\alpha\beta} x_\beta$ for all $\alpha < \beta$. The functor $\underleftarrow{\mathrm{Lim}}$ is representable in this category; for let Z be the integers, and let $\underline{Z}$ be the inverse system in which $Z_\alpha = Z$ and $z_{\alpha\beta} = 1$; then $\mathrm{Hom}(\underline{Z}, \underline{G}) \cong \underleftarrow{\mathrm{Lim}}\, \underline{G}$. Moreover, this category has enough injectives. In fact, let I be an injective abelian group; let $\underline{I}_\gamma$ be the inverse system in which

$$G_\alpha = \begin{cases} I & \text{if } \gamma \leq \alpha \\ 0 & \text{otherwise} \end{cases}, \qquad g_{\alpha\beta} = \begin{cases} 1 & \text{if } \gamma \leq \alpha \\ 0 & \text{otherwise} \end{cases}.$$

Then I_γ is injective, and we get enough injectives by taking products of objects like $\underline{I}_\gamma$. We can therefore do homological algebra; in particular, we have the functors

$$\underleftarrow{\mathrm{Lim}}^i\, \underline{G} = \mathrm{Ext}^i(\underline{Z}, \underline{G}) .$$

We have $\underleftarrow{\mathrm{Lim}}^0\, \underline{G} = \underleftarrow{\mathrm{Lim}}\, \underline{G}$.

Frequently we have $I = \{1, 2, 3, \ldots\}$. In this case we have an alternative construction of $\underleftarrow{\mathrm{Lim}}^i$. Given $\underline{G}$, define a cochain complex C by

$$C_0 = C_1 = \prod_i G_n , \qquad \delta\{x_n\} = \{x_n - g_{n,n+1} x_{n+1}\} \quad \text{for } \{x_n\} \in C_0 .$$
$$C_r = 0 \quad \text{for } r > 1 ,$$

Let H^i be the ith cohomology group of C. Then it is immediate that $H^0 = \underleftarrow{\mathrm{Lim}}\, \underline{G}$. To show that $H^i \cong \underleftarrow{\mathrm{Lim}}^i\, \underline{G}$, it is sufficient to make the following remarks.

(i) Let $0 \longrightarrow \underline{G}' \xrightarrow{i} \underline{G} \xrightarrow{j} \underline{G}'' \longrightarrow 0$ be a sequence which is exact in the category of inverse systems, that is, $0 \longrightarrow G'_\alpha \xrightarrow{i_\alpha} G_\alpha \xrightarrow{j_\alpha} G''_\alpha \longrightarrow 0$ is exact for each α. Then we obtain an exact sequence of chain complexes

$$0 \longrightarrow C' \longrightarrow C \longrightarrow C'' \longrightarrow 0,$$

and hence an exact cohomology sequence

$$0 \longrightarrow H'^0 \longrightarrow H^0 \longrightarrow H''^0 \longrightarrow H'^1 \longrightarrow H^1 \longrightarrow H''^1 \longrightarrow 0 .$$

(ii) We have constructed enough injectives with the property that all their maps $g_{\alpha\beta}$ are epi. If all the maps $g_{\alpha\beta}$ are epi, it follows that H^1 is zero. So H^1 vanishes on enough injectives.

It follows that $\underleftarrow{\mathrm{Lim}}^i\, \underline{G} \cong H^i$, and in particular $\underleftarrow{\mathrm{Lim}}^i\, \underline{G} = 0$ for $i \geq 2$, assuming $I = \{1, 2, 3, \ldots\}$. For a general I we would not have this.

Exercise (i). Let $I = \{1, 2, 3, \ldots\}$, and let $\underline{G}$ be an inverse system in which the maps $g_{n,m}$ are mono; thus we may regard G_1 as a topological group, topologized by giving the decreasing sequence of subgroups $\operatorname{Im} g_{1n}$. Then $\underleftarrow{\operatorname{Lim}}^0 \underline{G} = 0$ if and only if G_1 is Hausdorff; $\underleftarrow{\operatorname{Lim}}^1 \underline{G} = 0$ if and only if G_1 is complete. (Here we use words so that "complete" does not imply "Hausdorff"; it means that each Cauchy sequence has a limit, perhaps not unique.)

Exercise (ii). Let $I = \{1, 2, 3, \ldots\}$. We say that $\underline{G}$ satisfies the Mittag-Leffler condition if for each n, there exists m such that $\operatorname{Im} g_{np} = \operatorname{Im} g_{nm}$ for $p \geq m$; that is, $\operatorname{Im} g_{np}$ converges. Show that if $\underline{G}$ satisfies the Mittag-Leffler condition then $\underleftarrow{\operatorname{Lim}}^1 \underline{G} = 0$.

The cochain complex used above is due to Milnor, "On axiomatic homology theory", Pacific J. Math. 12(1962), 337-341. He made the following use of it. Let E^* be a generalized cohomology theory satisfying the wedge axiom; this axiom says that the canonical map

$$\tilde{E}^*\left(\bigvee_\alpha X_\alpha\right) \longrightarrow \prod_\alpha \tilde{E}^*(X_\alpha)$$

is an isomorphism. (One can use E^* instead of $\tilde{E}^*$ if one uses the disjoint union instead of the wedge.) Suppose given an increasing sequence of CW-pairs (X_n, A_n), and set

$$X = \bigcup_n X_n, \qquad A = \bigcup_n A_n .$$

PROPOSITION 8.1 (Milnor). There is an exact sequence

$$0 \longrightarrow \underset{n}{\underleftarrow{\operatorname{Lim}}}^1 E^{q-1}(X_n, A_n) \longrightarrow E^q(X, A) \longrightarrow \underleftarrow{\operatorname{Lim}}^0 E^q(X_n A_n) \longrightarrow 0 .$$

Sketch proof. First consider the absolute case. Replace X by the telescope $\bigcup_n [n, n+1] \times X_n$. Set $U = \bigcup_n [2n, 2n+1] \times X_{2n}$, $V = \bigcup_n [2n+1, 2n+2] \times X_{2n+1}$, so that U consists of the even-numbered cylinders, V of the odd-numbered cylinders. Using the wedge axiom, show that the part

$$E^q(U) \oplus E^q(V) \longrightarrow E^q(U \cap V)$$

of the Mayer-Vietoris sequence coincides, up to isomorphism, with the cochain complex

$$\prod_1^\infty E^q(X_n) \longrightarrow \prod_1^\infty E^q(X_n)$$

considered above. When you have a sound proof for the absolute case, relativize it.

Proposition 8.1 is evidently valid for spectra as well as spaces.

Sketch of applications. It may happen that we wish to construct a morphism $f:X \longrightarrow E$, and can construct morphisms $f_n:X_n \longrightarrow E$, where $\{X_n\}$ is an increasing sequence of subspectra whose union is X. Suppose that $f_n|X_{n-1} = f_{n-1}$. Then 8.1 assures us that there is a morphism $f:X \longrightarrow E$ whose restriction to each X_n is f_n. (In fact, so much is easy to prove directly by using the homotopy extension property.) However, it is difficult to check that morphisms constructed in this way have any good properties, unless one has a uniqueness statement; one needs to know that f is determined by giving $f|X_n$ for all n. By 8.1, it is sufficient to prove that $\varprojlim^1 [X_n, E]_1 = 0$.

For some applications it is important to know how inverse limits work in spectral sequences. Suppose, for example, that we take a generalized cohomology theory E^* satisfying the wedge axiom and a CW-complex X containing an increasing sequence of subcomplexes $\emptyset = X_{-1} \subset X_0 \subset X_1 \subset \ldots \subset X_n \subset \ldots \subset X$. Suppose also that $\varprojlim^0 E^*(X, X_n) = 0$, $\varprojlim^1 E^*(X, X_n) = 0$. (For example, we might have $X = \bigcup_n X_n$). Applying E^*, we obtain a half-plane spectral sequence whose term $E^{p,q}$ is $E^{p+q}(X_p, X_{p-1})$. In what sense does this spectral sequence converge? We may be interested in three conditions.

(i) Observe that $E^{p,q}_{r+1} \longrightarrow E^{p,q}_r$ is mono for $r > p$. So we can ask that the map $E^{p,q}_\infty \longrightarrow \underset{\overleftarrow{r}}{\mathrm{Lim}}\, E^{p,q}_r$ should be iso.

(ii) Similarly, we can ask that $\underset{\overleftarrow{r}}{\mathrm{Lim}}^1\, E^{p,q}_r = 0$.

(iii) Let $F^{p,q}$ be the filtration quotients of $E^{p+q}(X)$, so that we have exact sequences

$$0 \longrightarrow E^{p,q}_\infty \longrightarrow F^{p,q} \longrightarrow F^{p-1,q+1} \longrightarrow 0$$

and $F^{-1,q} = 0$. We can ask that the map $E^n(X) \longrightarrow \underset{\longleftarrow}{\mathrm{Lim}}^0\, F^{p,n-p}$ should be iso.

THEOREM 8.2. Condition (ii) is equivalent to (i) plus (iii).

In practice we verify condition (ii) (see exercise (i)). We then use 8.2 to deduce that conditions (i) and (iii) hold.

We can also generalize 8.1. For convenience I consider the absolute case. Let X be any CW-complex which is the union of a directed set of subcomplexes X_α. Then we have a spectral sequence

$$\underset{\overleftarrow{\alpha}}{\mathrm{Lim}}^p\, E^q(X_\alpha) \underset{p}{\Longrightarrow} E^{p+q}(X) .$$

This spectral sequence is convergent in the sense that 8.2 holds.

9. PRODUCTS

There are four external products we need: an external product in homology, an external product in cohomology and two slant products. Perhaps I should give some motivation for the slant products. The first thing to say is that I need one of them for the duality theorems. The second is to point to the case of ordinary homology. There the Eilenberg-Zilber theorem gives one chain equivalences

$$C_*(X) \otimes C_*(Y) \xrightarrow{\mu} C_*(X \times Y) \xrightarrow{\Delta} C_*(X) \otimes C_*(Y) .$$

So if we have a cycle u in X and a cycle v in Y, then μ gives us a cycle $\mu(u \otimes v)$ on $X \times Y$, whence the external homology product

$$H_*(X) \otimes H_*(Y) \xrightarrow{\mu_*} H_*(X \times Y) .$$

Also we can dualize Δ : if u is a cocycle in X and v is a cocycle in Y, then $\Delta^*(u \otimes v)$ is a cocycle in $X \times Y$, whence the external product in cohomology

$$H^*(X) \otimes H^*(Y) \xrightarrow{\Delta^*} H^*(X \times Y)$$

But you could also consider $\mu(x \otimes y)$ as a function of x with y fixed, and then dualize it, so as to get

$$C^*(X \times Y) \longrightarrow C^*(X) \quad \text{depending on } y,$$

that is,

$$C^*(X \times Y) \otimes C_*(Y) \longrightarrow C^*(X),$$

whence

$$H^*(X \times Y) \otimes H_*(Y) \longrightarrow H^*(X) .$$

Similarly, if we had a cocycle $C_*(X) \xrightarrow{u} Z$, we could form

$$C_*(X \times Y) \xrightarrow{\Delta} C_*(X) \otimes C_*(Y) \xrightarrow{u \otimes 1} C_*(Y) ,$$

and so get

$$H^*(X) \otimes H_*(X \times Y) \longrightarrow H_*(Y).$$

If anything, these products are even more obvious with spectra. Suppose I want to define products in generalized theories, say

$$E^*(X) \otimes F^*(Y) \longrightarrow G^*(X \wedge Y)$$

where X and Y are spectra, or

$$\widetilde{E}^*(X) \otimes \widetilde{F}^*(Y) \longrightarrow \widetilde{G}^*(X \wedge Y)$$

where X and Y are complexes with base-point. Then I should assume given a pairing, i.e., a map $\mu: E \wedge F \longrightarrow G$ of spectra. But then I might as well consider the case $G = E \wedge F$, because everything follows from it by naturality.

(i) The external product in cohomology is a map

$$E^p(X) \otimes F^q(Y) \longrightarrow (E \wedge F)^{p+q}(X \wedge Y)$$

defined as follows. If

$$f \in E^p(X) = [X, E]_{-p} \quad , \quad g \in F^q(Y) = [Y, F]_{-q}$$

then

$$f \wedge g \in [X \wedge Y, E \wedge F]_{-p-q} = (E \wedge F)^{p+q}(X \wedge Y) .$$

The result is written $f \bar{\wedge} g$ to distinguish it from the external product in homology.

(ii) The external product in homology is a map

$$E_p(X) \otimes F_q(Y) \longrightarrow (E \wedge F)_{p+q}(X \wedge Y) .$$

To define it, suppose

$$f \in E_p(X) = [S, E \wedge X]_p \quad , \quad g \in F_q(Y) = [S, F \wedge Y]_q \, ,$$

and form

$$S \xrightarrow{f \wedge g} E \wedge X \wedge F \wedge Y \xrightarrow{1 \wedge c \wedge 1} E \wedge F \wedge X \wedge Y .$$

This gives

$$f \underline{\wedge} g \in (E \wedge F)_{p+q}(X \wedge Y) ,$$

the external product in homology.

In order to see the slant products, one way is to suppose X and Y are finite complexes. Suppose given an element of $E^*(X \wedge Y)$, represented by a map

$$S \xrightarrow{f} E \wedge (X \wedge Y)^* = E \wedge X^* \wedge Y^* ,$$

and suppose given an element of $F_*(Y)$, represented by a map

$$S \xrightarrow{g} F \wedge Y .$$

Then we can form

$$S \xrightarrow{f \wedge g} E \wedge (X^* \wedge Y^*) \wedge F \wedge Y \xrightarrow{1 \wedge c \wedge 1} E \wedge F \wedge (X^* \wedge Y^*) \wedge Y \xrightarrow{1 \wedge 1 \wedge 1 \wedge e} E \wedge F \wedge X^* ;$$

this gives an element of $(E \wedge F)^*(X)$. Similarly, suppose given an element of $E^*(X)$, represented by a map

$$S \xrightarrow{f} E \wedge X^* ,$$

and an element of $F_*(X \wedge Y)$, represented by a map

$$S \xrightarrow{g} F \wedge X \wedge Y .$$

Then we get

$$S \xrightarrow{f \wedge g} E \wedge X^* \wedge F \wedge X \wedge Y \xrightarrow{1 \wedge c \wedge 1 \wedge 1} E \wedge F \wedge X^* \wedge X \wedge Y \xrightarrow{1 \wedge 1 \wedge e \wedge 1} E \wedge F \wedge Y;$$

this gives an element of $(E \wedge F)_*(Y)$.

It follows from §5 that these constructions are equivalent to the following ones, which work whether X and Y are finite or not.

(i) The first slant product is a map

$$E^p(X \wedge Y) \otimes F_q(Y) \longrightarrow (E \wedge F)^{p-q}(X) .$$

If $f: X \wedge Y \longrightarrow E$ represents an element of $E^p(X \wedge Y)$ and $g: S \longrightarrow F \wedge Y$ represents an element of $F_q(Y)$, we form

$$X \xrightarrow{1 \wedge g} X \wedge F \wedge Y \xrightarrow{1 \wedge c} X \wedge Y \wedge F \xrightarrow{f \wedge 1} E \wedge F .$$

The result is written f/g.

(ii) The second slant product

$$E^p(X) \otimes F_q(X \wedge Y) \longrightarrow (E \wedge F)_{-p+q}(Y)$$

is defined by taking

$$X \xrightarrow{f} E \quad \text{and} \quad S \xrightarrow{g} F \wedge X \wedge Y$$

and forming

$$S \xrightarrow{g} F \wedge X \wedge Y \xrightarrow{c \wedge 1} X \wedge F \wedge Y \xrightarrow{f \wedge 1 \wedge 1} E \wedge F \wedge Y .$$

The result is written $f \backslash g$.

<u>Note:</u> The following conventions are helpful.

(i) Fractions have the same variance as the numerator, and the opposite variance from the denominator.

(ii) Pay strict attention to the order of writing things on the page.

(a) Keep the cohomology variables (which are like functions) on the left of the homology variables (which are like arguments). That way both f/g and $f\backslash g$ mean composites in which you first apply g and afterwards apply f.

(b) If you have a class in $E^*(X \wedge Y)$ and want to "divide off" a homology class on one factor, by (a) you put the homology class on the right, so let it be a class in $F_*(Y)$ rather than $F_*(X)$. If you have a class in $F_*(X \wedge Y)$ and you want to divide it into a cohomology class on one factor, then by (a) you want to put the cohomology class on the left, so let it be a class in $E^*(X)$ rather than $E^*(Y)$.

Of course, once we have the external products for spectra, we get them for CW-complexes with base-point by specializing to suspension spectra. We then get them for relative groups by turning the handle. Note that if X, A and Y, B are pairs, then

$$X/A \wedge Y/B = X \times Y/(A \times Y \cup X \times B).$$

So for the relative groups we have the following products:

$$E^p(X, A) \otimes F^q(Y, B) \xrightarrow{\bar{\times}} (E \wedge F)^{p+q}(X \times Y, A \times Y \cup X \times B),$$

$$E_p(X, A) \otimes F_q(Y, B) \xrightarrow{\underline{\times}} (E \wedge F)_{p+q}(X \times Y, A \times Y \cup X \times B),$$

$$E^p(X \times Y, A \times Y \cup X \times B) \otimes F_q(Y, B) \xrightarrow{/} (E \wedge F)^{p-q}(X, A),$$

$$E^p(X, A) \otimes F_q(X \times Y, A \times Y \cup X \times B) \xrightarrow{\backslash} (E \wedge F)_{-p+q}(Y, B).$$

These products have various properties, of which we consider first naturality. I will do this for the case of spectra, because there we have to provide for maps of degree r. But first we need a remark about induced homomorphisms in cohomology. Let $f: X \longrightarrow Y$ be a morphism of degree $-p$, and let $g: Y \longrightarrow E$ be a morphism of degree $-q$, i.e., an element of $E^q(Y)$. Then the obvious thing to do is to define

$$f^*: E^q(Y) \longrightarrow E^{p+q}(X)$$

by

$$(g)f^* = gf .$$

But we usually write f^* on the left, and so we take care to introduce the proper sign:

$$f^*(g) = (-1)^{pq} gf .$$

For the next proposition assume for parts (i) to (iv) that we have morphisms

$$f: X \longrightarrow X' \quad \text{and} \quad g: Y \longrightarrow Y' .$$

PROPOSITION 9.1. (i) If $u \in E^*(X')$, $v \in F^*(Y')$, then

$$(u \bar{\wedge} v)(f \wedge g)^* = (-1)^{|f|\,|v|} uf^* \bar{\wedge}\, vg^*$$

or equivalently

$$(f \wedge g)^*(u \bar{\wedge} v) = (-1)^{|g|\,|u|} (f^* u) \bar{\wedge} (g^* v) .$$

(ii) If $u \in E_*(X)$, $v \in F_*(Y)$ then

$$(f \wedge g)_*(u \underline{\wedge} v) = (-1)^{|g|\,|u|} (f_* u) \underline{\wedge} (g_* v) .$$

(iii) If $u \in E^*(X' \wedge Y')$, $v \in F_*(Y)$ then

$$(u(f \wedge g)^*)/v = (-1)^{|f|(|g| + |v|)} (u/g_* v) f^*$$

or equivalently

$$((f \wedge g)^* u)/v = (-1)^{|g|\,|u|} f^*(u/g_* v) .$$

(iv) If $u \in E^*(X')$, $v \in F_*(X \wedge Y)$ then

$$u \backslash (f \wedge g)_* v = (-1)^{|g|(|u| + |f|)} g_*((uf^*) \backslash v)$$

or equivalently

$$u \backslash (f \wedge g)_* v = (-1)^{|g|\,|u| + |g|\,|f| + |f|\,|u|} g_*((f^* u) \backslash v) .$$

(v) With respect to morphisms of E and F, all the naturality statements are the same. Suppose given morphisms $e: E \longrightarrow E'$, $f: F \longrightarrow F'$. Then

$$(e \wedge f)_*(u \quad v) = (-1)^{|f|\,|u|}(e_* u) \quad (f_* v),$$

where the absence of a product symbol indicates that any of the four products may be used.

The proofs are elementary diagram-chasing.

PROPOSITION 9.2. All these products are biadditive.

We have two commutativity statements.

PROPOSITION 9.3. (i) Suppose $u \in E^p(X)$, $v \in F^q(Y)$. Then

$$v \bar{\wedge} u = (-1)^{pq} c_* c^* (u \bar{\wedge} v) .$$

(ii) Suppose $u \in E_p(X)$, $v \in F_q(Y)$. Then

$$v \underline{\wedge} u = (-1)^{pq} (c \wedge c)_* (u \underline{\wedge} v).$$

Of course, if we are going to apply maps $\mu: E \wedge F \longrightarrow G$ and $\mu': F \wedge E \longrightarrow G$ such that

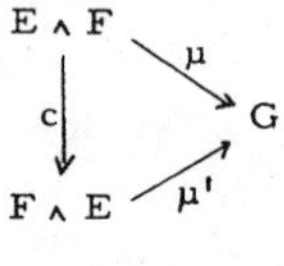

$$\begin{array}{ccc} E \wedge F & & \\ & \searrow^{\mu} & \\ c \downarrow & & G \\ & \nearrow_{\mu'} & \\ F \wedge E & & \end{array}$$

is a commutative diagram, then this absorbs the effect of $c: E \wedge F \longrightarrow F \wedge E$.

We have eight associativity statements. The first statement is obvious: suppose

$$u \in E^p(X), \quad v \in F^q(Y), \quad w \in G^r(Z) .$$

Then we have

$$(u \bar{\wedge} v) \bar{\wedge} w = u \bar{\wedge} (v \bar{\wedge} w) \in (E \wedge F \wedge G)^{p+q+r}(X \wedge Y \wedge Z) .$$

If we were using pairings of spectra, we would suppose that they made the following diagram commutative.

$$\begin{array}{ccc} E \wedge F \wedge G & \xrightarrow{\lambda \wedge 1} & H \wedge G \\ \downarrow{\scriptstyle 1\wedge\mu} & & \downarrow{\scriptstyle \nu} \\ E \wedge K & \xrightarrow{\pi} & L \end{array}$$

(Here, of course, H, K, and L are some spectra fitting into such a diagram.) Then we would obtain

$$(u \bar{\wedge} v) \bar{\wedge} w = u \bar{\wedge} (v \bar{\wedge} w) \in L^{p+q+r}(X \wedge Y \wedge Z).$$

The associativity law for the external homology product is entirely similar. With our conventions, the other six appear as very natural rules for manipulating fractions. For example, suppose

$$x \in E^*(X), \quad v \in F^*(Y \wedge Z), \quad z \in G_*(Z) .$$

Then $x \bar{\wedge} v \in (E \wedge F)^*(X \wedge Y \wedge Z)$, $(x \bar{\wedge} v)/z \in (E \wedge F \wedge G)^*(X \wedge Y)$. On the other hand, $v/z \in (F \wedge G)^*(Y)$, $x \bar{\wedge}(v/z) \in (E \wedge F \wedge G)^*(X \wedge Y)$. We have $(x \bar{\wedge} v)/z = x \bar{\wedge} (v/z)$.

THEOREM 9.4.

(i) If $x \in E^p(X)$, $y \in F^q(Y)$, $z \in G^r(Z)$

then

$$(x \bar{\wedge} y) \bar{\wedge} z = x \bar{\wedge} (y \bar{\wedge} z) \in (E \wedge F \wedge G)^{p+q+r}(X \wedge Y \wedge Z) .$$

(ii) If $x \in E^p(X)$, $u \in F^q(Y \wedge Z)$, $z \in G_r(Z)$

then

$$x \bar{\wedge} (u/z) = (x \bar{\wedge} u)/z \in (E \wedge F \wedge G)^{p+q-r}(X \wedge Y) .$$

(iii) If $v \in E^p(X \wedge Z)$, $y \in F^q(Y)$, $u \in G_r(Y \wedge Z)$

then

$$v/(y \backslash u) = [(1 \wedge c)^*(v \bar{\wedge} y)]/u \in (E \wedge F \wedge G)^{p+q-r}(X) .$$

(iv) If $t \in E^p(X \wedge Y \wedge Z)$, $z \in F_q(Z)$, $y \in G_r(Y)$

then

$$(t/z)/y = t/(c_*(z \underline{\wedge} y)) \in (E \wedge F \wedge G)^{p-q-r}(X) .$$

(v) If $y \in E^p(Y)$, $x \in F^q(X)$, $t \in G_r(X \wedge Y \wedge Z)$

then

$$y\backslash(x\backslash t) = (c^*(y \bar{\wedge} x))\backslash t \in (E \wedge F \wedge G)_{-p-q+r}(Z) .$$

(vi) If $w \in E^p(X \wedge Y)$, $y \in F_q(Y)$, $v \in G_r(X \wedge Z)$

then

$$(w/y)\backslash v = w\backslash[(c \wedge 1)_*(y \underline{\wedge} v)] \in (E \wedge F \wedge G)_{-p+q+r}(Z)$$

(vii) If $x \in E^p(X)$, $w \in F_q(X \wedge Y)$, $z \in G_r(Z)$

then

$$(x\backslash w) \underline{\wedge} z = x\backslash(w \underline{\wedge} z) \in (E \wedge F \wedge G)_{-p+q+r}(Y \wedge Z) .$$

(viii) If $x \in E_p(X)$, $y \in F_q(Y)$, $z \in G_r(Z)$

then

$$(x \underline{\wedge} y) \underline{\wedge} z = x \underline{\wedge} (y \underline{\wedge} z) \in (E \wedge F \wedge G)_{p+q+r}(X \wedge Y \wedge Z).$$

The proofs, as usual, are done by diagram-chasing.

Now we recall that the sphere spectrum S acts as a unit for the smash-product. It follows that we can identify

$$E_p(S) = [S, E]_p$$

with

$$E^{-p}(S) = [S, E]_p .$$

PROPOSITION 9.5. Suppose s is of this sort, say $s \in [S, E]_*$, and $y \in F_*(Y)$. Then

$$s\backslash y = s \underline{\wedge} y \in (E \wedge F)_*(Y) .$$

Suppose t is of this sort, say $t \in [S, F]_*$ and $x \in E^*(X)$. Then

$$x/t = x \bar{\wedge} t \in (E \wedge F)^*(X).$$

Suppose the result is of this sort, say $x \in E^*(X)$, $y \in F_*(X)$. Then

$$x\backslash y = x/y \in [S, E \wedge F]_* .$$

The proof is trivial diagram-chasing. The third case gives the Kronecker product $\langle x, y\rangle$. The explicit definition is as follows. Suppose given $X \xrightarrow{x} E$, $S \xrightarrow{y} F \wedge X$. Form

$$S \xrightarrow{y} F \wedge X \xrightarrow{c} X \wedge F \xrightarrow{x \wedge 1} E \wedge F.$$

The naturality properties of the Kronecker product are obvious and well known.

PROPOSITION 9.6. Suppose given $f: X \longrightarrow X'$ (of any degree), $x \in E^*(X')$, $y \in F_*(X)$. Then

$$\langle x'f^*, y\rangle = \langle x', f_* y\rangle ,$$

or equivalently

$$\langle f^* x', y\rangle = (-1)^{|f|\,|x'|} \langle x', f_* y\rangle .$$

We know that in the classical case the two slant products are obtained from the two more usual products by dualizing; in other words, they are related to them by the Kronecker product. We now state this for the generalized case.

PROPOSITION 9.7. (i) Suppose $u \in E^p(X \wedge Y)$, $y \in F_q(Y)$, $x \in G_r(X)$. Then

$$\langle u/y, x\rangle = \langle u, c_*(y \underline{\wedge} x)\rangle \in [S, E \wedge F \wedge G]_{-p+q+r} .$$

(ii) Suppose $y \in E^p(Y)$, $x \in G^q(X)$, $u \in G_r(X \wedge Y)$. Then

$$\langle y, x\backslash u\rangle = \langle c^*(y \bar{\wedge} x), u\rangle \in [S, E \wedge F \wedge G]_{-p-q+r} .$$

Proof. $\langle u/y, x\rangle$ may be viewed as either $(u/y)/x$ or $(u/y)\backslash x$. So part (i) follows by substituting into the appropriate associativity relation, number (iv) or (vi) on the list. Similarly for part (ii), using (v) or (iii).

These formulae are useful as a heuristic guide. For example, suppose you know some formula for the product $y \bar{\wedge} x$, and want to know the corresponding formula for the product $x\backslash u$. I really have in mind something like a coboundary formula, but I haven't yet done quite enough to use this case as an illustration, so let me consider a naturality formula. It's rather trivial, but it will do as an illustration of the method. Suppose $y \in E^p(Y)$, $x \in F^q(X)$, $u \in G_{p+q}(X' \wedge Y')$, $g: Y' \longrightarrow Y$, $f: X' \longrightarrow X$. We write down

$$\begin{array}{ccc}
< (y \bar{\wedge} x)(g \wedge f)^*, \ c_* u > & = & (-1)^{|x|\,|g|} < yg^* \bar{\wedge} xf^*, \ c_* u > \\
\| & & \| \\
(-1)^{|f|\,|g|} < y \bar{\wedge} x, \ c_*(f \wedge g)_* u > & & (-1)^{|x|\,|g|} < yg^*, xf^* \backslash u > \\
\| & & \| \\
(-1)^{|f|\,|g|} < y, x \backslash (f \wedge g)_* u > & & (-1)^{|x|\,|g|} < y, g_*(xf^* \backslash u) > .
\end{array}$$

If we knew that pairing with y were non-singular, we would have

$$(-1)^{|f|\,|g|}\, x \backslash (f \wedge g)_* u = (-1)^{|x|\,|g|}\, g_*((xf^*) \backslash u) .$$

But this argument is indeed a valid proof, because we can take

$$y = 1 \in Y^0(Y) .$$

PROPOSITION 9.8. Suppose $x^* \in E^p(X)$, $y^* \in F^q(Y)$, $x_* \in G_r(X)$, $y_* \in H_s(Y)$. Then

$$< x^* \bar{\wedge} y^*, x_* \bar{\wedge} y_* > = (-1)^{qr} (1 \wedge c \wedge 1)_* < x^*, x_* > < y^*, y_* > .$$

Here $1 \wedge c \wedge 1: E \wedge G \wedge F \wedge H \longrightarrow E \wedge F \wedge G \wedge H$.

Proof. Apply 9.7, commutativity, and associativity law (ii) or (vii).

Now we would like to write down the properties of our products for boundary and coboundary maps. One of them is immediate, that for the Kronecker product. We simply observe that the boundary or coboundary is induced by a map

$$X/A \longrightarrow A \quad \text{(of degree } -1) ;$$

we have a naturality formula for the Kronecker product valid for morphisms of any degree, so we get the following formula.

If $a \in E^p(A)$, $u \in F_q(X, A)$ then $\langle a, \partial u\rangle = \langle a\delta, u\rangle = (-1)^p \langle \delta a, u\rangle$, where we make the same sign conventions as before about $f^* a$.

In order to see what to expect in the other cases, let's go back to the classical case, and suppose given

$$u \in C_*(X), \quad \partial u \in C_*(A), \quad v \in C_*(Y), \quad \partial v \in C_*(B).$$

Then we expect to have

$$\partial(uv) = (\partial u)v + (-1)^{|u|} u(\partial v) \in C_*(A \times Y \cup X \times B).$$

However, the separate terms $(\partial u)v$ and $u(\partial v)$ do not define elements of $H_*(A \times Y \cup X \times B)$, so we have to work instead in

$$H_*(A \times Y \cup X \times B, A \times B) = H_*(A \times Y, A \times B) \oplus H_*(X \times B, A \times B).$$

Here $(\partial u)v$ defines an element in the first summand and $u(\partial v)$ in the second.

Additional motivation can be obtained if we consider the possibility of arguments using the five lemma. We have the exact sequence

$$E_p(A) \longrightarrow E_p(X) \longrightarrow E_p(X, A) \longrightarrow E_{p-1}(A).$$

If we tensor it with $F_q(Y, B)$ we get the left-hand column of the following diagram.

$$
\begin{array}{lcl}
\text{I} \quad E_p(A) \otimes F_q(Y,B) & \xrightarrow{\;\times\;} & (E \wedge F)_{p+q}(A \times Y, A \times B) \xrightarrow{\;\cong\;} (E \wedge F)_{p+q}(A \times Y \cup X \times B, X \times B) \\
\qquad \downarrow & & \\
\quad\;\; E_p(X) \otimes F_q(Y,B) & \xrightarrow{\;\times\;} & (E \wedge F)_{p+q}(X \times Y, X \times B) \longleftarrow (E \wedge F)_{p+q}(A \times Y \cup X \times B, X \times B) \\
\qquad \downarrow & & \qquad \downarrow \\
\text{II} \quad E_p(X, A) \otimes F_q(Y,B) & \xrightarrow{\;\times\;} & (E \wedge F)_{p+q}(X \times Y, A \times Y \cup X \times B) \xrightarrow{\;\partial\;} (E \wedge F)_{p+q-1}(A \times Y \cup X \times B, X \times B) \\
\qquad \downarrow \partial \otimes 1 & & \\
\quad\;\; E_{p-1}(A) \otimes F_q(Y,B) & \xrightarrow{\;\times\;} & (E \wedge F)_{p+q-1}(A \times Y, A \times B) \xrightarrow{\;\cong\;} (E \wedge F)_{p+q-1}(A \times Y \cup X \times B, X \times B)
\end{array}
$$

The oblique isomorphisms identify the second column with the exact sequence of a triple. The section of the diagram labeled I is commutattive by the naturality of $\times$, and we would like to know that the section labeled II is also commutative. So we wish to obtain commutative diagrams of the following form.

$$\begin{array}{ccc}
E_p(X,A) \otimes F_q(Y,B) & \xrightarrow{\partial \otimes 1} & E_{p-1}(A) \otimes F_q(Y,B) \\
 & & \downarrow \underline{\times} \\
\underline{\times} \downarrow & & (E \wedge F)_{p+q-1}(A \times Y, A \times B) \\
 & & \downarrow \cong \\
(E \wedge F)_{p+q}(X \times Y, A \times Y \cup X \times B) & \xrightarrow{\partial} & (E \wedge F)_{p+q-1}(A \times Y \cup X \times B, X \times B)
\end{array}$$

$$\begin{array}{ccc}
E_p(X,A) \otimes F_q(Y,B) & \xrightarrow{1 \otimes \partial} & E_p(X,A) \otimes F_{q-1}(B) \\
 & & \downarrow \underline{\times} \\
\underline{\times} \downarrow & & (E \wedge F)_{p+q-1}(X \times B, A \times B) \\
 & & \downarrow \cong \\
(E \wedge F)_{p+q}(X \times Y, A \times Y \cup X \times B) & \xrightarrow{\partial} & (E \wedge F)_{p+q-1}(A \times Y \cup X \times B, A \times Y)
\end{array}$$

Here we need a convention about signs. If $\theta: G \longrightarrow G'$ and $\varphi: H \longrightarrow H'$ are homomorphisms of graded groups, their tensor product is defined by

$$(\theta \otimes \varphi)(g \otimes h) = (-1)^{|\varphi|\,|g|}\, \theta g \otimes \varphi h .$$

In particular, $1 \otimes \partial$ is defined by $(1 \otimes d)(u \theta v) = (-1)^{|u|}\, u \otimes \partial v$.

Of course we propose to obtain our commutative diagrams by applying the results we already have to geometrical diagrams.

LEMMA 9.9. The following diagrams are commutative.

(i)

$$\begin{array}{ccccccc}
\dfrac{A \wedge Y \cup X \wedge B}{X \wedge B} & \longrightarrow & \dfrac{X \wedge Y}{X \wedge B} & \longrightarrow & \dfrac{X \wedge Y}{A \wedge Y \cup X \wedge B} & \xrightarrow{J} & \dfrac{A \wedge Y \cup X \wedge B}{X \wedge B} \\
 & & & & & & \uparrow \cong \\
 & & & & \cong \uparrow & & \dfrac{A \wedge Y}{A \wedge B} \\
 & & & & & & \uparrow \\
 & & & & \dfrac{X}{A} \wedge \dfrac{Y}{B} & \xrightarrow{j \wedge 1} & A \wedge \dfrac{Y}{B}
\end{array}$$

(ii)

$$
\begin{array}{ccccccc}
\dfrac{A \wedge Y \cup X \wedge B}{A \wedge Y} & \longrightarrow & \dfrac{X \wedge Y}{A \wedge B} & \longrightarrow & \dfrac{X \wedge Y}{A \wedge Y \cup X \wedge B} & \xrightarrow{J'} & \dfrac{A \wedge Y \cup X \wedge B}{A \wedge Y} \\
 & & & & & & \uparrow \cong \\
 & & & & \cong \uparrow & & \dfrac{X \wedge B}{A \wedge B} \\
 & & & & & & \uparrow \cong \\
 & & & & \dfrac{X}{A} \wedge \dfrac{Y}{B} & \xrightarrow{1 \wedge j'} & \dfrac{X}{B} \wedge B
\end{array}
$$

Notes. The diagrams are valid as they stand for spectra. The maps J, j, J', j' are the appropriate maps from the cofibre sequences, and they have degree -1. They may be replaced by maps of degree zero into the appropriate terms $S(\frac{A \wedge Y \cup X \wedge B}{X \wedge B})$, $S(\frac{A \wedge Y}{A \wedge B})$, etc., except that $1 \wedge j'$ in (ii) has to be replaced by

$$\frac{X}{A} \wedge \frac{Y}{B} \xrightarrow{1 \wedge j'} \frac{X}{A} \wedge S^1 \wedge B \xrightarrow{c \wedge 1} S^1 \wedge \frac{X}{A} \wedge B .$$

With this interpretation the diagrams are valid if X, Y, etc. are CW-complexes. For the case of spectra, the two ways of writing the diagrams are equivalent, because we have the canonical equivalence $Z \sim S^1 \wedge Z$ of degree 1.

It is sufficient to prove the commutativity of one of the diagrams, say the first; the other then follows by applying c (and checking that J corresponds to J'). But it is trivial to check the first diagram for CW-complexes by constructing the appropriate maps of $(X \cup CA) \wedge (Y \cup CB)$. The construction commutes with suspension on the right, and so passes to CW-spectra.

We now get the following eight commutative diagrams by applying Proposition 9.1 to the diagrams in Lemma 9.9. The morphisms J, j and sign conventions are as above.

THEOREM 9.10. The following diagrams are commutative.

$$\begin{array}{ccc}
E^p(A) \otimes F^q\left(\frac{Y}{B}\right) & \xrightarrow{j^* \otimes 1} & E^{p+1}\left(\frac{X}{A}\right) \otimes F^q\left(\frac{Y}{B}\right) \\
\Big\downarrow \bar{\wedge} & & \Big\downarrow \bar{\wedge} \\
(E \wedge F)^{p+q}\left(A \wedge \frac{Y}{B}\right) & & (E \wedge F)^{p+q+1}\left(\frac{X}{A} \wedge \frac{Y}{B}\right) \\
\Big\downarrow \cong & & \Big\| \\
(E \wedge F)^{p+q}\left(\frac{A \wedge Y \cup X \wedge B}{X \wedge B}\right) & \xrightarrow{J^*} & (E \wedge F)^{p+q+1}\left(\frac{X \wedge Y}{A \wedge Y \cup X \wedge B}\right)
\end{array}$$

$$\begin{array}{ccc}
E^p\left(\frac{X}{A}\right) \otimes F^q(B) & \xrightarrow{1 \otimes j^*} & E^p\left(\frac{X}{A}\right) \otimes F^{q+1}\left(\frac{Y}{B}\right) \\
\Big\downarrow \bar{\wedge} & & \Big\downarrow \bar{\wedge} \\
(E \wedge F)^{p+q}\left(\frac{X}{A} \wedge B\right) & & (E \wedge F)^{p+q+1}\left(\frac{X}{A} \wedge \frac{Y}{B}\right) \\
\Big\downarrow \cong & & \Big\| \\
(E \wedge F)^{p+q}\left(\frac{A \wedge Y \cup X \wedge B}{A \wedge Y}\right) & \xrightarrow{J^*} & (E \wedge F)^{p+q+1}\left(\frac{X \wedge Y}{A \wedge Y \cup X \wedge B}\right)
\end{array}$$

$$\begin{array}{ccc}
E^p\left(\frac{A \wedge Y \cup X \wedge B}{X \wedge B}\right) \otimes F_q\left(\frac{Y}{B}\right) & \xrightarrow{J^* \otimes 1} & E^{p+1}\left(\frac{X \wedge Y}{A \wedge Y \cup X \wedge B}\right) \otimes F_q\left(\frac{Y}{B}\right) \\
\Big\downarrow \cong \otimes 1 & & \Big\| \\
E^p\left(A \wedge \frac{Y}{B}\right) \otimes F_q\left(\frac{Y}{B}\right) & & E^{p+1}\left(\frac{X}{A} \wedge \frac{Y}{B}\right) \otimes F_q\left(\frac{Y}{B}\right) \\
\Big\downarrow / & & \Big\downarrow / \\
(E \wedge F)^{p-q}(A) & \xrightarrow{j^*} & (E \wedge F)^{p-q+1}\left(\frac{X}{A}\right)
\end{array}$$

$$\begin{array}{ccc}
E^p\left(\frac{A \wedge Y \cup X \wedge B}{A \wedge Y}\right) \otimes F_q\left(\frac{Y}{B}\right) & \xrightarrow{J^* \otimes 1} & E^{p+1}\left(\frac{X \wedge Y}{A \wedge Y \cup X \wedge B}\right) \otimes F_q\left(\frac{Y}{B}\right) \\
\Big\downarrow \cong \otimes 1 & & \Big\| \\
E^p\left(\frac{X}{A} \wedge B\right) \otimes F_q\left(\frac{Y}{B}\right) & & E^{p+1}\left(\frac{X}{A} \wedge \frac{Y}{B}\right) \otimes F_q\left(\frac{Y}{B}\right) \\
\Big\downarrow 1 \otimes j_* & & \Big\downarrow / \\
E^p\left(\frac{X}{A} \wedge B\right) \otimes F_{q-1}(B) & \xrightarrow{/} & (E \wedge F)^{p-q+1}\left(\frac{X}{A}\right)
\end{array}$$

$$\begin{array}{ccc}
E^p(A) \otimes F_q\left(\frac{X \wedge Y}{A \wedge Y \cup X \wedge B}\right) & \xrightarrow{1 \otimes J_*} & E^p(A) \otimes F_{q-1}\left(\frac{A \wedge Y \cup X \wedge B}{X \wedge B}\right) \\
\Vert & & \uparrow 1 \otimes \cong \\
E^p(A) \otimes F_q\left(\frac{X}{A} \wedge \frac{Y}{B}\right) & & E^p(A) \otimes F_{q-1}\left(A \wedge \frac{Y}{B}\right) \\
j^* \otimes 1 \downarrow & & \downarrow \backslash \\
E^{p+1}\left(\frac{X}{A}\right) \otimes F_q\left(\frac{X}{A} \wedge \frac{Y}{B}\right) & \xrightarrow{\backslash} & (E \wedge F)_{-p+q-1}\left(\frac{Y}{B}\right)
\end{array}$$

$$\begin{array}{ccc}
E^p\left(\frac{X}{A}\right) \otimes F_q\left(\frac{X \wedge Y}{A \wedge Y \cup X \wedge B}\right) & \xrightarrow{1 \otimes J_*} & E^p\left(\frac{X}{A}\right) \otimes F_{q-1}\left(\frac{A \wedge Y \cup X \wedge B}{A \wedge Y}\right) \\
\Vert & & \uparrow 1 \otimes \cong \\
E^p\left(\frac{X}{A}\right) \otimes F_q\left(\frac{X}{A} \wedge \frac{Y}{B}\right) & & E^p\left(\frac{X}{A}\right) \otimes F_{q-1}\left(\frac{X}{A} \wedge B\right) \\
\backslash \downarrow & & \downarrow \backslash \\
(E \wedge F)_{-p+q}\left(\frac{Y}{B}\right) & \xrightarrow{j_*} & (E \wedge F)_{-p+q-1}(B)
\end{array}$$

$$\begin{array}{ccc}
E_p\left(\frac{X}{A}\right) \otimes F_q\left(\frac{Y}{B}\right) & \xrightarrow{j_* \otimes 1} & E_{p-1}(A) \otimes F_q\left(\frac{Y}{B}\right) \\
\hat{\ } \downarrow & & \downarrow \hat{\ } \\
(E \wedge F)_{p+q}\left(\frac{X}{A} \wedge \frac{Y}{B}\right) & & (E \wedge F)_{p+q-1}\left(A \wedge \frac{Y}{B}\right) \\
\Vert & & \downarrow \cong \\
(E \wedge F)_{p+q}\left(\frac{X \wedge Y}{A \wedge Y \cup X \wedge B}\right) & \xrightarrow{J_*} & (E \wedge F)_{p+q-1}\left(\frac{A \wedge Y \cup X \wedge B}{X \wedge B}\right)
\end{array}$$

$$\begin{array}{ccc}
E_p\left(\frac{X}{A}\right) \otimes F_q\left(\frac{Y}{B}\right) & \xrightarrow{1 \otimes j_*} & E_p\left(\frac{X}{A}\right) \otimes F_{q-1}(B) \\
\Delta \downarrow & & \downarrow \Delta \\
(E \wedge F)_{p+q}\left(\frac{X}{A} \wedge \frac{Y}{B}\right) & & (E \wedge F)_{p+q-1}\left(\frac{X}{A} \wedge B\right) \\
\Vert & & \downarrow \cong \\
(E \wedge F)_{p+q}\left(\frac{X \wedge Y}{A \wedge Y \cup X \wedge B}\right) & \xrightarrow{J_*} & (E \wedge F)_{p+q-1}\left(\frac{A \wedge Y \cup X \wedge B}{A \wedge Y}\right)
\end{array}$$

By an immediate translation, we obtain commutative diagrams for the boundary and coboundary in relative groups of pairs.

THEOREM 9.11. The following diagrams are commutative.

$$\begin{array}{ccc}
E^p(A) \otimes F^q(Y,B) & \xrightarrow{\delta \otimes 1} & E^{p+1}(X,A) \otimes F^q(Y,B) \\
\downarrow{\scriptstyle \bar{x}} & & \\
(E \wedge F)^{p+q}(A \times Y, A \times B) & & \downarrow{\scriptstyle \bar{x}} \\
\uparrow{\scriptstyle \cong} & & \\
(E \wedge F)^{p+q}(A \times Y \cup X \times B, X \times B) & \xrightarrow{\delta} & (E \wedge F)^{p+q+1}(X \times Y, A \times Y \cup X \times B)
\end{array}$$

$$\begin{array}{ccc}
E^p(X,A) \otimes F^q(B) & \xrightarrow{1 \otimes \delta} & E^p(X,A) \otimes F^{q+1}(Y,B) \\
\downarrow{\scriptstyle \bar{x}} & & \\
(E \wedge F)^{p+q}(X \times B, A \times B) & & \downarrow{\scriptstyle \bar{x}} \\
\downarrow{\scriptstyle \cong} & & \\
(E \wedge F)^{p+q}(A \times Y \cup X \times B, A \times Y) & \xrightarrow{\delta} & (E \wedge F)^{p+q+1}(X \times Y, A \times Y \cup X \times B)
\end{array}$$

$$\begin{array}{ccc}
E^p(A \times Y \cup X \times B, X \times B) \otimes F_q(Y,B) & \xrightarrow{\delta \otimes 1} & E^{p+1}(X \times Y, A \times Y \cup X \times B) \otimes F_q(Y,B) \\
\downarrow{\scriptstyle \cong \otimes 1} & & \\
E^p(A \times Y, A \times B) \quad F_q(Y,B) & & \downarrow{\scriptstyle /} \\
\downarrow{\scriptstyle /} & & \\
(E \wedge F)^{p-q}(A) & \xrightarrow{\delta} & (E \wedge F)^{p-q+1}(X,A)
\end{array}$$

$$\begin{array}{ccc}
E^p(A \times Y \cup X \times B, A \times Y) \otimes F_q(Y,B) & \xrightarrow{\delta \otimes 1} & E^{p+1}(X \times Y, A \times Y \cup X \times B) \otimes F_q(Y,B) \\
\downarrow{\scriptstyle \cong \otimes 1} & & \\
E^p(X \times B, A \times B) \otimes F_q(Y,B) & & \downarrow{\scriptstyle /} \\
\downarrow{\scriptstyle 1 \otimes \partial} & & \\
E^p(X \times B, A \times B) \otimes F_{q-1}(B) & \xrightarrow{/} & (E \wedge F)^{p-q+1}(X,A)
\end{array}$$

$$\begin{array}{ccc}
E^p(A) \otimes F_q(X \times Y, A \times Y \cup X \times B) & \xrightarrow{1 \otimes \partial} & E^p(A) \otimes F_{q-1}(A \times Y \cup X \times B, X \times B) \\
 & & \uparrow{\scriptstyle 1 \otimes \cong} \\
\downarrow{\scriptstyle \delta \otimes 1} & & E^p(A) \otimes F_{q-1}(A \times Y, A \times B) \\
 & & \downarrow{\scriptstyle \backslash} \\
E^{p+1}(X,A) \otimes F_q(X \times Y, A \times Y \cup X \times B) & \xrightarrow{\backslash} & (E \wedge F)_{-p+q-1}(Y,B)
\end{array}$$

$$\begin{array}{ccc}
E^p(X,A)\otimes F_q(X\times Y, A\times Y\cup X\times B) & \xrightarrow{1\otimes\partial} & E^p(X,A)\otimes F_{q-1}(A\times Y\cup X\times B, A\times Y) \\
 & & \uparrow{\scriptstyle 1\otimes\cong} \\
\Big\downarrow{\scriptstyle \backslash} & & E^p(X,A)\otimes F_{q-1}(X\times B, A\times B) \\
 & & \Big\downarrow{\scriptstyle \backslash} \\
(E\wedge F)_{-p+q}(Y,B) & \xrightarrow{\quad\partial\quad} & (E\wedge F)_{-p+q-1}(B)
\end{array}$$

$$\begin{array}{ccc}
E_p(X,A)\otimes F_q(Y,B) & \xrightarrow{\partial\otimes 1} & E_{p-1}(A)\otimes F_q(Y,B) \\
 & & \Big\downarrow{\scriptstyle \times} \\
\Big\downarrow{\scriptstyle \times} & & (E\wedge F)_{p+q-1}(A\times Y, A\times B) \\
 & & \Big\downarrow{\scriptstyle \cong} \\
(E\wedge F)_{p+q}(X\times Y, A\times Y\cup X\times B) & \xrightarrow{\partial} & (E\wedge F)_{p+q-1}(A\times Y\cup X\times B, X\times B)
\end{array}$$

$$\begin{array}{ccc}
E_p(X,A)\otimes F_q(Y,B) & \xrightarrow{1\otimes\partial} & E_p(X,A)\otimes F_{q-1}(B) \\
 & & \Big\downarrow{\scriptstyle \times} \\
\Big\downarrow{\scriptstyle \times} & & (E\wedge F)_{p+q-1}(X\times B, A\times B) \\
 & & \Big\downarrow{\scriptstyle \cong} \\
(E\wedge F)_{p+q}(X\times Y, A\times Y\cup X\times B) & \xrightarrow{\partial} & (E\wedge F)_{p+q-1}(A\times Y\cup X\times B, A\times Y)
\end{array}$$

Unfortunately, we need still more diagrams. Let's return to our original formula in the classical case,

$$\partial(uv) = (\partial u)v + (-1)^{|u|}u(\partial v).$$

We have written a relation between $\partial(uv)$ and $(\partial u)v$ by working in a group where we can ignore $u(\partial v)$, and a relation between $\partial(uv)$ and $u(\partial v)$ by working in a group where we can ignore $(\partial u)v$. It remains to write a relation between

$$(\partial u)v \quad \text{and} \quad u(\partial v)$$

by working in a group where we can ignore $\partial(uv)$. And in this case the answer is obvious. We have to say that the following diagram commutes up to a sign -1.

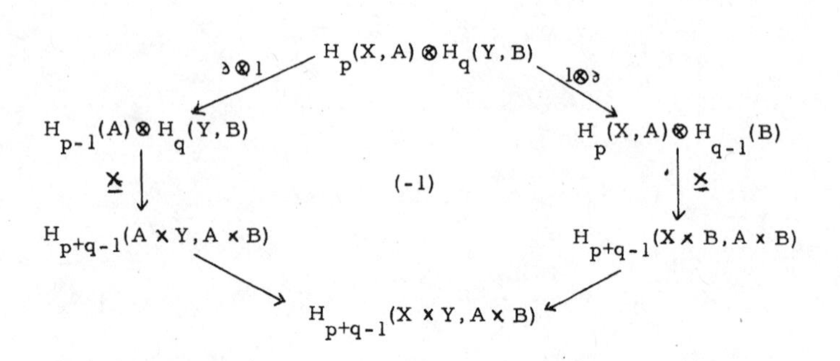

We can easily prove such a result for the generalised case. Consider the following diagram.

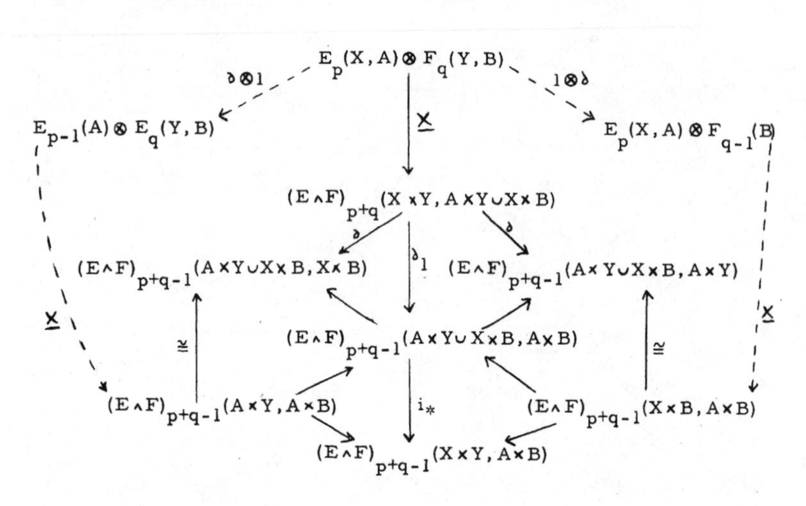

The diagram displays

$$(E \wedge F)_{p+q-1}(A \times Y \cup X \times B, A \times B)$$

as the direct sum

$$(E \wedge F)_{p+q-1}(A \times Y, A \times B) \oplus (E \wedge F)_{p+q-1}(X \times B, A \times B).$$

The composite $i_* \partial_1$ is zero, so the two paths from $(E \wedge F)_{p+q}(X \times Y, A \times Y \cup X \times B)$ to $(E \wedge F)_{p+q-1}(X \times Y, A \times B)$ around the outside of the lower hexagon gives maps whose sum is zero. This is the Eilenberg-Steenrod hexagon lemma. Of course, we know the result geometrically by 6.9. Now fill in the rest of the diagram by 9.11.

Proceeding in this way for the four products we obtain four more diagrams listed in the following theorem.

THEOREM 9.12. (i) The following diagram is commutative up to a sign -1.

$$
\begin{array}{ccccc}
 & & E^p(A)\otimes F^q(B) & & \\
 & \swarrow^{\delta\otimes 1} & & \searrow^{1\otimes\delta} & \\
E^{p+1}(X,A)\otimes F^q(B) & & & & E^p(A)\otimes F^{q+1}(Y,B) \\
\downarrow \overline{\times} & & & & \downarrow \overline{\times} \\
(E\wedge F)^{p+q+1}(X\times B, A\times B) & & (-1) & & (E\wedge F)^{p+q+1}(A\times Y, A\times B) \\
\cong\uparrow & & & & \uparrow\cong \\
(E\wedge F)^{p+q+1}(A\times Y\cup X\times B, A\times Y) & & & & (E\wedge F)^{p+q+1}(A\times Y\cup X\times B, X\times B) \\
 & \searrow & & \swarrow & \\
 & & (E\wedge F)^{p+q+1}(A\times Y\cup X\times B) & &
\end{array}
$$

(ii)

$$
\begin{array}{ccccc}
 & & E^p(A\times Y, A\times B)\otimes F_q(Y,B) & \xrightarrow{\ /\ } & (E\wedge F)^{p-q}(A) \\
 & \nearrow^{i^*} & \Big\downarrow{\scriptstyle \partial} & & \Big\downarrow{\scriptstyle \delta} \\
E^p(X\times Y, A\times B) & & & & \\
 & \searrow_{j^*} & & & \\
 & & E^p(X\times B, A\times B)\otimes F_{q-1}(B) & \xrightarrow{\ /\ } & (E\wedge F)^{p-q+1}(X,A)
\end{array}
$$

If $u \in E^p(X\times Y, A\times B)$ and $y \in F_q(Y,B)$, then

$$\delta((i^*u)/y) = (-1)^{p+1}(j^*u)/(\partial y) .$$

(iii)

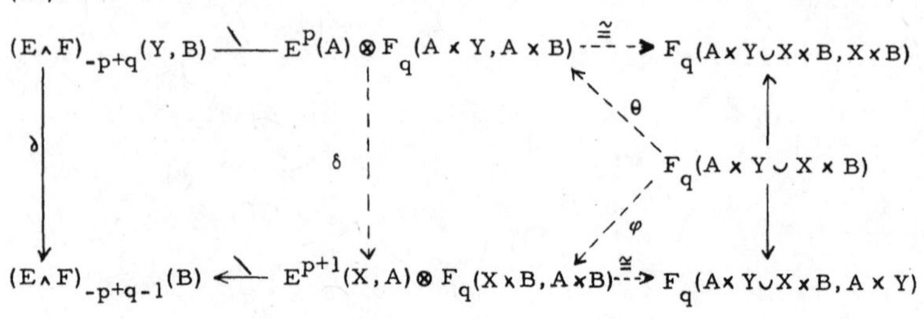

If $a \in E^p(A)$ and $u \in F_q(A\times Y \cup X\times B)$ then

$$\partial(a\backslash(\theta u)) = -(\delta a)\backslash(\varphi u) .$$

(iv) The following diagram is commutative up to a sign -1.

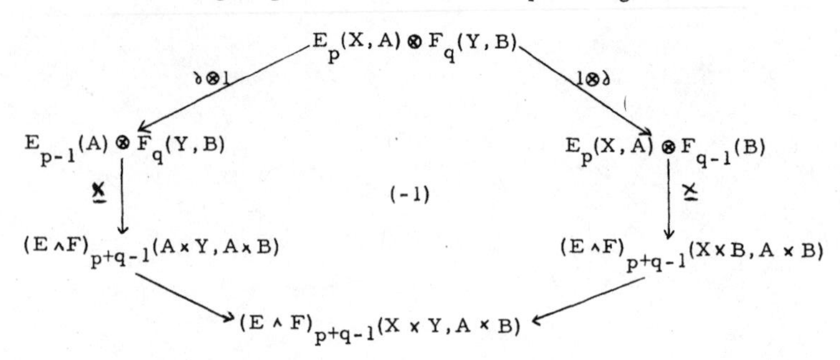

<u>Internal Products</u>

Following the idea of Lefschetz, these products are introduced by considering the diagonal map

$$\Delta : X \longrightarrow X\times X .$$

Here X is a CW-complex, Given $u \in E^p(X,A)$, $v \in F^q(X,B)$ we have defined

$$u \,\bar{\times}\, v \in (E\wedge F)^{p+q}(X\times X, A\times X \cup X\times B)$$

and we can define

$$u \cup v = \Delta^*(u \,\bar{\times}\, v) \in (E\wedge F)^{p+q}(X, A\cup B).$$

Similarly, given $u \in E^p(X, A)$, $v \in F_q(X, A \cup B)$, we can form

$$\Delta_* v \in F_q(X \times X, A \times X \cup X \times B)$$

and define

$$u \cap v = u \backslash \Delta_* v \in (E \wedge F)_{-p+q}(X, B).$$

Conversely, the $\bar{\times}$ and $\backslash$ products can be recovered from $\cup$ and $\cap$. Let $p_1 : X \times Y \longrightarrow X$, $p_2 : X \times Y \longrightarrow Y$ be the projections on the two factors.

PROPOSITION 9.13. If $u \in E^p(X, A)$, $v \in F^q(Y, B)$ then

$$u \bar{\times} v = (p_1^* u) \cup (p_2^* v) \in (E \wedge F)^{p+q}(X \times Y, A \times Y \cup X \times B) .$$

If $u \in E^p(X, A)$, $v \in F_q(X \times Y, A \times Y \cup X \times B)$ then

$$u \backslash v = p_{2*}((p_1^* u) \cap v) \in (E \wedge F)_{-p+q}(Y, B) .$$

The proof is immediate, by naturality.

Since we can recover the Kronecker product from either slant product, we can recover it from the cap product.

PROPOSITION 9.14. If $u \in E^p(X, A)$, $v \in F_q(X, A)$ then

$$\langle u, v \rangle = \epsilon_*(u \cap v) \in \pi_{-p+q}(E \wedge F) ,$$

where $\epsilon : X \longrightarrow pt.$ is the constant map.

All the properties of the internal products can be deduced from those of the external ones, by naturality. The list of associativity properties, however, will look less symmetrical than in the case of the external products.

10. DUALITY IN MANIFOLDS

In the classical case, to have a duality theorem relating $H_r(M;A)$ and $H^{m-r}(M;A)$ we need to assume M is orientable, and then we can take A to be any abelian group. Otherwise, we can suppose that M is non-orientable; then either we must use twisted coefficients, or we must suppose that A is a module over Z_2. The point is that an orientable manifold has classes in Z-homology and cohomology which enter into the statements and the proofs; and even a non-orientable manifold has such classes if we use homology and cohomology with coefficients in Z_2.

To generalize this situation, G. W. Whitehead introduced the notion of a ring-spectrum and a module-spectrum. The idea is that if M is orientable with respect to E_* and E^*, where E is a ring-spectrum, then the duality theorem will hold for F_* and F^*, where F is any module-spectrum over E.

Examples: To illustrate the situation above, take

$$E = H, \quad F = HA \quad \text{for any abelian group } A;$$

or

$$E = HZ_2, \quad F = HA \quad \text{for any } Z_2\text{-module } A.$$

A spectrum E is said to be a ring-spectrum if it has given maps $\mu: E \wedge E \longrightarrow E$, $\eta: S \longrightarrow E$ of degree 0 such that the following diagrams commute.

$$\begin{array}{ccc} E \wedge E \wedge E & \xrightarrow{\mu \wedge 1} & E \wedge E \\ {\scriptstyle 1 \wedge \mu}\downarrow & & \downarrow{\scriptstyle \mu} \\ E \wedge E & \xrightarrow{\mu} & E \end{array} \qquad \begin{array}{ccc} S \wedge E & \xrightarrow{\eta \wedge 1} & E \wedge E \\ \wr & & \downarrow \mu \\ E & \xrightarrow{1} & E \\ \wr & & \uparrow \mu \\ E \wedge S & \xrightarrow{1 \wedge \eta} & E \wedge E \end{array}$$

Let E be a ring-spectrum. We say a spectrum F is a <u>module-spectrum over E</u> if it has given a map $\nu: E \wedge F \longrightarrow F$ of degree 0 such that the following diagrams commute.

$$
\begin{array}{ccc}
E \wedge E \wedge F & \xrightarrow{\mu \wedge 1} & E \wedge F \\
{\scriptstyle 1 \wedge \nu}\downarrow & & \downarrow{\scriptstyle \nu} \\
E \wedge F & \xrightarrow{\nu} & F
\end{array}
\qquad
\begin{array}{ccc}
S \wedge F & \xrightarrow{\eta \wedge 1} & E \wedge F \\
\wr\!\!\!\downarrow & & \downarrow{\scriptstyle \nu} \\
F & \xrightarrow{1} & F
\end{array}
$$

A ring-spectrum E is said to be <u>commutative</u> if the following diagram commutes.

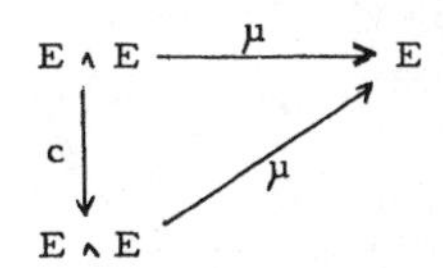

If E is a ring-spectrum, we can use the product map $\mu: E \wedge E \to E$ to obtain products with values in E_* or E^* instead of $(E \wedge E)_*$ or $(E \wedge E)^*$. For example, we obtain a cup-product

$$E^p(X, A) \otimes E^q(X, B) \longrightarrow E^{p+q}(X, A \cup B).$$

Similarly for an action map $\nu: E \wedge F \longrightarrow F$.

Practically all the examples of spectra which I have mentioned are, in fact, ring-spectra. I will only illustrate the case $E = H$. We have

$$\pi_r(H \wedge H) = H_r(H) = \begin{cases} 0 & (r < 0) \\ Z & (r = 0) \end{cases}$$

so that by the Hurewicz theorem,

$$H_0(H \wedge H) = Z .$$

Alternatively, by the Kunneth theorem

$$H_0(H \wedge H) \cong H_0(H) \otimes H_0(H) \cong Z \otimes Z \cong Z .$$

By the universal coefficient theorem,

$$H^0(H \wedge H) = \mathrm{Hom}(Z, Z) = Z .$$

Therefore I can take a map $\mu: H \wedge H \longrightarrow H$ realizing the product map $Z \otimes Z \longrightarrow Z$ of π_0.

Alternatively, realize $H \wedge H$ with no stable cells of dimension $d < 0$, map the cells of dimension 0 in the indicated way, and similarly for the cells of dimension 1. Now the map extends over the higher stable cells of $H \wedge H$, because the higher homotopy groups of H are zero. For the same reason, the map is unique up to homotopy.

For similar reasons, if R is a ring then HR is a ring-spectrum; if M is an R-module then HM is a module-spectrum over HR.

So far our generalised homology and cohomology theories have been defined on CW-pairs X, A. Now we would like to extend them to other categories of pairs.

We begin with the singular extension of E_* and E^*. Take any pair X, A and let X', A' be a weakly equivalent CW-pair. Define the singular E-homology and E-cohomology groups of X, A to be

$$E_p(X, A) = E_p(X', A'),$$
$$E^p(X, A) = E^p(X', A').$$

The result is independent of the choice of X', A', up to a canonical isomorphism.

All the properties of E_p and E^p carry over very well, except for excision. Here one has to be careful. Let $U \cup V$ be a space which comes as the union of two subspaces U and V intersecting in $U \cap V$. Then we can certainly take a CW-complex W' equipped with a weak equivalence

$$W' \xrightarrow{w} U \cap V,$$

and we can enlarge W' on the one hand to a CW-complex U' admitting a weak equivalence

$$U' \xrightarrow{u} U$$

extending w, and on the other hand to a CW-complex V' admitting a weak equivalence

$$V' \xrightarrow{v} V$$

also extending w. Then we can put them together to get

$$U' \cup_{W'} V' \longrightarrow U \cup V .$$

But this map is not a weak equivalence in general. For example, take subsets of the real numbers; let $U = Q$, $V = R - Q$; then W' will be empty, U' will be a countable discrete space, V' will be an uncountable discrete space, and $U' \cup_{W'} V'$ will be an uncountable discrete space, which is not weakly equivalent to R.

However, if we assume that $\mathrm{Int}\, U \cup \mathrm{Int}\, V = U \cup V$, then $U' \cup_{W'} V' \longrightarrow U \cup V$ is a weak equivalence, and all is well. So the excision axiom holds with this extra hypothesis, which is actually the standard one for ordinary (singular) homology and cohomology.

I must also comment on the behaviour of singular homology for limits. Let X, Y be a pair containing a directed family of subpairs X_α, Y_α. Then we can form

$$\underset{\alpha}{\underrightarrow{\mathrm{Lim}}}\ E_p(X_\alpha, Y_\alpha) \longrightarrow E_p(X, Y) .$$

PROPOSITION 10.1. In order that this map be an isomorphism, it is sufficient that for any compact pair $K, L \subset X, Y$ we can find an α such that $X_\alpha, Y_\alpha \supset K, L$.

The proof is easy.

Now I want to define a Čech-type cohomology theory for compact pairs K, L which happen to come embedded in some topological manifold M, possibly not compact, possibly with boundary. The definition is as follows. Let U, V run over open pairs in M with $U \supset K$, $V \supset L$. These form a directed set; if $U_i \supset K$, $V_i \supset L$ for $i = 1, 2$ then $U_1 \cap U_2 \supset K$, $V_1 \cap V_2 \supset L$. So I define

$$\check{E}^*(K, L) = \underset{(U, V)}{\underrightarrow{\mathrm{Lim}}} \; E^*(U, V) .$$

(The notation E^*, when applied to an arbitrary topological pair X, A will mean singular E-cohomology.) Of course we always have a map

$$E^*(U, V) \longrightarrow E^*(K, L);$$

this passes to the limit, and gives up a map

$$\check{E}^*(K, L) \longrightarrow E^*(K, L) .$$

In general this map need not be an isomorphism. However, there are cases when it is.

Example (i). Suppose M is a compact topological manifold, possibly with boundary. Then

$$\check{E}^*(M) \longrightarrow E^*(M)$$

is an isomorphism.

In fact, the pair $M, \emptyset$ qualifies as an open pair containing the compact pair $M, \emptyset$, and is terminal.

Example (ii). Suppose K is a point x. Then

$$\check{E}^*(x) \longrightarrow E^*(x)$$

is an isomorphism.

In fact, any point x lies in a coordinate neighborhood, so we can choose a cofinal system of open pairs $U_\alpha, \emptyset \supset x, \emptyset$ with U_α contractible. Then

$$E^*(U_\alpha) \longrightarrow E^*(x)$$

is an isomorphism for all α.

Next I would like to know that $\check{E}^*(K, L)$ is a topological invariant of the pair (K, L), and does not depend on the embedding in M.

LEMMA 10.2 (i). Suppose given compact pairs $K_1, L_1 \subset M_1$ and $K_2, L_2 \subset U_2, V_2 \subset M_2$, where U_2, V_2 is an open pair, and a continuous map

$$f: K_1, L_1 \longrightarrow K_2, L_2 .$$

Then f can be extended so as to map some open pair $U_1, V_1 \supset K_1, L_1$ in M_1 into U_2, V_2.

(ii) Suppose given a homotopy

$$h: I \times K_1, I \times L_1 \longrightarrow K_2, L_2 ,$$

and extensions f^0 of h^0, f^1 of h^1 which map (possibly different) open pairs U^0, V^0 and U^1, V^1 into an open pair $U_2, V_2 \supset K_2, L_2$. Then there exists an open pair U, V with $K_1, L_1 \subset U, V \subset U^0 \cap U^1, V^0 \cap V^1$ and a homotopy

$$h: I \times U, I \times V \longrightarrow U_2, V_2$$

extending $f^0 | U, V$, $f^1 | U, V$ and h.

<u>Proof.</u> Standard but repeated use of compactness, plus the Tietze extension theorem; we rely heavily on the fact that M_2 is a manifold.

COROLLARY 10.3. A map $f: K_1, L_1 \longrightarrow K_2, L_2$ induces $f^*: \check{E}^*(K_2, L_2) \longrightarrow \check{E}^*(K_1, L_1)$ depending only on the homotopy class of f, and satisfying $1^* = 1$, $(fg)^* = g^* f^*$.

The exactness properties of $\check{E}^*$ are fine, since direct limits preserve exactness.

Example. Let $M, \partial M$ be a pair consisting of a compact topological manifold with boundary and its boundary. Then

$$\check{E}^*(M, \partial M) \longrightarrow E^*(M, \partial M)$$

is an isomorphism.

Proof. Consider the following commutative diagram.

$$\begin{array}{ccccccccc}
\cdots \longrightarrow & \check{E}^*(M, \partial M) & \longrightarrow & \check{E}^*(M) & \longrightarrow & \check{E}^*(\partial M) & \xrightarrow{\delta} & \check{E}^*(M, \partial M) & \longrightarrow \cdots \\
& \downarrow & & \downarrow \cong & & \downarrow \cong & & \downarrow & \\
\cdots \longrightarrow & E^*(M, \partial M) & \longrightarrow & E^*(M) & \longrightarrow & E^*(\partial M) & \xrightarrow{\delta} & E^*(M, \partial M) & \longrightarrow \cdots
\end{array}$$

The rows are exact, and the two arrows marked are isomorphisms by a previous example. The result follows by the 5-lemma.

N.B. This way of saying things relies on the previous proof that $\check{E}^*(\partial M)$ is independent of the embedding of ∂M in M, but it does not need the construction of a collar for ∂M inside M.

The excision properties of $\check{E}^*$ are excellent, because $\check{E}^*$ was defined using only open pairs in M.

PROPOSITION 10.4. If U, V are any compact sets in M, then

$$\check{E}^*(U \cup V, V) \longrightarrow \check{E}^*(U, U \cap V)$$

is an isomorphism.

This follows from the definitions by a bit of general topology (compact Hausdorff spaces again.)

I also have to comment on the behaviour of $\check{E}^*$ for limits.

PROPOSITION 10.5. Let K_α, L_α be a downward-directed set of compact pairs in M, with intersection K, L. Then

$$\underset{\overrightarrow{\alpha}}{\text{Lim}}\ \check{E}^*(K_\alpha, L_\alpha) \longrightarrow \check{E}^*(K, L)$$

is an isomorphism.

Again, this is easy modulo a bit of general topology. One must show that given any open pair $U, V \supset K, L$ there is an α with $K_\alpha, L_\alpha \subset U, V$.

Experience in Manchester and Cambridge suggests I had better give some exposition about orientations. Suppose $E \xrightarrow{p} B$ is an n-plane bundle and E^0 is the complement of the zero cross-section. Then for each point $x \in B$, I have the fibre $E_x = p^{-1}x$; let $E_x^0 = E_x \cap E^0$. I can form

$$H_n(E_x, E_x^0) \cong Z$$
$$H^n(E_x, E_x^0) \cong Z .$$

Now since $E \xrightarrow{p} B$ is a bundle, locally it is a product; and if x and y are close together we can easily tell which element in $H_n(E_x, E_x^0)$ corresponds to which in $H_n(E_y, E_y^0)$. That is, we get a bundle over B, with fibre Z, and with structure group Z_2 acting on Z by $n \longrightarrow -n$. A similar situation occurs in cohomology.

One may say that the original n-plane bundle was orientable if the Z-bundle $\bigcup_x H_n(E_x, E_x^0)$ is trivial. The definition may be given equally well in terms of homology or cohomology; we have

$$H^n(E_x, E_x^0) = \mathrm{Hom}(H_n(E_x, E_x^0), Z) \quad ,$$

so the two bundles are trivial or non-trivial together.

If we are given orientations consistently on each fibre, that amounts to saying there is given a continuous section

$$\lambda: B \longrightarrow \bigcup_x H_n(E_x, E_x^0)$$

which assigns to each point $x \in B$ a generator

$$\lambda(x) \in H_n(E_x, E_x^0) \cong Z .$$

The same goes for cohomology. But I would like a statement more global than that. In this case it is clear that cohomology rather than homology is required. Suppose, for example, that B had an infinity of path-components; then a singular homology class could only have a non-zero component in a finite number of them, but this difficulty does not arise in cohomology. We can ask if there is an element

$$\omega \in H^n(E, E^0)$$

such that for each $x \in B$, the induced homomorphism

$$i_x^*: H^n(E, E^0) \longrightarrow H^n(E_x, E_x^0)$$

has $i_x^* \omega = \lambda(x)$ for any given section λ.

Now in ordinary homology, the answer is yes: if you are given a section λ, there exists a cohomology class ω such that $i_x^* \omega = \lambda(x)$ for each x, and ω is unique. However, the proof makes essential use of the dimension axiom. For a generalised cohomology theory the corresponding result is not true. There is an n-plane bundle $E \longrightarrow B$ and a section $\lambda: B \longrightarrow \bigcup_x KO^n(E_x, E_x^0)$ such that there exists no $\omega \in KO^n(E, E^0)$ with $i_x^* \omega = \lambda(x)$ for all x; in another such example the required ω exists but is not unique.

It seems best to choose our definitions so as to avoid the difficulty. First I consider the meaning to be assigned to the word "generator." Let F be a ring-spectrum; then $F_*(R^n, R^n - 0) = \tilde{F}_*(S^n)$ and $F^*(R^n, R^n - 0) = \tilde{F}^*(S^n)$ are modules over $\pi_*(F)$. In fact each is a free module on one generator, because we have canonical classes

$$\gamma_n \in F_n(R^n, R^n - 0) \quad , \quad \gamma^n \in F^n(R^n - 0) .$$

I will say that $\varphi \in F^*(R^n, R^n - 0)$ is a <u>generator</u> if $\{\varphi\}$ is a $\pi_*(F)$-base for $F^*(R^n, R^n - 0)$. φ is a generator if and only if $\varphi = u\gamma^n$, where u is a unit in $\pi_*(F)$. φ need not lie in $F^n(R^n, R^n - 0)$, because we may have units of non-zero degree in $\pi_*(F)$; e.g., this occurs if $F = K$.

The property which I need of generators is the following. Let G be a module-spectrum over F. Then the map

$$G_*(R^n, R^n - 0) \longrightarrow \pi_*(G)$$

given by

$$y \longmapsto \langle \varphi, y \rangle$$

is an isomorphism. In fact, it is trivially so if $\varphi = \gamma^n$, and the general case differs only by a unit in $\pi_*(F)$.

We say $\omega \in F^*(E, E^0)$ is an orientation for E if

$$i_x^* \omega \in F^*(E_x, E_x^0)$$

is a generator for each $x \in B$.

Of course, the question of constructing an orientation for a vector-bundle, or of constructing orientations for some class of bundles, is non-trivial. However it can be done in several cases which are important in the applications. For example, complex n-plane bundles can be oriented over K^* or MU^*; Spin bundles can be oriented over KO^*; and so on. I will not give the constructions here.

We have defined orientations as they apply to n-plane bundles; but we want the notion as well for topological manifolds, which might not have a tangent bundle in the same sense as smooth manifolds. But it is well known what one substitutes for the tangent bundle. That is, one replaces E by $M \times M$, where M is a topological manifold, say without boundary. One replaces E^0 by the complement of the diagonal, $M \times M - \Delta$. One replaces the fibres E_x by the fibres $x \times M$ of the projection $p_1: M \times M \longrightarrow M$. One replaces E_x^0 by $x \times M - x \times x$. Since M is a topological manifold without boundary, x has a neighborhood U in M which is homeomorphic to a neighborhood of 0 in R^n, by a homeomorphism mapping x onto 0. Then $F^*(M, M-x) \cong F^*(U, U-x)$ by excision and so is isomorphic to $F^*(R^n, R^n - 0)$.

By an orientation over F^* for the tangent bundle of a topological manifold M, we will therefore mean a class $\omega \in F^*(M \times M, M \times M - \Delta)$ such that

$$i_x^* \omega \in F^*(x \times M, x \times M - x \times x)$$

is a generator for each x.

If M happens to have a boundary, there are two things we can do. The first starts from the observation that for a smooth manifold M, the tangent bundle to M contains over ∂M tangent vectors which point out, as well as tangent vectors which point in. To copy this in the topological case, one adds an open collar on the boundary; that is, one forms

$$M' = M \cup [0,1) \times \partial M .$$

This is a topological manifold without boundary, and it has a fully satisfactory topological tangent bundle, and one can ask for an orientation.

The other thing we can do is to use the same form of words as before, and ask for a class

$$F^*(M \times M, M \times M - \Delta) ,$$

but only demand that $i_x^* \omega$ be a generator for $x \in M - \partial M$.

Evidently the first sort of orientation restricts to the second, but I will not go into the relations between them.

Having completed the discussion of bundles, we go back to using E for a ring-spectrum.

Suppose given an orientation class

$$\omega \in E^d(M \times M, M \times M - \Delta) ,$$

where E is a ring-spectrum. Let F be a module-spectrum over E.

We define a duality map, which ultimately will be a map of the following form. Let K, L be a compact pair in M. Then $M - L$, $M - K$ is an open pair in M. The duality map will be a homomorphism

$$D: F_p(M - L, M - K) \longrightarrow \check{F}^{d-p}(K, L)$$

where the left-hand side, as before, indicates singular F-homology.

We will define the map D in a number of steps. Let $U, V \supset K, L$ be an open pair in M and V', U' another open pair with $U \cap U' = \emptyset$, $V \cap V' = \emptyset$. Then we have

$$U \times U' \subset M \times M - \Delta$$

$$V \times V' \subset M \times M - \Delta.$$

Therefore we can form

$$i^* \omega \in E^d(U \times V', U \times U' \cup V \times V').$$

So given $x \in F_p(V', U')$ we can form

$$D(x) = (i^* \omega)/x \in F^{d-p}(U, V).$$

I claim D is natural for inclusion maps. First, suppose $U'' \subset U$, $V'' \subset V$. Then surely $U'' \cap U' = \emptyset$, $V'' \cap V' = \emptyset$. The following diagram commutes.

$$\begin{array}{ccc} & F_p(V', U'') & \\ \swarrow & & \searrow \\ F^{d-p}(U, V) & \longrightarrow & F^{d-p}(U'', V'') \end{array}$$

Next suppose $V''', U''' \subset V', U'$. Again $U''' \cap U = \emptyset$, $V''' \cap V = \emptyset$. The following diagram commutes.

$$\begin{array}{ccc} F_p(V''', U''') & \longrightarrow & F_p(V', U') \\ \searrow & & \swarrow \\ & F^{d-p}(U, V) & \end{array}$$

Both facts are immediate from 9.1.

Again, I claim that D commutes with boundary maps, up to a suitable sign. More precisely, suppose we have

$$U \supset V \supset W \qquad\qquad U' \subset V' \subset W'$$

with $U \cap U' = \emptyset$, $V \cap V' = \emptyset$, $W \cap W' = \emptyset$. Then the diagram

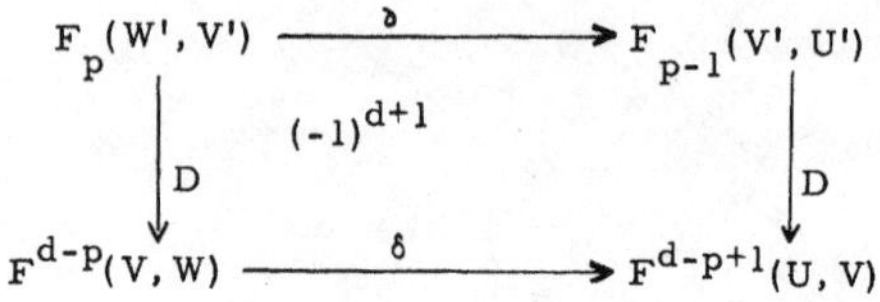

$$\begin{array}{ccc} F_p(W', V') & \xrightarrow{\partial} & F_{p-1}(V', U') \\ \Big\downarrow D & (-1)^{d+1} & \Big\downarrow D \\ F^{d-p}(V, W) & \xrightarrow{\delta} & F^{d-p+1}(U, V) \end{array}$$

commutes up to the sign $(-1)^{d+1}$. For we can easily reduce it to the case $W = \emptyset$, $U' = \emptyset$, by the following diagram.

$$\begin{array}{ccccccc} & F_p(W', V') & \xrightarrow{\partial} & F_{p-1}(V', \emptyset) & \longrightarrow & F_{p-1}(V', U') & \\ \swarrow & & \searrow & & \searrow & & \swarrow \\ F^{d-p}(V, W) & \longrightarrow & F^{d-p}(V, \emptyset) & \xrightarrow{\delta} & F^{d-p+1}(U, V) & & \end{array}$$

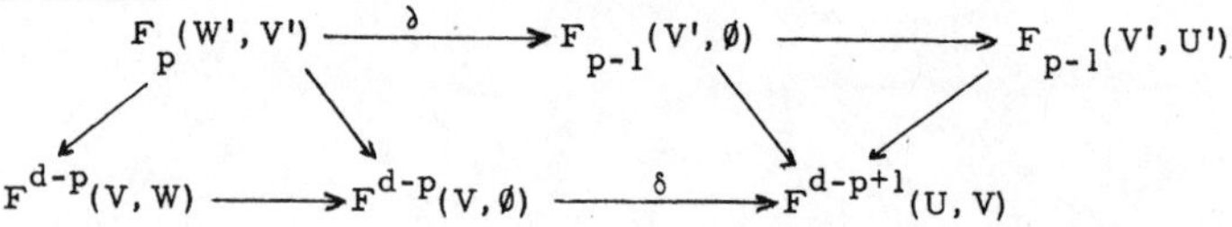

Now since $\omega \in E^d(M \times M, M \times M - \Delta)$ and $V \cap V' = \emptyset$, the class ω comes via $E^d(U \times W', V \times V')$, and by 9.12(ii) we have

$$\delta((i^* \omega)/x) = (-1)^{d+1} (j^* \omega)/\partial x .$$

Now we can start to pass to limits. Let us take a compact pair $K, L \subset M$ and consider the complementary open pair $M-L$, $M-K$. We vary the pair U, V over open pairs containing K, L. We vary V', U' over open pairs contained in $M-L, M-K$, of course arranging that $U \cap U' = \emptyset$, $V \cap V' = \emptyset$. Now $\check{F}^*(K, L) = \varinjlim F^*(U, V)$, so with a class in $\check{F}^*(U, V)$ for any U, V we get an image in $\check{F}^*(K, L)$. Now I claim that for any $x \in F_p(V', U')$, its image in $\check{F}^{d-p}(K, L)$ is independent of the choice of the pair U, V, provided, of course, that there exists a pair $U, V \supset K, L$ with $U \cap U' = \emptyset$, $V \cap V' = \emptyset$. And this is immediate, by the following diagram.

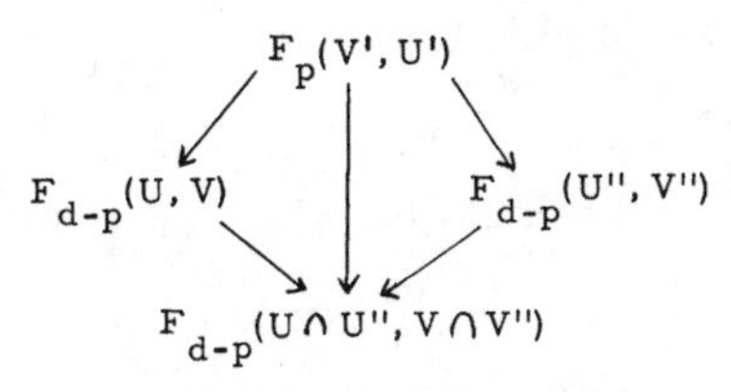

So now we have a well defined function

$$D: F_p(V', U') \longrightarrow \check{F}^{d-p}(K, L)$$

which of course gives

$$D: \varinjlim_{(V', U')} F_p(V', U') \longrightarrow \check{F}^{d-p}(K, L) .$$

(In fact, if $V''', U''' \subset V', U'$ and there exists a pair U, V with $U \cap U' = \emptyset$, $V \cap V' = \emptyset$, then the following diagram commutes.)

$$\begin{array}{ccc} F_p(V''', U''') & \longrightarrow & F_p(V', U') \\ & \searrow \quad \swarrow & \\ & F^{d-p}(U, V) & \end{array}$$

But I claim we have

$$\varinjlim_{(V', U')} F_p(V', U') = F_p(M-L, M-K) .$$

For this we need only check, by general topology, that the available pairs (V', U') satisfy 10.1.

At this stage, then, we have a transformation

$$D: F_p(M-L, M-K) \longrightarrow \check{F}^{d-p}(K, L)$$

which is natural in the sense that it commutes with the homomorphisms induced by inclusion maps, and, up to a sign $(-1)^{d+1}$, with the boundary maps.

THEOREM 10.6. D is an isomorphism if $K \cap \partial M \subset L$.

We build up the proof by stages, We always assume our pairs K, L have $K \cap \partial M \subset L$.

Remark 10.7. D is an isomorphism if K is a point x, and $L = \emptyset$.

Proof. Our assumption is $x \cap \partial M \subset L = \emptyset$, so $x \notin \partial M$. I claim the following diagram is commutative.

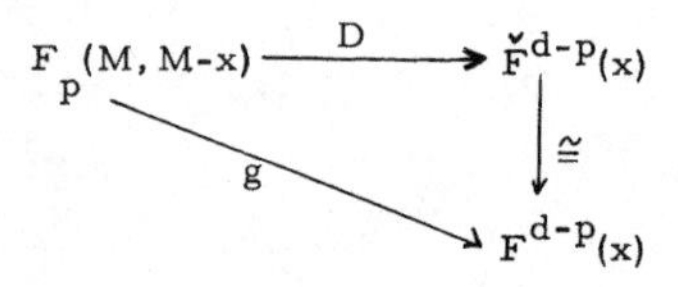

Here $g(y) = \langle i_x^* \omega, y \rangle$. Our assumption is that $i_x^* \omega$ is a generator, so the Kronecker product with it is an isomorphism.

The commutativity of the diagram follows easily from naturality. We can begin by supposing that we start from a class in $F_p(M, M-x)$ which comes from $y \in F_p(M, M-B)$, where B is a small closed ball in a coordinate neighborhood. (This uses 10.1.) If we apply D and the map into $F^{d-p}(x)$, we obtain

$$j^*((i^*\omega)/y) \; ,$$

where $j: x \longrightarrow \operatorname{Int} B$ is the injection. We have

$$j^*((i^*\omega)/y) = ((j \times 1)^* i^* \omega)/y = i_x^* \omega / y = \langle i_x^* \omega, y \rangle \; .$$

Remark 10.8. Suppose K is a rectilinear simplex in a coordinate neighborhood and L is one face of K (which may be K but must not be $\emptyset$). Then D is an isomorphism between groups which are zero.

Note: coordinate neighborhoods near ∂M are supposed to map ∂M into a linear subspace of R^n, e.g., R^{n-1}.

Proof. (i) $\check{F}^*(K, L) = 0$. In fact, we can even show this without appealing to the homotopy invariance of $\check{F}^*$; just surround K, L by a cofinal system of open convex neighborhoods U, V, for which $F^*(U, V) = 0$.

(ii) Also $F_*(M-L, M-K) = 0$. This is seen geometrically; see the accompanying figure. We can write K as the join $K = L*K'$. If $K' = \emptyset$, $K = L$ then the result is trivially true. Since $K' \cap \partial M = \emptyset$, we can draw a slightly larger simplex K'' slightly farther away from L so that $L*K''$ is n-dimensional and contains $L*K'$, while $L*K' \cap \partial(L*K'') = L$.

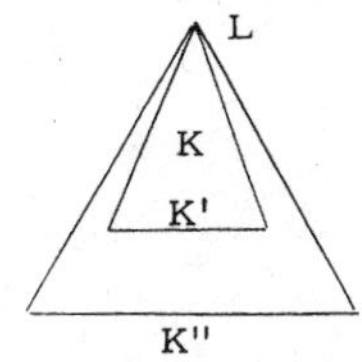

Then $(L*K'') - L$ is homeomorphic to $L \times (0, 1] \times K'' \cup K''$:

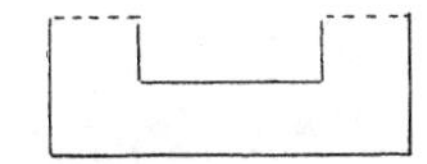

Now clearly $L*K'' - L*K' \longrightarrow L*K'' - L$ is a homotopy equivalence, by maps and homotopies keeping $\partial(L*K'') - L$ fixed throughout. These maps and homotopies extend over M by keeping everything fixed outside $L*K''$.

<u>Remark 10.9.</u> Suppose K is a rectilinear simplex in a coordinate neighborhood and $L = \emptyset$. Then D is an isomorphism.

<u>Proof.</u> Since $L = \emptyset$, we have $K \cap \partial M = \emptyset$. Let x be one vertex of K. Then we have the following commutative diagram.

$$\begin{array}{ccccccc} 0 = F_p(M-x, M-K) & \to & F_p(M, M-K) & \to & F_p(M, M-x) & \to & F_{p-1}(M-x, M-K) = 0 \\ & & \downarrow & & \downarrow \cong & & \\ 0 = \check{F}^{d-p}(K, x) & \longrightarrow & \check{F}^{d-p}(K) & \longrightarrow & \check{F}^{d-p}(x) & \longrightarrow & \check{F}^{d-p+1}(K, x) = 0 \end{array}$$

The four groups marked zero are so by 10.8. The map marked as an isomorphism is so by 10.7.

Remark 10.10. Suppose K_1, L_1 and K_2, L_2 are compact pairs in M with $K_1 \cap \partial M \subset L_1$, $K_2 \cap \partial M \subset L_2$, and $K_1 \cap L_2 = L_1 \cap K_2$. If D is an isomorphism for (K_1, L_1), (K_2, L_2) and $(K_1 \cap K_2, L_1 \cap L_2)$, then it is an isomorphism for $(K_1 \cup K_2, L_1 \cup L_2)$.

Proof. Consider the diagram of Mayer-Vietoris sequences on the following page. The Mayer-Vietoris sequences are slightly more general than those considered in Eilenberg and Steenrod but none the worse for that. The second row works because $K_1 \cap L_2 = L_1 \cap K_2$; this is the condition that the Mayer-Vietoris sequence may be replaced by one in which the subspaces remain fixed (namely at $L_1 \cup L_2$). The first row works for the dual reason that

$$(M-K_1) \cup (M-L_2) = (M-L_1) \cup (M-K_2);$$

this is the condition that the Mayer-Vietoris sequence may be replaced by one in which the total spaces remain fixed (namely at $(M-L_1) \cap (M-L_2)$) and the subspaces vary. We have the excision necessary for the first row because all the subspaces are open, and for the second because excision always holds for Čech **F**-cohomology on compact spaces.

The result follows from the five lemma.

Remark 10.11. Suppose K, L is a finite simplicial pair linearly embedded in a coordinate neighborhood. Then D is an isomorphism.

Proof. By barycentric subdivision we can suppose that for each simplex σ of K, $\sigma \cap L$ is either 0 or 1 faces of σ. For such pairs we argue by induction over the number of simplices in K. If this number is zero the result is trivial; if this number is one it is true by 10.8 and 10.9. The inductive step is immediate from 10.10.

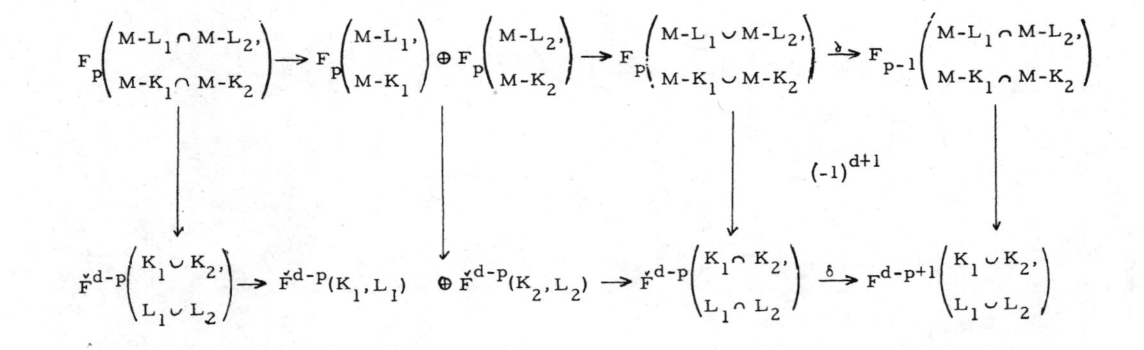

F_p(M-L_1 ∩ M-L_2, M-K_1 ∩ M-K_2) → F_p(M-L_1, M-K_1) ⊕ F_p(M-L_2, M-K_2) → F_p(M-L_1 ∪ M-L_2, M-K_1 ∪ M-K_2) ∂→ F_{p-1}(M-L_1 ∩ M-L_2, M-K_1 ∩ M-K_2)
(-1)^{d+1}
F̌^{d-p}(K_1 ∪ K_2, L_1 ∪ L_2) → F̌^{d-p}(K_1, L_1) ⊕ F̌^{d-p}(K_2, L_2) → F̌^{d-p}(K_1 ∩ K_2, L_1 ∩ L_2) δ→ F^{d-p+1}(K_1 ∪ K_2, L_1 ∪ L_2)

Remark 10.12. Suppose K, L is any compact pair in a single coordinate neighbourhood. Then D is an isomorphism.

Proof. Pass to direct limits from finite simplicial neighbourhoods U, V.

Proof of Theorem 10.6. Each point of K is in the interior of a compact neighbourhood which is in a single coordinate neighbourhood. Hence, K can be covered by finitely many such subsets. Now argue by induction on the number of such subsets; if the number is one, 10.12 gives us the result; the inductive step is immediate from 10.10.

COROLLARY 10.13. (Poincaré duality). Let M be a compact topological manifold without boundary, oriented over E^*. Then we have an isomorphism

$$D: F_p(M) \longrightarrow F^{d-p}(M)$$

which may be given by

$$D(y) = \omega / y.$$

Now we observe that we can make $E^*(M)$ act on $F_*(M)$ via the cap product, and on $F^*(M)$ via the cup product. We could like to know that D is a map of modules, up to sign, provided that E is a commutative ring-spectrum. Actually this is not quite general enough for what follows; in any case, it helps to keep the details in order if we assume our spectra are distinct as long as we can. So I suppose given two module-spectra G, G' over E, and a pairing $\mu: F \wedge G \longrightarrow G'$, where F is not necessarily a module-spectrum over E. I also assume the pairing is right-linear over E, in the sense that the following diagram is commutative.

$$\begin{array}{ccccc} E\wedge F\wedge G & \xrightarrow{1\wedge\mu} & E\wedge G' & \xrightarrow{\nu'} & G' \\ \downarrow{\scriptstyle c\wedge 1} & & & & \downarrow{\scriptstyle 1} \\ F\wedge E\wedge G & \xrightarrow{1\wedge\nu} & F\wedge G & \xrightarrow{\mu} & G' \end{array}$$

Example. Take E and F to be E; take G and G' to be F; and assume E is a commutative ring-spectrum.

PROPOSITION 10.14. If $u \in F^p(M)$, the following diagram is commutative up to a sign $(-1)^{dp}$.

$$\begin{array}{ccc} G_q(M) & \xrightarrow{u\cap} & G'_{-p+q}(M) \\ \downarrow{\scriptstyle D} & (-1)^{dp} & \downarrow{\scriptstyle D} \\ G^{d-q}(M) & \xrightarrow{u\cup} & G'^{d+p-q}(M) \end{array}$$

That is, $D(u \cap v) = (-1)^{dp} u \cup (Dv)$, $v \in G_q(M)$.

Proof. $D(u \cap v) = \omega/(u \cap v)$

$u \cup (Dv) = u \cup (\omega/v)$

using the pairings from the first and second rows of the diagram. Now we want the following associativity formulae.

LEMMA 10.15. If

$$\omega \in E^d(X \times Y, A \times Y \cup X \times B), \quad u \in F^p(Y, C), \quad v \in G_q(Y, B \cup C)$$

then

$$\omega/(u \cap v) = (\omega \cup p_2^* u)/v \in (E \wedge F \wedge G)^{d+p-q}(X, A).$$

If

$$u \in F^p(X, A), \quad \omega \in E^d(X \times Y, B \times Y \cup X \times C), \quad v \in G_q(Y, C)$$

then

$$u \cup (\omega/v) = (p_1^* u \cup \omega)/v \in (F \wedge E \wedge G)^{p+d-q}(X, A \cup B).$$

The proof is immediate from the associativity formulae we have, by naturality.

This gives

$$D(u \cap v) = (\omega \cup p_2^* u)/v$$

$$u \cup (Dv) = (p_1^* u \cup \omega)/v$$

where we are still using the pairing from the second row of the diagram for the second formula. However, because the diagram of pairings is commutative we can write

$$u \cup (Dv) = (-1)^{dp} (\omega \cup p_1^* u)/v$$

using the pairing from the top row of the diagram. Now it is sufficient to prove

$$\omega \cup p_1^* u = \omega \cup p_2^* u .$$

Consider the maps

$$p_1 \colon M \times M \longrightarrow M$$

$$p_2 \colon M \times M \longrightarrow M .$$

They have the same restriction to Δ; a fortiori they are homotopic on Δ. By 10.2 (ii) there is an open neighborhood U of Δ in M and a homotopy $h \colon U \longrightarrow M$ between $p_1 | U$ and $p_2 | U$. Hence, if we apply

$$i^* \colon F^p(M \times M) \longrightarrow F^p(U),$$

we have

$$i^* p_1^* u = i^* p_2^* u \in F^p(U) .$$

But by excision,

$$(E \wedge F)^{d+p}(M \times M, M \times M - \Delta) \longrightarrow (E \wedge F)^{d+p}(U, U - \Delta)$$

is an isomorphism. The classes

$$\omega \cup p_1^* u \quad , \quad \omega \cup p_2^* u$$

restrict to

$$(i^* \omega) \cup (i^* p_1^* u) \quad , \quad (i^* \omega) \cup (i^* p_2^* u) ,$$

that is, they restrict to the same thing. Therefore they already were equal in

$$(E \wedge F)^{d+p}(M \times M, M \times M - \Delta).$$

This proves 10.14.

Applying 10.13 to the case $F = E$, we see that there is a class $[M] \in E_d(M)$ such that

$$D([M]) = 1 \in E^0(M).$$

This is called the <u>fundamental class</u> of M (corresponding to the given orientation).

The usual way to present the Poincaré duality isomorphism is to say that it is the homomorphism

$$F^p(M) \longrightarrow F_{d-p}(M)$$

given by $x \longmapsto x \cap [M]$. Of course the pairing being considered is

$$F \wedge E \xrightarrow{c} E \wedge F \xrightarrow{\nu} F \ .$$

PROPOSITION 10.16. This homomorphism is the inverse of D, up to a sign $(-1)^{dp}$; it is therefore an isomorphism.

<u>Proof.</u> In 10.14, take E and G to be E; take F and G' to be F. The resulting diagram is commutative even without the assumption that E is a commutative ring-spectrum. Then

$$D(u \cap [M]) = (-1)^{dp} u \cup D([M]) = (-1)^{dp} u.$$

The relative version of Poincaré duality is called Lefschetz duality. It asserts that we have the following diagram, commutative up to sign.

$$\begin{array}{ccccccccc}
F_p(\partial M) & \longrightarrow & F_p(M) & \longrightarrow & F_p(M, \partial M) & \xrightarrow{\partial} & F_{p-1}(\partial M) & \longrightarrow & \cdots \\
\cong \downarrow & \pm & \cong \downarrow & \pm & \cong \downarrow & \pm & \cong \downarrow & & \\
F^{d-1-p}(\partial M) & \xrightarrow{\delta} & F^{d-p}(M, \partial M) & \longrightarrow & F^{d-p}(M) & \longrightarrow & F^{d-p}(\partial M) & \longrightarrow & \cdots
\end{array}$$

I will omit the proof. It involves discussing the relation between an orientation on M and one on ∂M, and also manipulation of collars.

11. APPLICATIONS IN K-THEORY

The material presented so far may have seemed rather theoretical. But topologists also like to do sums and see how things work out in concrete cases, so I ought to show you some examples. I choose to present some examples from complex K-theory.

First we recall some facts we need about complex K-theory. This has a geometric interpretation; a complex vector-bundle ξ over X represents an element of $K^0(X)$. (See §6.) Similarly, a formal linear combination of bundles, such as $\xi - \eta$, gives an element of $K^0(X)$. The Whitney sum of bundles gives addition in $K^0(X)$; the tensor product of bundles gives multiplication in $K^0(X)$.

We need to know the K-cohomology of a few simple spaces. Over $BU(1) = CP^\infty$ we have the universal $U(1)$-bundle, which gives a line-bundle, i.e., a complex vector bundle with fibres of dimension 1. Call this line bundle ξ. Define $x = \xi - 1 \in \tilde{K}^0(CP^\infty)$. Use the same symbol x for the restriction of this class to CP^n.

PROPOSITION 11.1. (Atiyah and Todd). $K^*(CP^n)$ is free over $\pi_*(K)$ with a base consisting of $1, x, x^2, \ldots, x^n$ $(x^{n+1} = 0)$. $K^*(CP^\infty) = \pi_*(K)[[x]]$.

We need a cohomology operation in K-theory.

PROPOSITION 11.2. There exists a function

$$\Psi^2 : K^0(X) \longrightarrow K^0(X)$$

such that

(i) Ψ^2 is natural,

(ii) Ψ^2 is a homomorphism of rings, and

(iii) if η is a line bundle, then $\Psi^2(\eta) = \eta^2$.

Now I have said something about orientations for particular vector-bundles. If we construct orientations for a whole class of vector-bundles, we would like them to have various properties. First, the orientations should be natural for maps of vector-bundles. Secondly, we would like good behaviour on Whitney sums. Suppose given two bundles ξ', ξ'' over X; form their Whitney sum $\xi = \xi' \oplus \xi''$. Let the total spaces be E, E', E'' and the complements of the zero cross-sections E_0, E'_0, E''_0 respectively. Then we have maps $p': E \longrightarrow E'$, $p'': E \longrightarrow E''$; over each $x \in X$ one projects the sum of two fibres onto either summand. Then

$$E_0 = ((p')^{-1}E'_0) \cup ((p'')^{-1}E''_0).$$

Let $\omega \in F^*(E, E_0)$, $\omega' \in F^*(E', E'_0)$, $\omega'' \in F^*(E'', E''_0)$ be the three orientations. We would like them to satisfy

$$\omega = ((p')^*\omega') \cup ((p'')^*\omega'').$$

Thirdly, we have a normalisation axiom. Consider the canonical line bundle ξ over $BU(1)$. I claim its Thom complex $MU(1)$ is equivalent to $BU(1)$. In fact, we have to consider the associated pair of bundles with fibres D^2 and S^1. But $S^1 = U(1)$; the associated S^1-bundle is the universal S^1-bundle, so it is contractible. Thus, when we form a Thom complex by collapsing it to a point, we do not change anything. But D^2 is contractible and the associated D^2-bundle is equivalent to $BU(1)$. Hence, $MU(1) \simeq BU(1)$.

PROPOSITION 11.3. There is an orientation ω for each complex vector-bundle ξ which satisfies the following axioms.

(i) Naturality.

(ii) The axiom on Whitney sums.

(iii) Normalization; for the universal bundle, $\omega \in \tilde{K}^0(MU(1))$ corresponds under the equivalence to $x \in \tilde{K}^0(BU(1))$.

Now we can construct various characteristic classes. The easiest is the Euler class. Suppose we have an orientation ω in F-cohomology for some class of bundles; let $\zeta: X \longrightarrow E$ be the zero cross-section. We define the Euler class of ξ by

$$\chi_F(\xi) = \zeta^* \omega .$$

Its formal properties are: naturality (if ω is natural);

$$\chi_F(\xi' + \xi'') = \chi_F(\xi')\chi_F(\xi''),$$

(if ω satisfies the axiom on Whitney sums); and normalisation (if ω satisfies the nomalisation axiom). For example, in the case of complex K-theory we have

$$\chi_K(\eta) = \eta - 1 \qquad \text{where } \eta \text{ is a line bundle.}$$

PROPOSITION 11.4. Suppose the bundle in question is the tangent bundle τ of a compact smooth manifold M^n, orientable for ordinary homology. Then

$$\chi_F(\tau) = f^* i^* \omega .$$

Here $i^*\omega$ is the restriction of the orientation ω to one fibre, so that it lies in

$$\tilde{F}^*(R^n, R^n - 0) \cong \tilde{F}^*(S^n) ,$$

and $f: M^n \longrightarrow S^n$ is a map of degree $\chi(M)$, this being the ordinary Euler characteristic for M.

<u>Proof.</u> By a result going back to Hopf, we can construct on M a field γ of tangent vectors with non-degenerate singularities, so that the number of singularities, when counted with appropriate signs, is $\chi(M)$. But now the zero section $\zeta: M \longrightarrow E(\tau)$ is homotopic to a section λ,

which crosses the zero-section transversely a total of $\chi(M)$ times. So $\zeta^*\omega = \lambda^*\omega$. But here the contribution comes from many small discs, each of which contributes $\pm i^*\omega$.

Given an orientation, we can also construct a Thom isomorphism. This allows us to copy Thom's treatment of the Stiefel-Whitney classes. We consider the following diagram.

$$\begin{array}{ccc} K^0(E, E_0) & \xrightarrow{\Psi^2} & K^0(E, E_0) \\ \uparrow{\scriptstyle \varphi_K} & & \uparrow{\scriptstyle \varphi_K} \\ K^0(X) & & K^0(X) \end{array}$$

We define

$$\rho_2(\xi) = \varphi_K^{-1} \Psi^2 \varphi_K(1) .$$

PROPOSITION 11.5. $\rho_2(\xi) \in K^0(X)$ is a characteristic class with the following properties.

(i) Naturality.

(ii) $\rho_2(\xi \oplus \eta) = \rho_2(\xi)\rho_2(\eta)$.

(iii) If η is a line bundle,

$$_2(\eta) = 1 + \eta .$$

PROPOSITION 11.6. ρ_2 extends to a function

$$\rho_2 : K^0(X) \longrightarrow K^0(X; Z[\tfrac{1}{2}]).$$

We need the denominators because $\rho_2(1) = 2$, so $\rho_2(-1) = \frac{1}{2}$.

Now we are ready to study the following problem. In terms of our knowledge of $K^*(CP^n)$, what is the fundamental class in $K_*(CP^n)$? If we look at our account of duality, it appears we should ask a prior question. Take $CP^n \times CP^n$ and embed CP^n in the diagonal Δ. We have an orientation

$$\omega \in K^0(CP^n \times CP^n, CP^n \times CP^n - \Delta) .$$

What is its image in $K^0(CP^n \times CP^n)$? Of course we require our answer in terms of the base we know in $K^*(CP^n \times CP^n)$.

PROPOSITION 11.7. $K^*(CP^n \times CP^n)$ is free over $\pi_*(K)$ with a base consisting of the products $x_1^i x_2^j$ for $0 \leq i \leq n$, $0 \leq j \leq n$ ($x_1^{n+1} = 0$, $x_2^{n+1} = 0$). Here x_1 and x_2 are the generators for the two factors--see 11.1.

The difficulty is that the construction of ω refers to a tubular neighbourhood of the diagonal, and it is not clear how to relate that to the whole of $M \times M$.

LEMMA 11.8. Consider

$$j^*: K^*(CP^n \times CP^n, CP^n \times CP^n - \Delta) \longrightarrow K^*(CP^n \times CP^n) .$$

If $k \in \operatorname{Im} j^*$, then $x_1 k = x_2 k$.

See the proof of 10.14.

LEMMA 11.9. The subgroup of elements $k \in K^0(CP^n \times CP^n)$ such that $(x_1 - x_2)k = 0$ has a Z-base $p_0, p_1, \ldots, p_n$, where

$$p_r = \sum_{i+j=n+r} x_1^i x_2^j .$$

The proof is a trivial calculation.

LEMMA 11.10. We have

$$j^*\omega = 1 \cdot p_0 + a_1 p_1 + a_2 p_2 + \ldots + a_n p_n , \qquad a_i \in Z.$$

Proof. By Lemmas 11.8 and 11.9 we have

$$j^*\omega = \sum_i a_i p_i .$$

Now consider the restriction of j^* to the diagonal. p_0 restricts to $(n+1)x^n$; p_i restricts to 0 for $i > 0$. But $j^*\omega$ restricts to the Euler class; $\chi(CP^n) = (n+1)$, and the orientation was chosen so that $1^* i^* \omega = x^n$. So $a_0 = 1$.

LEMMA 11.11. $j^*\omega$ satisfies

$$\Psi^2(j^*\omega) = (\rho_2\tau)(j^*\omega)$$

where $\rho_2(\tau) = \frac{1}{2}(2 + x)^{n+1}$.

Proof. The first equation is immediate from the definition of ρ_2. For the second,

$$\tau + 1 = (n + 1)\xi$$

$$\rho_2(\xi) = 1 + \xi = 2 + x,$$

$$\rho_2(1) = 2 ;$$

so

$$\rho_2(\tau) = \frac{1}{2}(2 + x)^{n+1} .$$

LEMMA 11.12. $j^*\omega$ is uniquely determined by 11.10 and 11.11.

Proof. Suppose as an inductive hypothesis that $a_1, \ldots, a_{i-1}$ are determined. Then

$$\Psi^2(a_i p_i) = 2^{n+i} a_i p_i + T_1 ,$$

where T_1 is a sum of terms in $p_{i+1}, \ldots, p_n$; so

$$\Psi^2(j^*\omega) = T_2 + 2^{n+i} a_i p_i + T_3 ,$$

where T_2 is a sum of known terms and T_3 is a sum of terms in $p_{i+1}, \ldots, p_n$. Similarly, $(\rho_2\tau)(j^*\omega)$ is the sum of known terms, terms in $p_{i+1}, \ldots, p_n$ and the term $2^n a_i p_i$. So we can find a_i by equating the coefficients of p_i.

LEMMA 11.13. We have

$$2(1 + x)\psi^2(p_0) = (2 + x)^{n+1} p_0 .$$

Proof. Calculating in $K^0(CP^\infty \times CP^\infty)$ we have

$$(x_1 - x_2)p_0 = x_1^{n+1} - x_2^{n+1} ,$$

therefore

$$\psi^2(x_1 - x_2)\psi^2 p_0 = \psi^2 x_1^{n+1} - \psi^2 x_2^{n+1} .$$

i.e.,

$$(2x_1 + x_1^2 - 2x_2 - x_2^2)\Psi^2 p_0 = (2x_1 + x_1^2)^{n+1} - (2x_2 + x_2^2)^{n+1}$$

$$= \sum_{i+j=n+1} \frac{(n+1)!}{i!j!} 2^i (x_1^{n+1+j} - x_2^{n+1+j}) .$$

Dividing by $x_1 - x_2$, which is not a zero-divisor in $K^0(CP^\infty \times CP^\infty)$, we have

$$(2 + x_1 + x_2)\Psi^2 p_0 = \sum_{i+j=n+1} \frac{(n+1)!}{i!\, j!} 2^i p_j .$$

Now restricting to $K^0(CP^n \times CP^n)$, we get

$$2(1 + x)\Psi^2 p_0 = \sum_{i+j=n+1} \frac{(n+1)!}{i!\, j!} 2^i x^j p_0 = (2 + x)^{n+1} p_0 .$$

This proves 11.13.

It follows that

$$\Psi^2((1 + x)p_0) = (1 + x)^2 \Psi^2 p_0 = \frac{1}{2}(2 + x)^{n+1}(1 + x)p_0 .$$

We conclude that the solution to our problem is:

THEOREM 11.14.

$$j^* \omega = (1 + x)p_0 = \sum_{i+j=n} x_1^i x_2^j + \sum_{i+j=n+1} x_1^i x_2^j .$$

As a corollary, we obtain the relation between the fundamental class $[CP^n]_K$ in K-homology and our base $\{x^i\}$.

THEOREM 11.15. $< x^i, [CP^n]_K > = (-1)^{n-i}$.

<u>Proof.</u> Suppose we choose a base $\{b_j\}$ in $K_0(CP^n)$ such that $< x^i, b_j > = \delta_{ij}$. Then

$$x_1^i x_2^j / b_k = x_1^i < x_2^j , b_k > = x_1^i \delta_{jk} .$$

Thus

$$j^* \omega / b_n = 1 + x_1 ,$$

$$j^* \omega / b_{n-1} = x_1 + x_1^2$$

$$j^*\omega/b_1 = x_1^{n-1} + x_1^n ,$$

$$j^*\omega/b_0 = x_1^n .$$

We require the class $[CP^n]_K$ such that $j^*\omega/[CP^n]_K = 1$. Clearly the answer is

$$[CP^n]_K = b_n - b_{n-1} + b_{n-2} - b_{n-3} \cdots + (-1)^n b_0 .$$

This proves the result.

THEOREM 11.16. If M is a weakly almost complex manifold, then

$$\text{Index}(M) = \langle \rho_2(\tau) , [M]_K \rangle .$$

Proof. The index is a homomorphism of rings from the cobordism ring of weakly almost complex manifolds, that is, $\pi_*(MU)$. It is therefore sufficient to prove the result for a set of generators of the Q-algebra $\pi_*(MU) \otimes Q$. But the complex projective spaces CP^n are such generators. For CP^n we have

$$\rho_2(\tau) = \frac{1}{2}(2 + x)^{n+1} .$$

So

$$\begin{aligned}
\langle \rho_2(\tau), [CP^n]_K \rangle &= \langle \tfrac{1}{2}(2 + x)^{n+1}, [CP^n]_K \rangle \\
&= \frac{1}{2} \sum_{i+j=n+1} \frac{(n+1)!}{i!\, j!} 2^i \langle x^j, [CP^n]_K \rangle \\
&= \frac{1}{2} [(\sum_{i+j=n+1} \frac{(n+1)!}{i!\, j!} 2^i (-1)^{n-j}) + 1] \\
&= \frac{1}{2} [(-1)^n (2-1)^{n+1} + 1] \\
&= \frac{1}{2} [1 + (-1)^n] \\
&= \begin{cases} 1 & (n \equiv 0 \quad (2)) \\ 0 & (n \equiv 1 \quad (2)) \end{cases} \\
&= \text{Index}\,(CP^n) .
\end{aligned}$$

12. THE STEENROD ALGEBRA AND ITS DUAL

One knows that in order to perform calculations in ordinary cohomology, it is very useful to have operations like the Steenrod squares.

In the general case, let E be a spectrum. Then to every element of $E^*(E)$ we can associate a natural transformation $E^*(X) \longrightarrow E^*(X)$ defined for all spectra X. Namely, given

$$X \xrightarrow{f} E \quad \text{and} \quad E \xrightarrow{g} E,$$

we form $X \xrightarrow{gf} E$. This gives a 1-1 correspondence between elements of $E^*(E)$ and such natural transformations (consider the case $X = E$).

Now $E^*(E)$ is of course a group; addition in it corresponds to adding operations

$$(g_1 + g_2)f = (g_1 f) + (g_2 f) .$$

But $E^*(E)$ is in fact a ring; multiplication in its corresponds to composing operations,

$$(g_1 g_2)f = g_1(g_2 f) .$$

Example. Suppose given a prime p; take $E = HZ_p$. Then $A^* = (HZ_p)^*(HZ_p)$ is the mod p Steenrod algebra, the algebra of stable cohomology operations on ordinary cohomology with Z_p coefficients. That it is an algebra over Z_p is clear from the fact that it contains Z_p.

It is a fact that A^* is generated by the Steenrod operations. If $p = 2$ these are the Steenrod squares

$$Sq^i: H^n(X, Y; Z_2) \longrightarrow H^{n+i}(X, Y; Z_2) .$$

If $p > 2$ these are the Steenrod powers

$$P^k: H^n(X, Y; Z_p) \longrightarrow H^{n+2k(p-1)}(X, Y; Z_p)$$

together with the Bockstein boundary

$$\beta_p : H^n(X, Y; Z_p) \longrightarrow H^{n+1}(X, Y; Z_p) .$$

The fact that A^* is generated by the Steenrod operations is not obvious, and should not be taken as a definition; it comes from the calculation of $(HZ_p)^*(HZ_p)$, which is due to Serre for $p = 2$, and to Cartan for $p > 2$.

Actually A^* has more structure than just the structure of an algebra. Before going into this, I want to comment on the work of Milnor [10]. Milnor showed that it is also good to look at the dual of the Steenrod algebra,

$$A_* = (HZ_p)_*(HZ_p) .$$

Here A_* and A^* are dual graded vector spaces over Z_p. Of course, if we did not know that A_* is finite-dimensional over Z_p in each degree we would only say

$$A^n = \mathrm{Hom}_{Z_p}(A_n, Z_p);$$

but of course we do know it.

Now HZ_p is a ring-spectrum; we have a map

$$\mu: HZ_p \wedge HZ_p \longrightarrow HZ_p .$$

So we get

$$\begin{aligned} A_* \otimes A_* = (HZ_p)_*(HZ_p) \otimes (HZ_p)_*(HZ_p) \\ \xrightarrow{\wedge} (HZ_p)_*(HZ_p \wedge HZ_p) \\ \xrightarrow{\mu_*} (HZ_p)_*(HZ_p) = A_* . \end{aligned}$$

So A_* also is an algebra.

The dual of the product map $\varphi: A_* \otimes A_* \longrightarrow A_*$ is of course a coproduct map $\psi = \varphi^*: A^* \longrightarrow A^* \otimes A^*$. The interpretation of this coproduct is as follows. Suppose

$$\psi(a) = \sum_i a'_i \otimes a''_i .$$

Then

$$a(x \barwedge y) = \sum_i (-1)^{|a''_i|\,|x|} (a'_i x) \barwedge (a''_i y) \quad \text{(Cartan formula)} .$$

There exists one and only one element $\sum_i a'_i \otimes a''_i$ such that this formula is true for all x and y. Of course the formula is then true for $x \bar{\times} y$ and $x \cup y$. For example,

$$Sq^k(xy) = \sum_{i+j=k} (Sq^i x)(Sq^j y) ,$$

so that

$$\psi\, Sq^k = \sum_{i+j=k} Sq^i \otimes Sq^j .$$

It can easily be shown that in this way A^* becomes a Hopf algebra. Dually, A_* becomes a Hopf algebra; its coproduct is the dual of the composition product in A^*.

More generally, let X be a space such that $(HZ_p)_*(X)$ is finite-dimensional in each degree. Then $(HZ_p)^*(X)$ is a module over A^*. The action is given by a map

$$A^* \otimes (HZ_p)^*(X) \longrightarrow (HZ_p)^*(X) .$$

The dual of this map is a coaction map

$$(HZ_p)_*(X) \longrightarrow A_* \otimes (HZ_p)_*(X) .$$

Thus $(HZ_p)_*(X)$ becomes a comodule over the coalgebra A_*. The assumption that $(HZ_p)_*(X)$ is locally finite-dimensional is in fact unnecessary, since the coaction map can be defined directly, as will be done below in a more general setting.

It turns out that the structure of A_* is very much easier to describe than the structure of A^*. One reason is that the product in A_* is commutative, whereas that in A^* is not ($Sq^1Sq^2 \neq Sq^2Sq^1$).

We give a description for the case $p = 2$. We start from RP^∞, which is an Eilenberg-Mac Lane space of type $(Z_2, 1)$. We have $(HZ_2)^*(RP^\infty) = Z_2[x]$, a polynomial algebra on one generator x of dimension 1 (the fundamental class). We may take in $(HZ_2)_*(RP^\infty)$ a base of elements $b_i \in (HZ_2)_i(RP^\infty)$ such that

$$\langle x^i, b_j \rangle = \delta_{ij} .$$

Since RP^∞ is term 1 in the HZ_2 spectrum, b_j yields some element in $(HZ_2)_{j-1}(HZ_2) = A_{j-1}$. It can easily be shown that this element is zero unless j is a power of 2. We define ξ_n to be the image of b_{2^n} in A_{2^n-1}. The element ξ_0 turns out to be the unit $1 \in A_0$.

THEOREM 12.1 (Serre-Milnor). If $p = 2$,

$$A_* = Z_2[\xi_1, \xi_2, \ldots] .$$

The proof is non-trivial, and is omitted here.

The construction of ξ_i yields the following description of ξ_i as a linear function on A^*.

PROPOSITION 12.2. The action of $a \in A^*$ on $(HZ_2)^1(RP^\infty)$ is given by

$$ax = \sum_{i \geq 0} \langle a, \xi_i \rangle x^{2^i} .$$

For x is a morphism from the suspension spectrum of RP^∞ to HZ_2 of degree -1, and

$$\langle ax, b_j \rangle = \langle x^* a, b_j \rangle = \langle a, x_* b_j \rangle = \begin{cases} 0 & \text{if } j \neq 2^r \text{ for some } r \\ \langle a, \xi_r \rangle & \text{if } j = 2^r . \end{cases}$$

From this it is rather easy to work out the effect of a on x^2, x^4, etc. We get:

PROPOSITION 12.3. $a(x^{2^i}) = \sum_{j \geq 0} \langle a, \xi_j^{2^i} \rangle x^{2^{i+j}} .$

It now becomes easy to work out the effect of a composite ba on x, which gives us $\langle ba, \xi_i \rangle$ and therefore $\psi \xi_i$.

PROPOSITION 12.4. $\psi \xi_k = \sum_{i+j=k} \xi_j^{2^i} \otimes \xi_i$.

We would now like to carry over some of this work to generalised homology theories. Let E be a ring-spectrum with multiplication μ. Then obviously the appropriate generalisation of A_* is $E_*(E)$. It turns out that this works quite well even in various cases where $E^*(E)$ works horribly badly. However, one needs an assumption and one must give a warning. The warning is that in the classical case A_* is an algebra over Z_p, but in the generalised case $E_*(E)$ is a bimodule over $\pi_*(E)$. There are two actions of $\pi_*(E)$ on $E_*(E)$, and one has to remember that they are different. The left action $\pi_*(E) \otimes E_*(E) \longrightarrow E_*(E)$ is obtained by using the morphism $E \wedge E \wedge E \xrightarrow{\mu \wedge 1} E \wedge E$; the right action $E_*(E) \otimes \pi_*(E) \longrightarrow E_*(E)$ is obtained by using the morphism $E \wedge E \wedge E \xrightarrow{1 \wedge \mu} E \wedge E$.

The assumption we have to make is that $E_*(E)$ is flat as a right module over $\pi_*(E)$. I say "as a right module", but if E is commutative, which is the usual case, it is equivalent to say that $E_*(E)$ is flat as a left module; this is seen by using $c: E \wedge E \longrightarrow E \wedge E$ to interchange the two sides.

The assumption is satisfied for the following cases: $E = KO$, K, MO, MU, MSp, S, and HZ_p. See [1], Lemma 28, p. 45.

With this assumption, we have the following lemma. Consider the morphism

$$(E \wedge E) \wedge (E \wedge X) \xrightarrow{1 \wedge \mu \wedge 1} E \wedge E \wedge X.$$

It induces a product map

$$E_*(E) \otimes_{\pi_*(E)} E_*(X) \longrightarrow [S, E \wedge E \wedge X]_* .$$

LEMMA 12.5. This product map is an isomorphism.

Proof. (i) If $X = S^p$, the result is trivial,

(ii) If we have a cofibering

$$X_1 \longrightarrow X_2 \longrightarrow X_3 \longrightarrow X_4 \longrightarrow X_5$$

and the result is true for X_1, X_2, X_4, and X_5, then it is true for X_3 (by the 5-lemma).

(iii) The result is true if X is any finite spectrum, by induction on the number of cells, using (i) and (ii).

(iv) The result is true if X is any spectrum, by passing to direct limits.

We can now define the coaction map we want. Consider the morphism

$$E \wedge X \simeq E \wedge S \wedge X \xrightarrow{1 \wedge i \wedge 1} E \wedge E \wedge X.$$

This induces

$$E_*(X) \xrightarrow{(1 \wedge i \wedge 1)_*} [S, E \wedge E \wedge X]_* .$$

Composing this with the inverse of the isomorphism in Lemma 12.5, we obtain a homomorphism

$$\psi_X: E_*(X) \longrightarrow E_*(E) \otimes_{\pi_*(E)} E_*(X) .$$

Specialising to the case $X = E$, we obtain the homomorphism

$$\psi_E: E_*(E) \longrightarrow E_*(E) \otimes_{\pi_*(E)} E_*(E) .$$

We also define a counit map

$$\varepsilon: E_*(E) \longrightarrow \pi_*(E) ,$$

which is simply the homomorphism induced by the product morphism

$$\mu: E \wedge E \longrightarrow E.$$

THEOREM 12.6. (i) $E_*(E)$ is a coalgebra with ψ_E as a coproduct map and ϵ as a counit map.

(ii) $E_*(X)$ is a comodule over $E_*(E)$ with ψ_X as the coaction map.

(iii) If $E = HZ_p$, then ψ_X, ψ_E and ϵ become the structure maps classically considered.

To give a complete proof of 12.6, one has to introduce a few more structure maps, which is very easy, and check their properties by diagram chasing. See [1], chapter 3.

13. A UNIVERSAL COEFFICIENT THEOREM

The theme for the next part of the course is the following. Let E be a fixed ring-spectrum. Suppose given $E_*(X)$ and $E_*(Y)$; what can be said about $[X, Y]_*$? In other words, given homological information, what can we say about homotopy?

I propose to treat this problem under a restrictive hypothesis; that is, I will assume that $E_*(X)$ is projective over $\pi_*(E)$. I do know how to avoid this hypothesis, but it involves extra work; one has to resolve both X and Y and mix the resolutions geometrically. The present hypothesis is sufficient for the applications to be given here. To see that the hypothesis is reasonable, consider two examples.

Example (i). Let $X = S$. Initially most people need to compute stable homotopy, that is, $[S, Y]_*$. Of course $E_*(S)$ is projective over $\pi_*(E)$ for any ring-spectrum E; in fact it is free on one generator.

(ii). Let $E = HZ_p$. In this case $\pi_*(E)$ is the field Z_p, so any module over it is projective; in particular, $(HZ_p)_*(X)$ is projective over Z_p for any X.

All the same, the correct level of generality will probably turn out to be the maximum level, so ultimately we will probably want to go beyond the case in which $E_*(X)$ is projective over $\pi_*(E)$.

To handle even this case, we need some results of the general type of universal coefficient theorems. The reader interested only in the case $X = S$ may without loss omit this section.

In the situation of the universal coefficient theorem, E is the ring-spectrum and F is a module-spectrum over E. $E_*(X)$ is given and the aim is to find information about $F_*(X)$ and $F^*(X)$.

LEMMA 13.1. Let E be a ring-spectrum, F a module-spectrum over E, and X any spectrum. If $E_*(X) = 0$, then $F_*(X) = 0$ and $F^*(X) = 0$.

<u>Proof.</u> $E_*(X) = 0$ is equivalent to $\pi_*(E \wedge X) = 0$, i.e., $E \wedge X$ is contractible. Now any morphism

$$S \xrightarrow{f} F \wedge X$$

can be factored as

$$\begin{array}{ccc} S \wedge F \wedge X & \xrightarrow{i \wedge 1 \wedge 1} & E \wedge F \wedge X \\ {\scriptstyle 1 \wedge f} \uparrow & & \downarrow {\scriptstyle \nu \wedge 1} \\ S & \xrightarrow{f} & F \wedge X \end{array} ,$$

and of course $E \wedge F \wedge X \simeq F \wedge (E \wedge X)$, so it is contractible; hence $f = 0$.

Similarly, any morphism $X \xrightarrow{f} F$ can be factored as

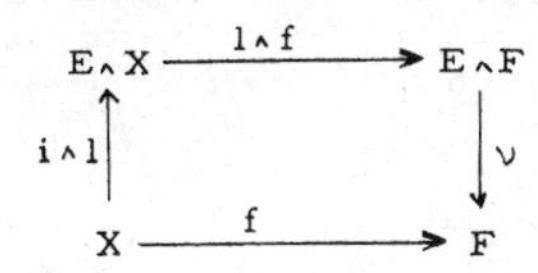

so $f = 0$.

Now observe that for any element $x^* \in F^*(X)$ we get a homomorphism

$$E_*(X) \longrightarrow \pi_*(F) .$$

One way to say it is that this map is

$$x_* \longmapsto \langle x^*, x_* \rangle$$

where we use the pairing

$$F \wedge E \xrightarrow{c} E \wedge F \xrightarrow{\nu} F .$$

Another way to say it is that if $X \xrightarrow{x^*} F$, we form

$$E_*(X) \xrightarrow{(x^*)_*} E_*(F) \xrightarrow{\nu_*} \pi_*(F) .$$

In any case, we get a homomorphism

$$F^*(X) \longrightarrow \mathrm{Hom}_{\pi_*(E)}(E_*(X), \pi_*(F)) .$$

We will be interested in spectra X which satisfy the following condition.

Condition 13.2. $F^*(X) \longrightarrow \mathrm{Hom}^*_{\pi_*(E)}(E_*(X), \pi_*(F))$ is an isomorphism for all module-spectra F over E.

I now introduce a condition on E.

Condition 13.3. E is the direct limit of finite spectra E_α for which $E_*(DE_\alpha)$ is projective over $\pi_*(E)$ and DE_α satisfies 13.2.

Here DE_α means the S-dual of E_α.

PROPOSITION 13.4. Condition 13.3 is satisfied by the following spectra E:

$$S, HZ_p, MO, MU, MSp, K, KO .$$

For the moment I postpone the proof of this proposition; it will be outlined below. Evidently one needs a lemma to say that DE_α satisfies 13.2, but one can impose very restrictive condition on DE_α.

The result we want is as follows.

PROPOSITION 13.5. Suppose E satisfies Condition 13.3 (e.g., E may be one of the examples listed in 13.4). Suppose $E_*(X)$ is projective over $\pi_*(E)$. Then 13.2 holds, i.e.,

$$F^*(X) \longrightarrow \mathrm{Hom}^*_{\pi_*(E)}(E_*(X), \pi_*(F))$$

is an isomorphism for all module-spectra F over E.

This is a special case of a more general result.

THEOREM 13.6. Suppose E satisfies Condition 13.3. Then there is a spectral sequence

$$\mathrm{Ext}^{p,*}_{\pi_*(E)}(E_*(X), \pi_*(F)) \underset{p}{\Longrightarrow} F^*(X)$$

whose edge-homomorphism is the homomorphism

$$F^*(X) \longrightarrow \mathrm{Hom}^*_{\pi_*(E)}(E_*(X), \pi_*(F))$$

considered above, and convergent in the sense that Theorem 8.2 holds.

<u>Proof of 13.5 from 13.6.</u> If $E_*(X)$ is projective over $\pi_*(E)$, then

$$\mathrm{Ext}^{p,*}_{\pi_*(E)}(E_*(X), \pi_*(E))$$

is zero for $p > 0$. Hence, the spectral sequence collapses to its edge-homomorphism. Note that we have enough convergence; condition (ii) of Theorem 8.2 is trivially satisfied, so (i) and (iii) of 8.2 hold.

We now prove some intermediate results necessary to prove Theorem 13.6.

The force of Condition 13.3 is that it allows us to make resolutions of the sort used by Atiyah in his paper on a Künneth theorem for K-theory. Recall that E is the direct limit of finite spectra E_α. The injection $E_\alpha \longrightarrow E$ corresponds to a cohomology class $i_\alpha \in E^0(E_\alpha)$ or to a homology class $g_\alpha \in E_0(DE_\alpha)$.

LEMMA 13.7. For any spectrum X and any class $e \in E_p(X)$ there is an E_α and a morphism $f\colon DE_\alpha \longrightarrow X$ of degree p such that

$e = f_*(g_\alpha)$.

Proof. Take a class $e \in E_p(X)$. Then there is a finite subspectrum $X' \overset{i}{\subset} X$ and a class $e' \in E_p(X')$ such that $i_*(E') = e$. We may interpret e' as a morphism $DX' \longrightarrow E$ of degree p; here I need the fact (not proved in §5) that $D^2Y \simeq Y$. By assumption, this morphism factors through some E_α, so that

$$\begin{array}{ccc} DX' & \longrightarrow & E \\ & \searrow^{\varphi} \quad \nearrow_{i_\alpha} & \\ & E_\alpha & \end{array}$$

and $\varphi^* i_\alpha = e'$ considered as an element of $E^{-p}(DX')$. Dualising back,

$$(D\varphi)_* g_\alpha = e' \in E_p(X') .$$

Take f to be

$$DE_\alpha \xrightarrow{D\varphi} X' \xrightarrow{i} X.$$

LEMMA 13.8. For any spectrum X there exists a spectrum of the form

$$W = \bigvee_\beta S^{p(\beta)} \wedge DE_{\alpha(\beta)}$$

and a morphism $g: W \longrightarrow X$ (of degree 0) such that

$$g_*: E_*(W) \longrightarrow E_*(X)$$

is an epimorphism.

Proof. Immediate from 13.7, by allowing the class e in 13.7 to run over a set of generators for $E_*(X)$.

Note that $W = \bigvee_\beta S^{p(\beta)} \wedge DE_{\alpha(\beta)}$ inherits from its factors the properties that $E_*(W)$ is projective and 13.2 holds, that is

$$F^*(W) \longrightarrow \mathrm{Hom}^*_{\pi_*(E)}(E_*(W), \pi_*(F))$$

is an isomorphism for all module-spectra F over E.

Proof of 13.6. We will construct a resolution of the following form, with the properties listed below.

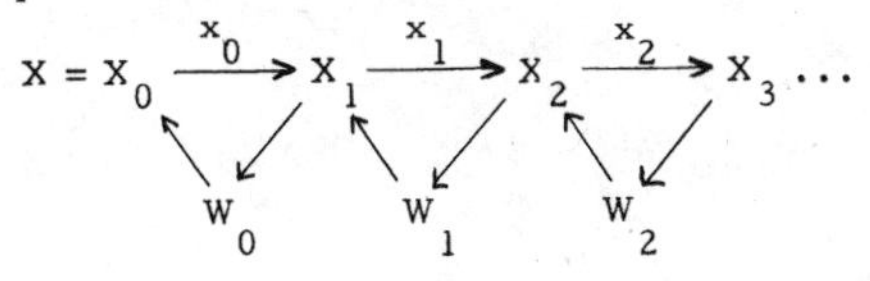

(i) The triangles

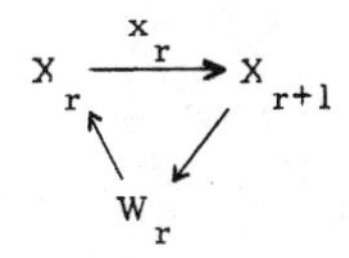

are cofibre triangles.

(ii) For each r,

$$(x_r)_*: E_*(X_r) \longrightarrow E_*(X_{r+1})$$

is zero.

(iii) For each r, $E_*(W_r)$ is projective over $\pi_*(E)$.

(iv) For each r, the map

$$F^*(W_r) \longrightarrow \mathrm{Hom}^*_{\pi_*(E)}(E_*(W_r), \pi_*(F))$$

is an isomorphism.

Let $X_0 = X$. Assume X_r is constructed. By 13.8, there exists a spectrum W_r and a morphism

$$g_r: W_r \longrightarrow X_r$$

as described in 13.8. Form a cofibering

$$W_r \xrightarrow{g_r} X_r \longrightarrow X_{r+1} \longrightarrow W_r$$

where the last morphism has degree -1. Without any essential loss of generality we may suppose by using a telescope that $X_0 \subset X_1 \subset X_2 \subset \ldots$; let X_∞ be their union. Since

$$E_*(W_r) \longrightarrow E_*(X_r)$$

is an epimorphism,

$$E_*(X_r) \longrightarrow E_*(X_{r+1})$$

is zero. Therefore

$$E_*(X_\infty) = \varinjlim_r E_*(X_r) = 0$$

By Lemma 13.1, we have $F^*(X_\infty) = 0$.

By applying F^*, we get a spectral sequence, convergent in the sense that Theorem 8.2 holds. It is convergent to $F^*(X_\infty, X_0) \cong F^*(X_0)$ and has E_1-term

$$E_1^{p,*} = F^*(W_p) .$$

Now we have arranged that

$$F^*(W_r) = \operatorname{Hom}_{\pi_*(E)}(E_*(W_r), \pi_*(F))$$

and

$$0 \longleftarrow E_*(X) \longleftarrow E_*(W_0) \longleftarrow E_*(W_1) \longleftarrow E_*(W_2) \quad \cdots$$

is a resolution of $E_*(X)$ by projective modules over $\pi_*(E)$. Moreover, the boundary d_1 in the spectral sequence is that induced by the boundary in this resolution. Therefore

$$E_2^{p,*} = \operatorname{Ext}^{p,*}_{\pi_*(E)}(E_*(X), \pi_*(F)),$$

as claimed.

It can be checked that the edge-homomorphism is the obvious map.

Now we start work on the proof of Proposition 13.4. We need the following lemma.

LEMMA 13.9. Suppose (i) X is a finite spectrum,

(ii) the spectral sequence

$$H_*(X; \pi_*(E)) \longrightarrow E_*(X)$$

is trivial, i.e., its differentials are zero, and

(iii) for each p, $H_p(X;\pi_*(E))$ is projective as a left module over $\pi_*(E)$.

Then $E_*(X)$ is projective and X satisfies Condition 13.2, i.e.,

$$F^*(X) \longrightarrow \mathrm{Hom}^*_{\pi_*(E)}(E_*(X), \pi_*(F))$$

is an isomorphism for all module-spectra F over E.

(The condition that X is finite is not essential, but is satisfied in the applications.)

In order to apply Lemma 13.9 to DE_α, we simply have to check that

(i) the spectral sequence

$$H^*(E_\alpha;\pi_*(E)) \longrightarrow E^*(E_\alpha)$$

is trivial, and

(ii) for each p, $H^p(E_\alpha;\pi_*(E))$ is projective over $\pi_*(E)$.

<u>Proof of 13.9</u> (from [1], Lecture 1, Prop. 17). Let $E^r_{p,q}(0)$ and $E_r^{p,q}(2)$ be respectively the spectral sequences

$$H^*(X;\pi_*(E)) \Longrightarrow E_*(X)$$

$$H^*(X;\pi_*(F)) \Longrightarrow F^*(X) .$$

It follows immediately from the assumptions on the spectral sequence $E^*_{**}(0)$ that $E_*(X)$ is projective.

The Kronecker product yields a homomorphism

$$E_r^{p*}(2) \longrightarrow \mathrm{Hom}_{\pi_*(E)}(E^r_{p*}(0), \pi_*(F)) .$$

This homomorphism sends d_r into $(d^r)^*$. (This assertion needs detailed proof from the definitions of the spectral sequences, but it can be done using only formal properties of the product and the fact that Hom is left exact.) Because of the assumption that the spectral sequence $E^*_{**}(0)$ is trivial, which is used here, the groups $\mathrm{Hom}_{\pi_*(E)}(E^r_{p*}(0), \pi_*(F))$, equipped with the boundaries $(d^r)^*$ (which happen to be zero) form a (trivial) spectral sequence $E_r^{p,q}(4)$. We now have a map of spectral sequences

$$E_r^{p,q}(2) \longrightarrow E_r^{p,q}(4).$$

For $r = 2$ it becomes the obvious map

$$H^p(X;\pi_*(F)) \longrightarrow \mathrm{Hom}^*_{\pi_*(E)}(H_p(X;\pi_*(E)),\pi_*(F)) .$$

Since we are assuming $H_p(X;\pi_*(E))$ is projective over $\pi_*(E)$, a theorem on ordinary homology shows that for $r = 2$ the map is an isomorphism. Therefore it is an isomorphism for all finite r, and the spectral sequence $E_r^{p,q}(2)$ is trivial. Since X is a finite spectrum, it is easy to deduce that the map

$$E_\infty^{p,*}(2) \longrightarrow \mathrm{Hom}^*_{\pi_*(E)}(E^\infty_{p*}(0),\pi_*(F))$$

is an isomorphism, because the limit is attained for some finite value of r.

Let us now introduce notation for the filtration quotient groups, say

$$G_{p*}(0) = \mathrm{Im}(E_*(X^p) \longrightarrow E_*(X))$$

$$G^{p*}(2) = \mathrm{Coim}(F^*(X) \longrightarrow F^*(X^p)).$$

The Kronecker product yields a homomorphism

$$G^{p*}(2) \longrightarrow \mathrm{Hom}^*_{\pi_*(E)}(G_{p*}(0),\pi_*(F)).$$

(Again, the verification uses formal properties of the product and the fact that Hom is left exact.) Consider the following diagram.

$$\begin{array}{ccc}
0 & & 0 \\
\downarrow & & \downarrow \\
E_\infty^{p*}(2) & \longrightarrow & \mathrm{Hom}^*_{\pi_*(E)}(E^\infty_{p*}(0),\pi_*(F)) \\
\downarrow & & \downarrow \\
G^{p*}(2) & \longrightarrow & \mathrm{Hom}^*_{\pi_*(E)}(G_{p*}(0),\pi_*(F)) \\
\downarrow & & \downarrow \\
G^{p-1\,*}(2) & \longrightarrow & \mathrm{Hom}^*_{\pi_*(E)}(G_{p-1\,*}(0),\pi_*(F) \\
\downarrow & & \downarrow \\
0 & & 0
\end{array}$$

The second column is exact because $E^{\infty}_{p*}(0)$ is projective. Induction over p, using the short five lemma, now shows that

$$G^{p*}(2) \longrightarrow \mathrm{Hom}^{*}_{\pi_*(E)}(G_{p*}(0), \pi_*(F))$$

is an isomorphism. Since X is a finite spectrum, in a finite number of steps we obtain the result that

$$F^*(X) \longrightarrow \mathrm{Hom}_{\pi_*(E)}(E_*(X), \pi_*(F))$$

is an isomorphism.

We now sketch the proof of 13.4. (See [1], pp. 29-30.)

(i) $E = S$, the sphere spectrum. Take $E_\alpha = S$; then 13.3 may be verified directly.

(ii) $E = HZ_p$. The hypotheses of 13.9 are satisfied by any X, and it is sufficient to let E_α run over any system of finite spectra whose limit is HZ_p.

(iii) $E = MO$. It is well known that

$$MO \simeq \bigvee_i S^{n(i)} HZ_2 \simeq \prod_i S^{n(i)} HZ_2 .$$

The hypotheses of 13.9 are satisfied by any X, and it is sufficient to let E_α run over any system of finite spectra whose limit is MO.

(iv) $E = MU$. We have $H^p(MU; \pi_q(MU)) = 0$ unless p and q are even. Therefore the spectral sequence

$$H^*(MU; \pi_*(MU)) \Longrightarrow MU^*(MU)$$

is trivial. Again, $H^p(MU; \pi_*(MU))$ is free over $\pi_*(MU)$. It is sufficient to let E_α run over a system of finite spectra which approximate MU in the sense that

$$i_*: H_p(E_\alpha) \longrightarrow H_p(MU)$$

is an isomorphism for $p \leq n$, while $H_p(E_\alpha) = 0$ for $p > n$.

(v) $E = MSp$. A simple adaptation of the method of S. P. Novikov [12, 13] from the unitary to the symplectic case shows that the spectral sequence

$$H^*(MSp; \pi_*(MSp)) \Longrightarrow MSp^*(MSp)$$

is trivial. Again, $H^p(MSp; \pi_*(MSp))$ is free over $\pi_*(MSp)$. The rest of the argument is as in (iv).

(vi) $E = K$. Recall that in the spectrum K every even term is the space BU. We have

$$H^p(BU; \pi_q(K)) = 0 \text{ unless } p \text{ and } q \text{ are even.}$$

Therefore the spectral sequence

$$H^*(BU; \pi_*(K)) \Longrightarrow K^*(BU)$$

is trivial. Again, $H^p(BU; \pi_*(K))$ is free over $\pi_*(K)$. It is sufficient to let E_α run over a system of finite spectra which approximate, as in (iv), the different spaces BU of the spectrum K.

(vii) $E = KO$. Recall that in the spectrum KO, every eighth term is the space BSp. I claim that the spectral sequence

$$H^*(BSp; \pi_*(KO)) \Longrightarrow KO^*(BSp)$$

is trivial. In fact, for each class $h \in H^{8p}(BSp(m))$ we can construct a real representation of $Sp(m)$ whose Chern character begins with h; for each class $h \in H^{8p+4}(BSp(m))$ we can construct a symplectic representation of $Sp(m)$ whose Chern character begins with h. The rest of the argument is as for (vi).

14. A CATEGORY OF FRACTIONS

We recall that our general object in these sections is to answer the following question. Suppose given $E_*(X)$ and $E_*(Y)$. What can we say about $[X, Y]_*$?

Now it is clear that we cannot say everything. For example, suppose $E = HZ_2$; given $(HZ_2)_*(X)$ and $(HZ_2)_*(Y)$ there is no hope of finding out anything about the odd torsion in $[X, Y]_*$.

More generally, we will say that a morphism $f: X \longrightarrow X'$ is an E-equivalence if the induced homomorphism

$$f_*: E_*(X) \longrightarrow E_*(X')$$

is an isomorphism. This can happen without f being an equivalence; for example, take $E = HZ_2$, $X = HZ_3$, $X' = pt$. Then it is clear that methods based on E-homology cannot tell X and X' apart.

It therefore seems best to introduce a new category in which one does not attempt to tell X and X' apart. In technical terms I have to start from the stable category and define a category of fractions.

(Added later.) I owe to A.K. Bousfield the remark that the procedure below involves very serious set-theoretical difficulties. Therefore it will be best to interpret this section not as a set of theorems, but as a programme, that is, as a guide to what one might wish to prove.

Let C be the stable category already constructed.

THEOREM 14.1. There exists a category F, called the category of fractions, and a functor

$$T: C \longrightarrow F$$

with the following properties.

(i) If $e: X \longrightarrow Y$ is an E-equivalence in C, then $T(e)$ is an actual equivalence in F, i.e., it has an inverse $T(e)^{-1}$.

(ii) T is universal with respect to this property; given a category G and a functor $U: C \longrightarrow G$ such that e an E-equivalence implies $U(e)$ is an equivalence in G, then there exists one and only one functor $V: F \longrightarrow G$ such that $U = VT$.

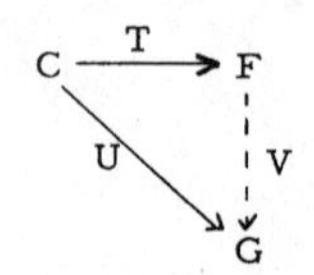

(iii) The objects of F are the same as the objects of C, and T is the identity on objects.

(iv) Every morphism in F from X to Y can be written $T(e)^{-1}T(f)$, where $f: X \longrightarrow Y'$ and $e: Y \longrightarrow Y'$ are in C and e is an E-equivalence.

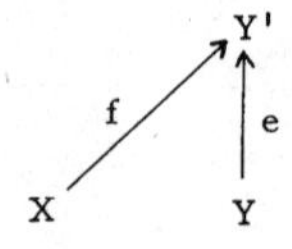

We have $T(e_1)^{-1}T(f_1) = T(e_2)^{-1}T(f_2)$ in F if and only if there exists a diagram of the following form in C.

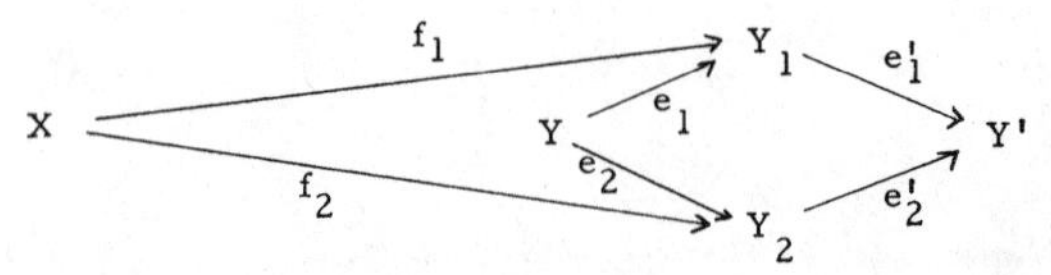

(v) Every morphism in F from X to Y can be written $T(f)\,T(e)^{-1}$, where $f: X' \longrightarrow Y$ and $e: X' \longrightarrow X$ are in C and e is an E-equivalence.

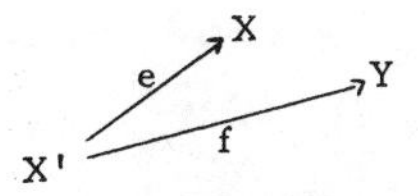

We have $T(f_1)T(e_1)^{-1} = T(f_2)T(e_2)^{-1}$ in F if and only if there exists a diagram of the following form in C.

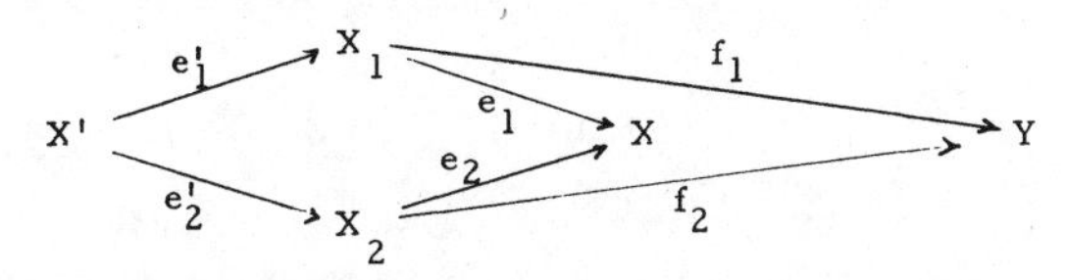

If one takes only parts (i), (ii), and (iii) the theorem is almost empty; such a category of fractions exists under negligible assumptions. (Added later: unfortunately there is no reason why the result should be a small category.) Our object, of course, is to construct F in such a way that we obtain a good hold on it. Parts (iv) and (v) essentially describe two ways of constructing F. We shall write $[X, Y]_*^E$ to mean the morphisms from X to Y in the category F. Besides constructing F, we must also give results calculating $[X, Y]_*^E$ in various cases which arise in the applications. When we construct the Adams spectral sequence, based on the homology theory E_* we will try to prove that it converges to $[X, Y]_*^E$.

Before proving 14.1, I will finish stating some results which help to show what F is.

We propose to get a hold on $[X, Y]^E$ by showing that if we keep Y fixed and vary X, then we get a functor of X which is representable in C. Then we give means for recognizing the representing object, and finally we construct the representing object in an elementary way in special cases.

PROPOSITION 14.2. The following conditions on Y are equivalent.

(i) $T: [X, Y]_* \longrightarrow [X, Y]^E_*$ is an isomorphism for all X.

(ii) If $E_*(X) = 0$, then $[X, Y]_* = 0$.

If these equivalent conditions hold, we say that Y is E-complete. This term can be justified by inspecting the special case $E = HZ_p$, which will be considered later.

As an example, we give:

COROLLARY 14.3. If Y is an E-module spectrum, then Y is E-complete and

$$T: [X, Y]_* \longrightarrow [X, Y]^E_* \text{ is an isomorphism.}$$

Proof. (from 14.2). Condition (ii) of 14.2 holds by 13.1.

THEOREM 14.4. (i) For any spectrum Y there is an E-equivalence $e: Y \longrightarrow Z$ such that Z is E-complete.

(ii) Such an E-equivalence is universal. That is, given any other E-equivalence $e': Y \longrightarrow Z'$, there exists a unique $f: Z' \longrightarrow Z$ such that $fe' = e$.

$$\begin{array}{ccc} & \overset{e'}{\nearrow} & Z' \\ Y & & \downarrow f \\ & \underset{e}{\searrow} & Z \end{array}$$

(iii) Therefore, Z is unique up to canonical equivalence.

(iv) For such a Z we have an isomorphism

$$[X, Z]_* \longrightarrow [X, Y]^E_* .$$

given by $f \longmapsto T(e)^{-1} T(f)$.

Notes. (iii) follows immediately from (ii). Since Z is defined up to canonical equivalence by Y, we may write it as a function of Y; we choose the notation $Z = Y^E$, so that

$$[X, Y^E]_* = [X, Y]^E_* .$$

We will call Y^E the E-completion of Y. Again, the term can be justified by considering the special case $E = HZ_p$. Note that $(Y^E)^E = Y^E$, so that the term "completion" is justified.

We say that X is <u>connective</u> if there exists $n_0 \in Z$ such that $\pi_r(X) = 0$ for $r < n_0$.

PROPOSITION 14.5. Suppose that E is a commutative ring-spectrum and $\pi_r(E) = 0$ for $r < 0$; suppose also that Y is connective. Then $[X, Y]_*^E$ depends only on the ring $\pi_0(E)$.

For example, $[X, Y]_*^E$ is the same whether $E = MUQ_p$ or $E = buQ_p$. The idea is that under these hypotheses, the difference between $[X, Y]_*^E$ and $[X, Y]_*$ is essentially arithmetical.

For the next result, we assume that E is a commutative ring-spectrum, that $\pi_r(E) = 0$ for $r < 0$, and Y is connective.

THEOREM 14.6. (i) Suppose $\pi_0(E)$ is a subring R of the rationals. Then

$$Y^E = YR.$$

(ii) Suppose $\pi_0(E) = Z_m$ and $\pi_r(Y)$ is finitely generated for all r. Then

$$Y^E = YI_m ,$$

where I_m is the ring of m-adic integers, $\underset{\overleftarrow{r}}{\mathrm{Lim}}\, Z_{m^r}$.

(iii) Suppose $\pi_0(E) = Z_m$ and the identity morphism $1: Y \longrightarrow Y$ satisfies $m^e \cdot 1 = 0$. Then

$$Y^E = Y.$$

Examples. (ia) Suppose $\pi_0(E) = Z$, then $Y^E = Y$ and

$$T: [X, Y]_* \rightarrow [X, Y]_*^E \text{ is an isomorphism.}$$

(ib) Suppose $\pi_0(E)$ is a subring R of the rationals and X is a finite spectrum. Then

$$[X, Y]_*^E = [X, YR]_* = [X, Y]_* \otimes R \quad \text{by 6.7.}$$

(iia) Suppose $\pi_0(E) = Z_m$, $\pi_r(Y)$ is finitely generated for all r and X is a finite spectrum. Then

$$[X, Y]_*^E = [X, YI_m]_* = [X, Y]_* \otimes I_m \quad \text{by 6.7.}$$

(iib) Take m to be a prime p, and take X = Y = S. Then

$$[S, S]_r^E = \begin{cases} 0 & (r < 0) \\ I_p & (r = 0) \\ \text{the p-component of } [S, S]_r & \text{if } r > 0 . \end{cases}$$

It is very plausible that the classical Adams spectral sequence should converge to these groups.

<u>Warnings.</u> (i) We have assumed that $\pi_r(E) = 0$ for $r < 0$. If we do not have this, the relationship between $[X, Y]_*$ and $[X, Y]_*^E$ may be much more distant. For example, take E = K; it can be shown that

$[S, S]_r^K \neq 0$ for infinitely many negative values of r.

(ii) Consider parts (ii) and (iii) of the theorem, in which $\pi_0(E) = Z_m$. Results of the form given do require some assumption on Y beyond the fact that it is connective. For example, take Y = S(Q/Z). It can be shown that

$$[S, Y]_1 = 0 \quad \text{and so} \quad [S, Y]_1 \otimes I_m = 0, \text{ but } [S, Y]_1^E = I_m .$$

If one takes m to be a prime p and checks the behaviour of the classical Adams spectral sequence based on $E = HZ_p$, one sees that it converges to $[S, Y]_1^E$, as indeed it must do by the theorem to be proved in the next section. So something which was previously a counterexample can now be used as evidence to support the theory.

The proof of Theorem 14.1 requires two lemmas.

LEMMA 14.7 (i) Suppose given a diagram

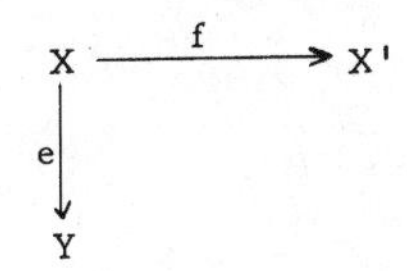

in which e is an E-equivalence. Then we can complete it to a commutative diagram

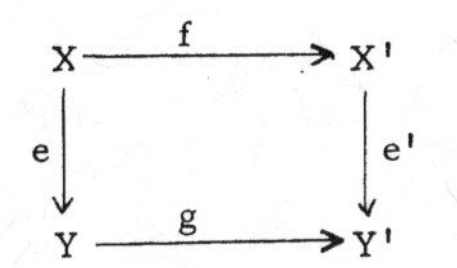

in which e' is an E-equivalence. If f is also an E-equivalence, so is g.

(ii) Suppose given a diagram

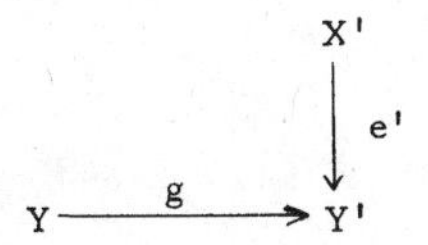

in which e' is an E-equivalence. Then we can complete it to a commutative diagram

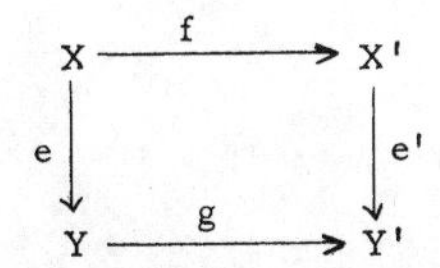

in which e is an E-equivalence. If g is also an E-equivalence, so is f.

<u>Proof.</u> (i) Let W be the fibre of $X \longrightarrow X'$, and let Y' be the cofibre of $W \longrightarrow Y$. The morphism e' is an E-equivalence by the five lemma. Part (ii) is similar.

LEMMA 14.8. (i) Suppose given

$$X' \xrightarrow{e} X \underset{g}{\overset{f}{\rightrightarrows}} Y$$

where e is an E-equivalence and $fe = ge$. Then we can construct

$$X \underset{g}{\overset{f}{\rightrightarrows}} Y \xrightarrow{e'} Y'$$

with e' an E-equivalence and $e'f = e'g$.

(ii) Suppose given

$$X \underset{g}{\overset{f}{\rightrightarrows}} Y \xrightarrow{e'} Y'$$

where e' is an E-equivalence and $e'f = e'g$. Then we can construct

$$X' \xrightarrow{e} X \underset{g}{\overset{f}{\rightrightarrows}} Y$$

with e an E-equivalence and $fe = ge$.

<u>Proof.</u> The proof is a manipulation with cofiberings using Verdier's axiom (6.8) and is left as an exercise.

Now, to construct F, let the objects of F be the same as the objects of C. To define morphisms in F, say $[X, Y]^E$, one makes a preliminary construction. Fix Y, and consider the category in which the objects are E-equivalences $Y \xrightarrow{e'} Y'$ and morphisms are diagrams of the following form.

$$\begin{array}{ccc} & \overset{e'}{\nearrow} & Y' \\ Y & & \downarrow \\ & \underset{e''}{\searrow} & Y'' \end{array}$$

Then 14.7 and 14.8 say that we get a directed category in the sense of Grothendieck. That is, given two objects A and B, there exists

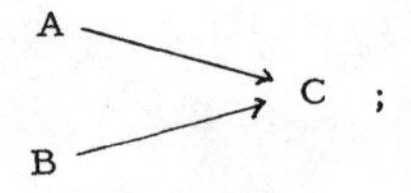

given two morphisms $A \underset{g}{\overset{f}{\rightrightarrows}} B$, there exists $A \underset{g}{\overset{f}{\rightrightarrows}} B \xrightarrow{h} C$ where $hf = hg$.

We define $[X, Y]_*^E = \underrightarrow{\text{Lim}}\, [X, Y']_*$, where the direct limit takes place over this directed category. An element of $\underrightarrow{\text{Lim}}\, [X, Y']$ is an equivalence class of diagrams

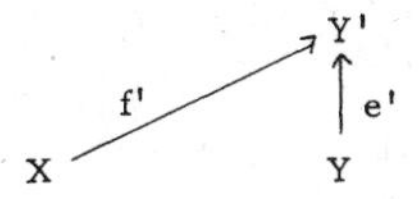

in which e' is an E-equivalence. Two such diagrams are equivalent if and only if there exists a diagram of the following form.

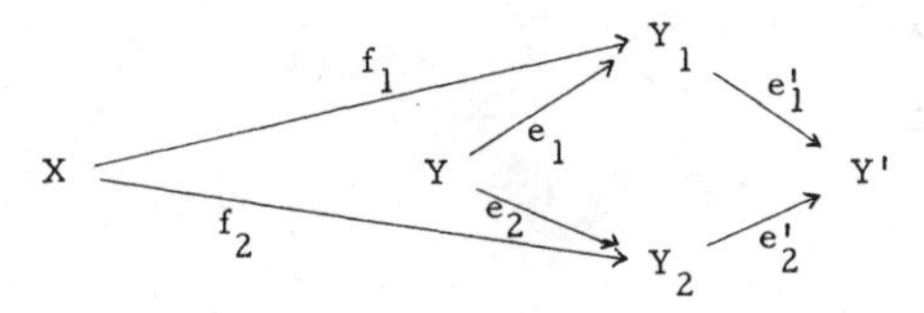

This is essentially the construction presented in (iv). To check that this is an equivalence relation one uses 14.7 (i).

To define composition in the category, suppose given the two diagram diagrams shown below with undotted arrows.

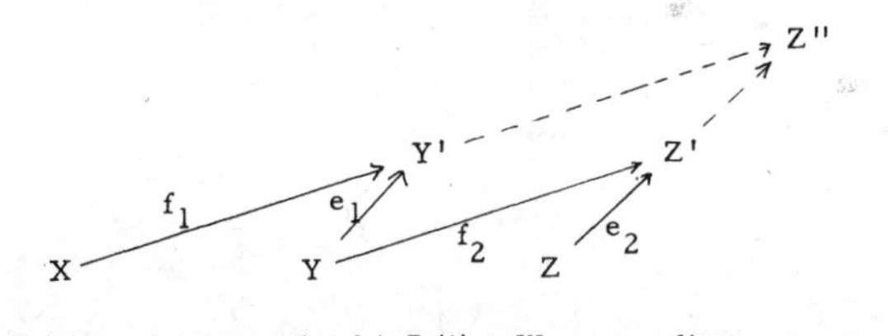

Add the dotted arrows by 14.7 (i). We get a diagram representing a morphism from X to Z in the new category. We check that the equivalence class of this diagram depends only on the equivalence classes of the factors, not on the choice of parallelogram (use 14.7 (i), 14.8 (i)).

We check the associativity law and the existence of identity morphisms. We now have a category F. We define $T: C \longrightarrow F$ as follows: if $f: X \longrightarrow Y$, let $T(f)$ be the class of the following diagram.

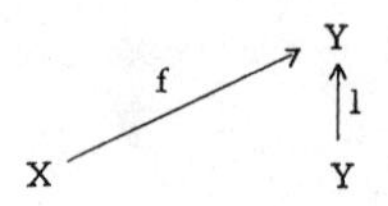

One checks that this is a functor. It is now almost trivial to verify properties (i)-(iv) of the theorem.

On the other hand, precisely the dual construction works using 14.7 (ii) and 14.8 (ii) to show that one can construct F so as to have properties (i)-(iii) and (v). But of course F is characterised by (i)-(iii), so it must have both properties (iv) and (v).

Now we turn to Proposition 14.2. First, suppose $E_*(X) = 0$. Then it is clear that the morphism pt. $\longrightarrow X$ is cofinal among E-equivalences $e': X' \longrightarrow X$. So we have $[X, Y]_* = 0$. If we assume that $T: [X, Y]_* \longrightarrow [X, Y]_*^E = 0$ is an isomorphism, then clearly we deduce that $[X, Y]_* = 0$. So condition (i) of 14.2 implies condition (ii). The proof that (ii) implies (i) will be given together with the proof of part of Theorem 14.4 to be considered below. This requires three lemmas, numbered 14.9, 14.10, and 14.11.

LEMMA 14.9. Let $A \longrightarrow B \longrightarrow C$ be a cofibering. Then

$$[A, Y]_*^E \longleftarrow [B, Y]_*^E \longleftarrow [C, Y]_*^E$$

and

$$[X, A]_*^E \longrightarrow [X, B]_*^E \longrightarrow [X, C]_*^E$$

are exact.

<u>Proof.</u> For any Y', the sequence

$$[A, Y']_* \longleftarrow [B, Y']_* \longleftarrow [C, Y']_*$$

is exact. The given sequence is obtained from such sequences by passing to a direct limit. But direct limits over a directed category preserve exactness. The same form of argument holds for the second sequence, using the fact that we can also define $[X, Y]_*^E$ by taking a direct limit of $[X', Y]_*$ as we vary X'.

LEMMA 14.10. The canonical map

$$[\bigvee_\alpha X_\alpha, Y]_*^E \longrightarrow \prod_\alpha [X_\alpha, Y]_*^E$$

is an isomorphism.

<u>Proof.</u> (i) Suppose given an element in $\prod_\alpha [X_\alpha, Y]^E$; each of its components is represented by a diagram

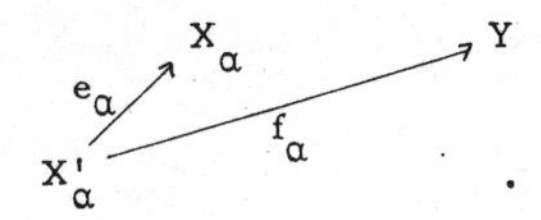

Then we can form the diagram

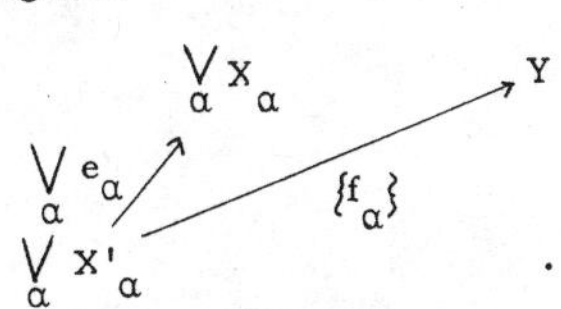

This gives an element of $[\bigvee_\alpha X_\alpha, Y]_*^E$ which maps in the required way.

(ii) Suppose given an element of $[\bigvee_\alpha X_\alpha, Y]^E$, say represented by

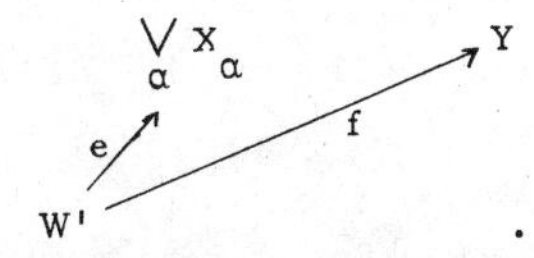

Suppose it restricts to zero in each $[X_\alpha, Y]_*^E$. This says that for each α we have a commutative diagram of the following form;

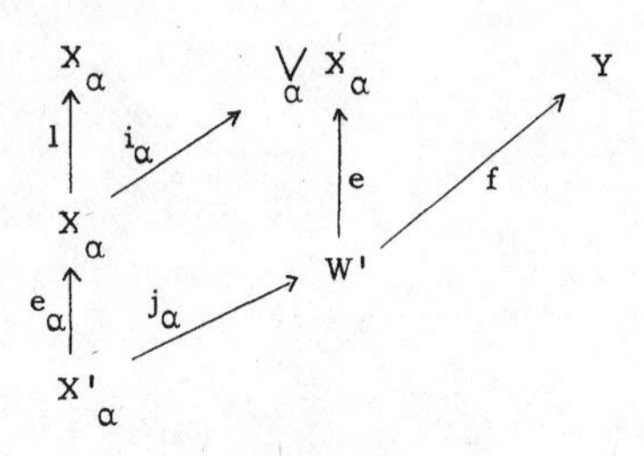

and moreover, $fj_\alpha = 0$. Then consider the following diagram.

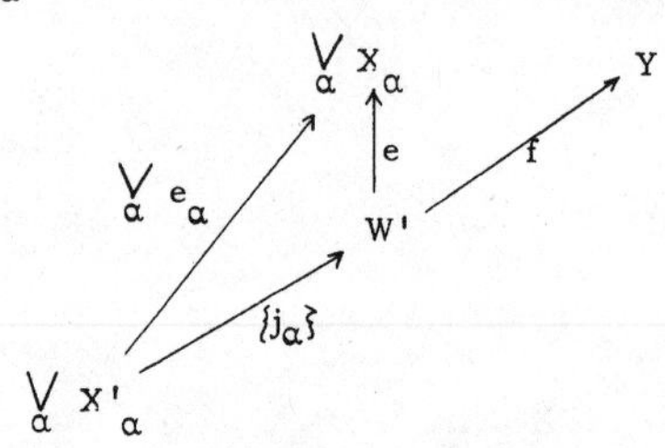

This shows that the diagram

$$\begin{array}{ccc} & \bigvee_\alpha X_\alpha & \longrightarrow Y \\ e \nearrow & & \nearrow f \\ W' & & \end{array}$$

gives the zero element of $[\bigvee_\alpha X_\alpha, Y]_*^E$.

Now we start the proof of 14.4. Consider $[X, Y]_*^E$. Hold Y fixed and vary X. By 14.9 and 14.10, we have the data for E.H. Brown's Theorem, and we deduce that $[X, Y]_*^E$ is a representable functor of X. That is, there is a spectrum Z and a natural transformation

$$U: [X, Z]_* \xrightarrow{\cong} [X, Y]_*^E.$$

Here Z satisfies condition (ii) of 14.2. For suppose $E_*(X) = 0$; then $[X, Y]_*^E = 0$, as we have remarked; so $[X, Z]_* = 0$, since U is an isomorphism.

Now consider $1 \in [Z, Z]$ and $U(1) \in [Z, Y]^E$. The latter is represented by a diagram

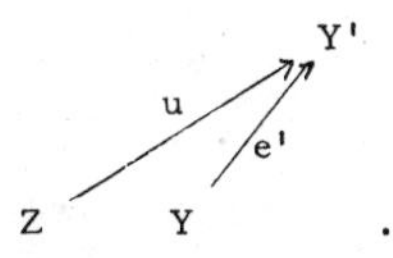

Extend this to a cofibre sequence

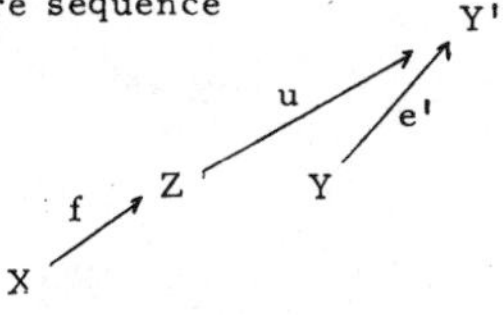

Then by naturality $U(f) = f^* U(1) = 0$. Since U is a monomorphism, $f = 0$. Therefore the morphism $Z \xrightarrow{u} Y'$ is equivalent to the injection $Z \longrightarrow Z \vee \mathrm{Susp}(X)$; we can replace the representative for $U(1)$ by the following diagram.

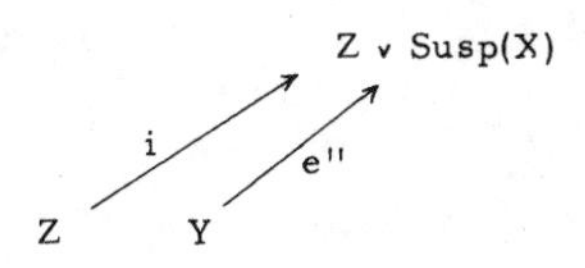

Now consider $1 \in [Y, Y]_*^E$; there exists $\epsilon: Y \longrightarrow Z$ such that $U(\epsilon) = 1 \in [Y, Y]^E$. That is, we have the following commutative diagram.

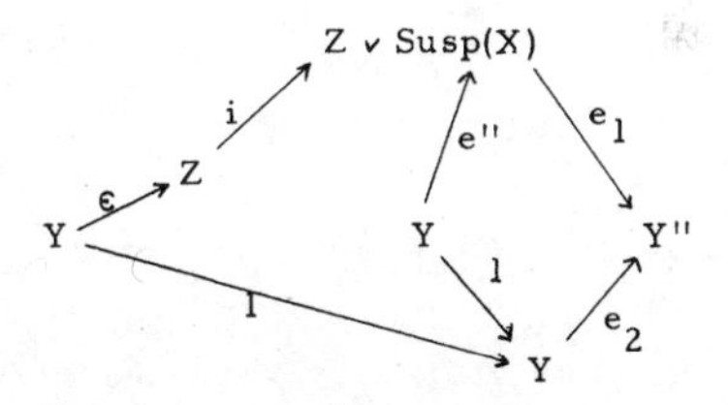

We conclude that $i_*: E_*(Z) \longrightarrow E_*(Z \vee \mathrm{Susp}(X))$ is an epimorphism. Therefore $E_*(\mathrm{Susp}(X)) = 0$. Hence, $i_*: E_*(Z) \longrightarrow E_*(Z \vee \mathrm{Susp}(X))$ and $\epsilon_*: E_*(Y) \longrightarrow E_*(Z)$ are isomorphisms.

Since we now know that $\epsilon: Y \longrightarrow Z$ is an E-equivalence, we allow ourselves to change its name to $e: Y \longrightarrow Z$. We have proved that any spectrum Y admits an E-equivalence $e: Y \longrightarrow Z$, where $E_*(X) = 0$ implies $[X, Z]_* = 0$.

We will now forget everything about Z except these two properties.

LEMMA 14.11. Suppose $e: Y \longrightarrow Z$ is an E-equivalence, and $E_*(X) = 0$ implies $[X, Z]_* = 0$. Then 14.4 (ii) and (iv) hold.

This will complete the proof of Proposition 14.2; for we take $e: Y \longrightarrow Z$ to be $1: Y \longrightarrow Y$, and deduce that

$$T: [X, Y]_* \longrightarrow [X, Y]_*^E$$

is an isomorphism. Moreover, it will obviously complete the proof of 14.4.

Proof of 14.11. We have to show that $e: Y \longrightarrow Z$ is universal. Suppose given an E-equivalence $e': Y \longrightarrow Z'$. Then up to equivalence we have $Z' = Y \cup_g CA$ for some $g: A \longrightarrow Y$; and here $E_*(A) = 0$, by the exact sequence of the cofibering $A \longrightarrow Y \xrightarrow{e'} Z'$.

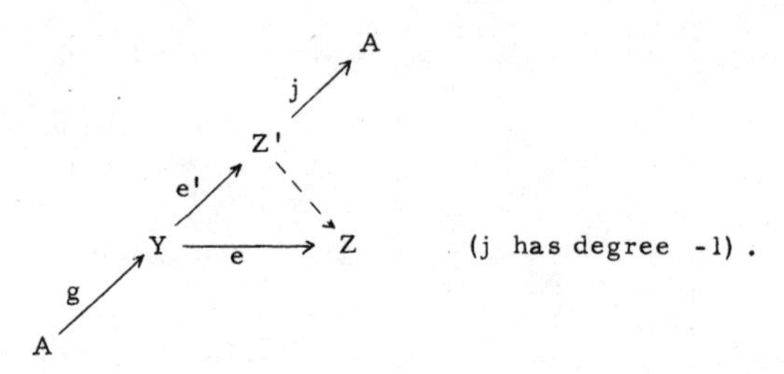

Then $eg = 0$ by the assumed property of Z, so e extends over $Y \cup_g CA$ and there is a map $f: Z' \longrightarrow Z$ with $fe' = e$. Also f is unique, because two choices differ by an element of $j^*[A, Z]_*$, and $[A, Z]_* = 0$ by the assumed property of Z.

This shows that $e: Y \longrightarrow Z$ is universal. Then clearly the single object $Y \xrightarrow{e} Z$ is cofinal in the directed category used to construct

$[X, Y]_*^E$; so we have an isomorphism

$$[X, Z]_* \longrightarrow [X, Y]_*^E ,$$

given by assigning to a morphism $f: X \longrightarrow Z$ the class of the diagram

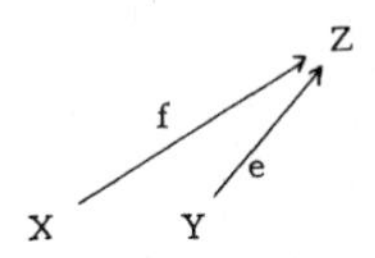

i.e., the element $T(e)^{-1}T(f) \in [X, Y]_*^E$. This completes the proof of 14.2, 14.3, and 14.4.

Now we start working toward the proof of 14.5.

LEMMA 14.12. Suppose $\pi_r(E) = 0$ for $r < 0$. Suppose a morphism $f: X \longrightarrow Y$ induces an isomorphism $E_*(X) \longrightarrow E_*(Y)$. Then it induces an isomorphism $H_*(X, \pi_0(E)) \longrightarrow H_*(Y; \pi_0(E))$.

Proof. First a remark. Let E be any spectrum, not necessarily a ring-spectrum, and not necessarily connective; then I claim

$$H \wedge E \simeq \bigvee_i S^i \wedge HG_i ,$$

where $G_i = H_i(E)$. In fact, for each i we can construct a Moore spectrum S^iG_i; then we can construct a morphism

$$S^iG_i \xrightarrow{a_i} H \wedge E$$

inducing the identity map

$$G_i = \pi_i(S^iG_i) \longrightarrow \pi_i(H \wedge E) = G_i .$$

Now we can form

$$H \wedge (S^iG_i) \xrightarrow{1 \wedge a_i} H \wedge H \wedge E \xrightarrow{\mu \wedge 1} H \wedge E .$$

Finally we form,

$$\bigvee_i H \wedge (S^iG_i) \xrightarrow{\{(\mu \wedge 1)(1 \wedge a_i)\}} H \wedge E .$$

This induces an isomorphism of homotopy groups, so it is an equivalence by the theorem of J.H.C. Whitehead.

Now we return to the lemma. Form the cofibering $X \xrightarrow{f} Y \rightarrow Z$. Then we have $E_*(Z) = 0$ and it is sufficient to deduce that $H_*(Z; \pi_0(E)) = 0$. Since $\pi_*(E \wedge Z) = 0$, $E \wedge Z$ is contractible. Therefore $H \wedge E \wedge Z$ is contractible. Now $\pi_r(E) = 0$ for $r < 0$, so by the Hurewicz theorem $G_0 = H_0(E) = \pi_0(E)$. We have just shown that HG_0 is a direct summand in $H \wedge E$, so $(HG_0) \wedge Z$ is contractible; that is, $H_*(Z; \pi_0(E)) = 0$. This proves the lemma.

LEMMA 14.13. Suppose E is a commutative ring-spectrum and $\pi_r(E) = 0$ for $r < 0$. Suppose X and Y are connective and $f: X \rightarrow Y$ induces an isomorphism $H_*(X; \pi_0(E)) \rightarrow H_*(Y, \pi_0(E))$. Then it induces an isomorphism $E_*(X) \rightarrow E_*(Y)$.

Proof. As before, we form the cofibering $X \xrightarrow{f} Y \rightarrow Z$. Then Z is connective; we have $H_*(Z; \pi_0(E)) = 0$, and it is sufficient to prove $E_*(Z) = 0$.

Since $\pi_0(E)$ is a commutative ring and $\pi_r(E)$ is a module over $\pi_0(E)$, we have the universal coefficient theorem in the form of the spectral sequence

$$\mathrm{Tor}^{\pi_0(E)}_{p*}(H_*(Z; \pi_0(E)), \pi_r(E)) \underset{p}{\Longrightarrow} H_*(Z; \pi_r(E)) .$$

This is a quarter-plane spectral sequence convergent in the naive sense. We see that $H_*(Z; \pi_r(E)) = 0$. We now consider the Atiyah-Hirzebruch spectral sequence

$$H_p(Z; \pi_q(E)) \underset{p}{\Longrightarrow} E_{p+q}(Z) .$$

This is a quarter-plane spectral sequence convergent in the naive sense. We conclude that $E_*(Z) = 0$.

Warning. This condition that X and Y are connective cannot be omitted (take $E = bu$, $X = pt.$, $Y = BUZ_p$ or vice versa).

Proof of 14.5. Recall that we wish to show that if E is a commutative ring-spectrum and $\pi_r(E) = 0$ for $r < 0$, then for any connective spectrum Y, $[X, Y]_*^E$ depends only on $\pi_0(E)$. More precisely, we show that $[X, Y]_*^E = [X, Y]_*^{E'}$, where $E' = H\pi_0(E)$.

(i) By 14.12, we have that any morphism $f: Y \longrightarrow Y'$ which induces an isomorphism in E-homology also induces an isomorphism in E'-homology.

(ii) Consider the directed category used in the construction of $[X, Y]_*^{E'}$. I claim that morphisms $f: Y \longrightarrow Y'$ which induce an isomorphism in E-homology are cofinal in those which induce an isomorphism in E'-homology. Once this is proved, 14.5 follows. We need a lemma.

LEMMA 14.14. Let Y be a connective spectrum, X any spectrum. Then any morphism $f: X \longrightarrow Y$ factors as

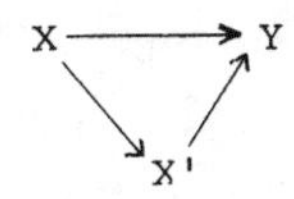

where X' is connective and $H_r(X') \begin{cases} \cong H_r(X) & (r \geq N) \\ = 0 & (r < N) \end{cases}$ for some $N \in Z$ depending only on Y.

Proof. Let N be such that $\pi_r(Y) = 0$ for $r < N + 1$. Then we can factor f through X/X^{N-1}; this spectrum is connective. However, it need not have the desired properties in homology. We have

$$H_r(X/X^{N-1}) \cong H_r(X) \qquad (r > N)$$
$$H_r(X/X^{N-1}) = 0 \qquad (r < N) ,$$

and in dimension N we have an exact sequence

$$0 \longrightarrow H_N(X) \longrightarrow H_N(X/X^{N-1}) \longrightarrow F \longrightarrow 0 ,$$

where F is free since it is a subgroup of $H_{N-1}(X^{N-1})$. By the Hurewicz theorem, we have

$$\pi_N(X/X^{N-1}) \cong H_N(X/X^{N-1}) .$$

Choose a set of elements

$$\theta_\alpha \in \pi_N(X/X^{N-1})$$

which project to a base of F, and form

$$X' = X/X^{N-1} \cup_{\theta_\alpha} CS^N .$$

X' is connective, and $X/X^{N-1} \longrightarrow Y$ factors through X'. We have

$$H_r(X') \begin{cases} \cong H_r(X) & (r \geq N) \\ = 0 & (r < N) . \end{cases}$$

Returning to (ii) above, suppose $f: Y \longrightarrow Y'$ induces an isomorphism in E'-homology. Form a cofibre sequence

$$A \longrightarrow Y \xrightarrow{f} Y' \longrightarrow \ldots .$$

Here $H_r(A;\pi_0(E)) = 0$, and so by the ordinary universal coefficient theorem,

$$H_r(A) \otimes_Z \pi_0(E) = 0 , \qquad \mathrm{Tor}_1^Z(H_r(A), \pi_0(E)) = 0.$$

By 14.14, we can factor $A \longrightarrow Y$ in the form

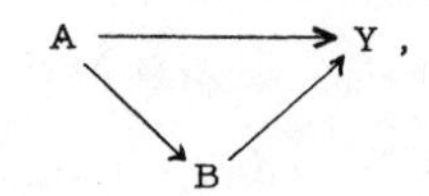

where B is connective and

$$H_r(A) \xrightarrow{\cong} H_r(B) \qquad \text{for } r \geq N$$

$$H_r(B) = 0 \qquad \text{for } r < N.$$

Then

$$H_r(B) \otimes_Z \pi_0(E) = 0, \quad \mathrm{Tor}_1^Z(H_r(A), \pi_0(E)) = 0$$

and so

$$H_r(B; \pi_0(E)) = 0$$

for all r. Now we can form the following diagram of cofiberings.

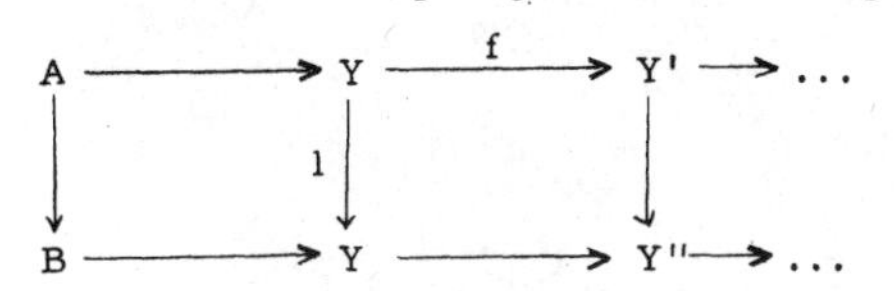

Here Y'' is connective, and $Y \longrightarrow Y''$ is an E'-equivalence, so it is an E-equivalence by 14.13. This completes the proof of (iii) above, and so completes the proof of (14.5).

Now we turn to Theorem 14.6. We have to take YR, or YI_m, or Y, according to the case, and show that it satisfies the conditions in 14.4. We have already shown that it will be sufficient to check 14.4 (i), that is to say that these spectra are E-equivalent to Y, under the hypotheses given for each case, and E-complete.

Consider the first condition. In case (i), suppose $\pi_0(E)$ is a subring R of the rationals Z. Consider the product

$$SR \wedge S(R/Z) .$$

By the Künneth theorem we have

$$H_*(SR \wedge S(R/Z)) = 0;$$

for $R \otimes_Z (R/Z) = 0$, $\mathrm{Tor}_1^Z(R, R/Z) = 0$, The spectrum is connective, so $SR \wedge S(R/Z)$ is contractible by the theorem of J. H. C. Whitehead.

Now we have a cofibering

$$Y \longrightarrow YR \longrightarrow Y \wedge S(R/Z) .$$

Here we have

$$HR_*(Y \wedge S(R/Z)) = \pi_*(H \wedge SR \wedge Y \wedge S(R/Z)) = 0,$$

for $H \wedge SR \wedge Y \wedge S(R/Z)$ is contractible. So

$$(HR)_*(Y) \longrightarrow (HR)_*(YR)$$

is an isomorphism.

We proceed similarly for case (ii), starting from the fact that $SZ_m \wedge S(I_m/Z)$ is contractible.

In case (iii) it is trivial that $Y \xrightarrow{1} Y$ is an E-equivalence.

Now we have to check the other condition of 14.4, namely that $E_*(X) = 0$ implies $[X, YR]_* = 0$, or $[X, YI_m]_* = 0$, or $[X, Y]_* = 0$ according to the case.

First suppose that we are in case (i), so that $\pi_0(E) = R$. Suppose that $f: X \longrightarrow YR$ is a map, and suppose I have already deformed it until all the stable n-cells map to the base-point. (The induction starts, because YR is connective.) I wish to keep it fixed on the (n-1)-cells and deform it until the n-cells and (n+1)-cells map to the base-point. There is an obstruction, and it lies in $H^{n+1}(X;\pi_{n+1}(YR))$. But R is a principal ideal ring, and $\pi_{n+1}(YR)$ is a module over R, so the ordinary universal coefficient theorem applies; we know $H_*(X;R) = 0$, so we can deduce

$$H^{n+1}(X;\pi_{n+1}(YR)) = 0.$$

So I can deform f as required. I continue by induction and conclude that $f = 0$. This shows that $[X, YR]_* = 0$.

Evidently the obstruction-theory argument will work just as well in case (ii), provided we prove that

$$H^{n+1}(X;\pi_{n+1}(YI_m)) = 0.$$

Here we have $\pi_{n+1}(YI_m) = \pi_{n+1}(Y) \otimes I_m$ by 6.7, and $\pi_{n+1}(Y)$ is a finitely-generated group. And in this case we start by knowing that

$$H^*(X;Z_m) = 0.$$

The exact sequence $0 \longrightarrow Z \xrightarrow{m} Z \longrightarrow Z_m \longrightarrow 0$ induces a long exact sequence in homology; it follows that $H_*(X) \xrightarrow{m} H_*(X)$ is an isomorphism, hence $H_*(X) \xrightarrow{m^t} H_*(X)$ is an isomorphism. Now consider

$$\mathrm{Hom}_Z(H_r(X), Z_{m^t}), \quad \mathrm{Ext}^1_Z(H_r(X), Z_{m^t}).$$

On the one hand multiplication by m^t is an isomorphism; on the other hand it is zero. Hence the groups must be zero. So by the ordinary universal coefficient theorem,

$$H^r(X;Z_{m^t}) = 0.$$

Now we have an exact sequence

$$0 \longrightarrow \underset{\longleftarrow}{\mathrm{Lim}}^1(H^*(X;Z_{m^t})) \longrightarrow H^*(X;I_m) \longrightarrow \underset{\longleftarrow}{\mathrm{Lim}}^0(H^*(X;Z_{m^t})) \longrightarrow 0.$$

Hence we have $H^*(X;I_m) = 0$. Finally, let G be any finitely-generated abelian group. We have a resolution

$$0 \longrightarrow F_1 \longrightarrow F_0 \longrightarrow G \longrightarrow 0$$

with F_0 and F_1 finitely-generated free. Therefore we have an exact sequence

$$0 \longrightarrow \prod_1^r I_m \longrightarrow \prod_1^s I_m \longrightarrow I_m \otimes G \longrightarrow 0.$$

This yields an exact cohomology sequence, from which we conclude that

$$H^*(X;I_m \otimes G) = 0.$$

We conclude that

$$H^{n+1}(X;\pi_{n+1}(YI_m)) = 0,$$

the obstruction-theory argument works, and

$$[X, YI_m]_* = 0 .$$

Finally we consider case (iii). Let $f:X \longrightarrow Y$ be a morphism. By Lemma 14.14, we can factor f as

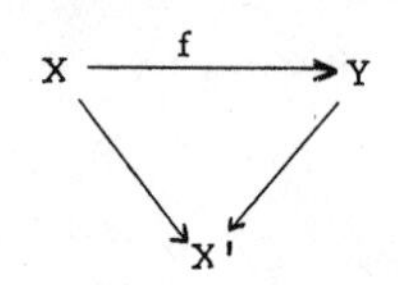

where X' is connective; $H_r(X') \cong H_r(X)$, $r \geq N$, and $H_r(X') = 0$ for $r < N$, for some $N \in Z$.

As above,

$$m: H_*(X) \longrightarrow H_*(X)$$

is an isomorphism; clearly the same is true for X'. Since X' is connective, the theorem of J.H.C. Whitehead shows that $m:X' \longrightarrow X'$ is an equivalence; so it has an inverse m^{-1}. Consider the following diagram.

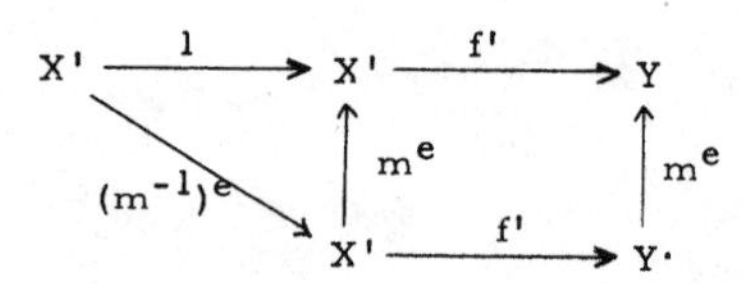

Since $m^e \cdot 1_Y: Y \longrightarrow Y$ is the zero morphism, we conclude $f' = 0$. Thus $[X', Y]_* = 0$. This completes the proof of 14.6.

Now we have some short lemmas which will be needed in the next section.

LEMMA 14.15. Suppose

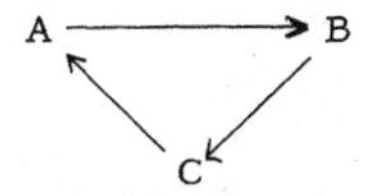

is a cofibre triangle and two of A, B, C are E-complete; then so is the third.

Proof. Suppose $E_*(X) = 0$. We have an exact sequence

$$[X, A]_* \longrightarrow [X, B]_* \longrightarrow [X, C]_* \longrightarrow [X, A]_* \longrightarrow \ldots$$

Two out of every three groups are zero, so the third must be zero also.

LEMMA 14.16. If $f: X \longrightarrow X'$ and $g: Y \longrightarrow Y'$ are E-equivalences, so is

$$f \wedge g: X \wedge Y \longrightarrow X' \wedge Y' .$$

This lemma says that smash products pass to the category of fractions.

Proof. We are given that $E \wedge X \xrightarrow{1 \wedge f} E \wedge X'$ and $E \wedge Y \xrightarrow{1 \wedge g} E \wedge Y'$ are equivalences. Then

$$E \wedge X \wedge Y' \xrightarrow{1 \wedge f \wedge 1} E \wedge X' \wedge Y'$$

and

$$E \wedge X \wedge Y \xrightarrow{1 \wedge 1 \wedge g} E \wedge X \wedge Y'$$

are equivalences; hence so is their composite; that is,

$$X \wedge Y \xrightarrow{f \wedge g} X' \wedge Y'$$

is an E-equivalence.

Now we introduce some arithmetical considerations. Let E be a commutative ring-spectrum such that $\pi_r(E) = 0$ for $r < 0$, and let $\theta: Z \longrightarrow \pi_0(E)$ be the unique homomorphism of rings. Let $S \subset Z$ be the set of n such that $\theta(n)$ is invertible in $\pi_0(E)$. Then S is multiplicatively closed. Let $R \subset Q$ be the localization of Z at S, i.e., the set of fractions n/m with $m \in S$. Then there exists a unique extension of θ to

$$\theta: R \longrightarrow \pi_0(E).$$

PROPOSITION 14.17. If Y is E-complete, then $\pi_r(Y)$ is an R-module. More generally, $[X, Y]_r$ is an R-module for any X.

Proof. Let $m \in S$; then m gives a morphism $Y \longrightarrow Y$, which must be an E-equivalence, since the induced map $E_*(Y) \longrightarrow E_*(Y)$ is multiplication by m, which is an invertible element of $\pi_0(E)$. So in $[Y, Y]_0^E$ the morphism m has an inverse. Therefore the canonical map

$$\varphi: Z \longrightarrow [Y, Y]_0^E$$

extends to give

$$\varphi: R \longrightarrow [Y, Y]_0^E .$$

So R acts on $[X, Y]_*^E$ for any X. If Y is E-complete, we have $[X, Y]_*^E = [X, Y]_*$.

15. THE ADAMS SPECTRAL SEQUENCE

Suppose given a ring-spectrum E and two spectra X, Y such that $E_*(X)$ is projective over $\pi_*(E)$. Our object in this section is to prove the following theorem.

THEOREM 15.1. Assume that X, Y and E satisfy the assumptions listed below. Then

(i) there exists a spectral sequence with the properties which follow,

(ii) its E_2 term is given by

$$E_2^{p,*} = \mathrm{Ext}^{p*}_{E_*(E)}(E_*(X), E_*(Y)), \quad \text{and}$$

(iii) the spectral sequence converges to $[X, Y]_*^E$ in the sense that a suitable analogue of Theorem 8.2 holds. More precisely, it may be obtained by applying the functor $[X, \ ]_*^E$ to a decreasing filtration

$$Y \simeq Y_0 \supset Y_1 \supset Y_2 \supset Y_3 \supset \ldots \supset Y_p \supset \ldots$$

such that

$$\underset{\overleftarrow{p}}{\mathrm{Lim}}^0 [X, Y_p]_*^E = 0$$
$$\underset{\overleftarrow{p}}{\mathrm{Lim}}^1 [X, Y_p]_*^E = 0 .$$

<u>Notes.</u> In (ii), Ext means Ext of comodules over the coalgebra $E_*(E)$. The rules for its calculation will be explained in due course.

<u>List of assumptions.</u> For part (i), none; no extra data is needed to construct the spectral sequence.

For part (ii), two assumptions.

(a) Either $X = S$, or E satisfies 13.3.

(b) $E_*(E)$ is flat as a right module over $\pi_*(E)$.

Both are satisfied for $E = S$, HZ_p, MO, MU, MSp, K, KO.

Of course the spectral sequence may be usable even if (ii) does not apply, if we can calculate the E_1 or E_2 term some other way.

For part (iii), three assumptions.

(a) Y is connective; that is, there exists $n_0 \in Z$ such that $\pi_r(Y) = 0$ for $r < n_0$.

(b) $\pi_r(E) = 0$ for $r < 0$, and

$$\mu_*: \pi_0(E) \otimes_Z \pi_0(E) \longrightarrow \pi_0(E)$$

is an isomorphism. (Examples: $\pi_0(E) = Z_m$; $\pi_0(E)$ is a subring of the rationals.)

Before proceeding, we observe that $H_*(E)$ is a ring, so $H_r(E)$ is a module over $H_0(E) = \pi_0(E)$. Let the subring R of the rationals Q be as in 14.17, so that we have a homomorphism $\theta: R \longrightarrow \pi_0(E)$; thus $H_r(E)$ becomes an R-module.

(c) $H_r(E)$ is finitely-generated over R for all r.

Examples. $E = S, H, HZ_p, MO, MU, MSp, bu, bo$ satisfy (b) and (c); indeed $H_r(E)$ is finitely-generated over Z. However, we might also wish to introduce suitable coefficients. For example, we might prefer some account of the Brown-Peterson spectrum in which $\pi_0(E)$ is Q_p, the integers localised at p. Then $R = Q_p$, and the groups $H_r(E)$ are finitely-generated over R but not over Z.

The basic construction is very easy. We start with $Y_0 = Y$. Suppose Y_p has been constructed. Let $W_p = E \wedge Y_p$. Then we can form the morphism

$$Y_p \simeq S \wedge Y_p \xrightarrow{i \wedge 1} E \wedge Y_p = W_p .$$

Construct a cofibering

$$Y_{p+1} \longrightarrow Y_p \longrightarrow W_p \longrightarrow Y_{p+1}$$

where $W_p \longrightarrow Y_{p+1}$ has degree -1. This completes the induction and constructs the following diagram.

$$\begin{array}{ccccccccc} Y = Y_0 & \longleftarrow & Y_1 & \longleftarrow & Y_2 & \longleftarrow & Y_3 & \longleftarrow & Y_4 \quad \cdots \\ & \searrow \quad \nearrow & & \searrow \quad \nearrow & & \searrow \quad \nearrow & & \searrow \quad \nearrow & \\ & W_0 & & W_1 & & W_2 & & W_3 & \end{array}$$

If we wish, we may use a telescope construction to replace Y_0 by an equivalent spectrum so that the morphisms actually become inclusions

$$Y_0 \supset Y_1 \supset Y_2 \supset Y_3 \supset \ldots ;$$

but this is not necessary.

Suppose we now apply the functor $[X, \ ]_*^E$. Using 14.9 we get a spectral sequence, and this is the spectral sequence required.

We can also write the spectra Y_p, W_p slightly differently. Let us form the cofibering

$$\bar{E} \longrightarrow S \xrightarrow{i} E \longrightarrow \bar{E} \ \ldots$$

where $E \longrightarrow \bar{E}$ has degree -1. Let

$$\bar{E}^p = \bar{E} \wedge \bar{E} \wedge \ldots \wedge \bar{E} \quad (p \text{ factors}).$$

Smashing with $\bar{E}^p \wedge Y$, we obtain a cofibering

$$\bar{E}^{p+1} \wedge Y \longrightarrow \bar{E}^p \wedge Y \longrightarrow E \wedge \bar{E}^p \wedge Y \longrightarrow \bar{E}^{p+1} \wedge Y$$

where again the last morphism shown has degree -1. So we may take

$$Y_p = \bar{E}^p \wedge Y, \qquad W_p = E \wedge \bar{E}^p \wedge Y .$$

This makes it trivial that a morphism $f: Y \longrightarrow Y'$ induces morphisms of the whole construction, and induces a homomorphism from the spectral sequence for Y to that for Y'.

Suppose now that $f: Y \longrightarrow Y'$ is an E-equivalence. Then all the induced morphisms $Y_p \longrightarrow Y'_p$, $W_p \longrightarrow W'_p$ are also E-equivalences (by 14.16) and induce isomorphisms of $[X, \ ]_*^E$. Thus an E-equivalence $f: Y \longrightarrow Y'$ induces an isomorphism of the whole spectral sequence.

It follows that we may suppose without loss of generality that Y is E-complete; for if not, replace it by its E-completion Y^E.

If Y is E-complete, then we easily see by induction over p that Y_p is E-complete; for W_p is E-complete since it is an E-module spectrum, and we use 14.15. So in this case everything in the construction is E-complete, and we could have used $[X, \ ]_*$ instead of $[X, \ ]_*^E$.

Now I had better proceed to part (ii) of the theorem, the calculation of the E_2 term. I ought to begin by recalling some facts from algebra, or perhaps from "coalgebra".

Let A be an algebra with multiplication μ over a ground-ring R, and let N be an R-module. Then we can construct $A \otimes_R N$, and it is an A-module with action map

$$A \otimes_R (A \otimes_R N) \xrightarrow{\mu \otimes 1} A \otimes_R N .$$

The most usual case is that in which N is R-free; then $A \otimes_R N$ is A-free. In general $A \otimes_R N$ is called an <u>extended</u> module, and it possesses the following important property, which generalises the characteristic property of a free module. Let M be an A-module with action map γ. Then we have an isomorphism

$$\mathrm{Hom}_A(A \otimes_R N, M) \xrightarrow{\theta} \mathrm{Hom}_R(N, M) .$$

It is given as follows. Suppose given

$$A \otimes_R N \xrightarrow{f} M ;$$

then θf is

$$N \cong R \otimes_R N \xrightarrow{\eta \otimes 1} A \otimes_R N \xrightarrow{f} M$$

where η is the unit map $R \longrightarrow A$. Suppose given $N \xrightarrow{g} M$; then $\theta^{-1}g$ is

$$A \otimes_R N \xrightarrow{1 \otimes g} A \otimes_R M \xrightarrow{\gamma} M.$$

In particular, if N is projective over R, then $A \otimes_R N$ is projective over A.

We also have the dual situation. Let C be a coalgebra with diagonal ψ over a ground-ring R. I emphasize that R is allowed to act differently on the two sides of C. Let N be an R-module. Then we can construct can construct $C \otimes_R N$, and it is a C-comodule with coaction map

$$C \otimes_R N \xrightarrow{\psi \otimes 1} C \otimes_R (C \otimes_R N).$$

It is called an extended comodule. It has the following property. Let M be a C-comodule with coaction map γ. Then we have an isomorphism

$$\mathrm{Hom}_C(M, C \otimes_R N) \xrightarrow{\theta} \mathrm{Hom}_R(M, N) .$$

It is given as follows. Suppose given

$$M \xrightarrow{f} C \otimes_R N;$$

then θf is

$$M \xrightarrow{f} C \otimes_R N \xrightarrow{\epsilon \otimes 1} R \otimes_R N \cong N,$$

where ϵ is the augmentation $C \to R$. Suppose given $M \xrightarrow{g} N$; then $\theta^{-1} g$ is

$$N \xrightarrow{\gamma} C \otimes_R N \xrightarrow{1 \otimes g} C \otimes_R N .$$

In particular, if N is injective over R, then $C \otimes_R N$ is injective over C.

There is a prescription of homological algebra for computing $\mathrm{Ext}_C^{**}(L, M)$, where L and M are comodules over the coalgebra C. However, it does not demand that we resolve M by absolute injectives. So long as L is projective over R it will be sufficient if we resolve M by relative injectives. More precisely, if L is projective over R we have to make a resolution

$$0 \to M \to M_0 \to M_1 \to M_2 \cdots$$

where each M_i is an extended comodule. Then we form

$$\mathrm{Hom}_C(L, M_0) \to \mathrm{Hom}_C(L, M_1) \to \mathrm{Hom}_C(L, M_2) \to \cdots$$

and the cohomology groups of this cochain complex are

$$\mathrm{Ext}_C^{**}(L, M) .$$

With this in mind, let us return to consider our geometrical situation. We have

$$W_p = E \wedge Y_p .$$

So of course we have

$$E_*(W_p) = E_*(E \wedge Y_p) \cong E_*(E) \otimes_{\pi_*(E)} E_*(Y_p) ;$$

this is by Lemma 12.5. It is rather trivial to check that this isomorphism throws the coaction map ψ_{W_p} onto $\psi_E \otimes 1$; so $E_*(E \wedge Y_p)$ is an extended comodule.

Again, consider our cofibering

$$Y_p \longrightarrow E \wedge Y_p \longrightarrow Y_{p+1}$$

where $E \wedge Y_p \longrightarrow Y_{p+1}$ has degree -1. When we smash with E we have

$$E \wedge Y_p \underset{1 \wedge i}{\overset{\mu \wedge 1}{\rightleftarrows}} E \wedge E \wedge Y_p \longrightarrow E \wedge Y_{p+1} \longrightarrow \cdots .$$

But $\mu \wedge 1$ is a left inverse for $1 \wedge i$, so we have the following short exact sequence, split as a sequence of modules over $\pi_*(E)$.

$$0 \longrightarrow E_*(Y_p) \longrightarrow E_*(E \wedge Y_p) \longrightarrow E_*(Y_{p+1}) \longrightarrow 0$$
$$\| \qquad\qquad$$
$$E_*(W_p)$$

Hence, the sequence

$$0 \longrightarrow E_*(Y) \longrightarrow E_*(W_0) \longrightarrow E_*(W_1) \longrightarrow E_*(W_2) \longrightarrow \cdots$$

is indeed a resolution of $E_*(Y)$ by extended comodules over $E_*(E)$.

Now I recall that the E_1 term of our spectral sequence is given by

$$\begin{aligned} E_1^{p*} &= [X, W_p]_*^E \\ &= [X, E \wedge Y_p]_*^E \\ &= [X, E \wedge Y_p]_* \quad \text{(since } E \wedge Y_p \text{ is E-complete).} \end{aligned}$$

The boundary d_1 is induced by the morphism

$$W_p \longrightarrow Y_{p+1} \longrightarrow W_{p+1}$$

where $W_p \longrightarrow Y_{p+1}$ has degree -1. We have the following commutative diagram.

$$\begin{array}{ccc} [X, E \wedge Y_p] & \xrightarrow{\alpha} & \mathrm{Hom}^*_{E_*(E)}(E_*(X), E_*(E \wedge Y_p)) \\ & \searrow^{\beta} & \Big\downarrow \theta \cong \\ & & \mathrm{Hom}^*_{\pi_*(E)}(E_*(X), E_*(Y_p)) \end{array}$$

Here $\alpha(f) = f_*$. The isomorphism θ comes because $E_*(E \wedge Y_p)$ is an extended comodule. The spectrum $E \wedge Y_p$ is a module-spectrum over E, and β is precisely the map which is asserted to be an isomorphism by 13.5, if we have the data for that, or trivially if $X = S$. We conclude that α is an isomorphism.

Now we have the following commutative diagram, in which the horizontal maps are induced by the morphisms $W_p \longrightarrow W_{p+1}$ (of degree -1), and Hom is $\mathrm{Hom}_{E_*(E)}$.

$$\begin{array}{ccccc} [X, W_{p-1}]_* & \xrightarrow{d_1} & [X, W_p]_* & \xrightarrow{d_1} & [X, W_{p+1}]_* \\ \Big\downarrow \cong & & \Big\downarrow \cong & & \Big\downarrow \cong \\ \mathrm{Hom}(E_*(X), E_*(W_{p-1})) & \longrightarrow & \mathrm{Hom}(E_*(X), E_*(W_p)) & \longrightarrow & \mathrm{Hom}(E_*(X), E_*(W_{p+1})) \end{array}$$

The cohomology groups of the top row are E_2^{p*} and those of the bottom row are

$$\mathrm{Ext}^{p*}_{E_*(E)}(E_*(X), E_*(Y)) .$$

This proves part (ii) of 15.1.

We now start work on part (iii). I recall we have assumed that

$$\pi_0(E) \otimes_Z \pi_0(E) \xrightarrow{\mu_*} \pi_0(E)$$

is an isomorphism. I claim it follows that for any module M over $\pi_0(E)$,

$$\pi_0(E) \otimes_Z M \xrightarrow{\nu} M$$

is an isomorphism. In fact, this follows from the following commutative diagram.

$$\begin{array}{ccc} \pi_0(E) \otimes_Z \pi_0(E) \otimes_{\pi_0(E)} M & \xrightarrow[\cong]{1 \otimes \nu} & \pi_0(E) \otimes_Z M \\ {\scriptstyle u \otimes 1} \downarrow \cong & & \downarrow \nu \\ \pi_0(E) \otimes_{\pi_0(E)} M & \xrightarrow[\cong]{\nu} & M \end{array}$$

Now I undertake to prove by induction over p that $\pi_r(E \wedge \bar{E}^p) = 0$ for $r < 0$. This is surely true for $p = 0$, by assumption. Suppose it true for p, and consider the following cofibering.

$$E \wedge S \wedge \bar{E}^p \xrightarrow{1 \wedge i \wedge 1} E \wedge E \wedge \bar{E}^p \longrightarrow E \wedge \bar{E}^{p+1} .$$

Here $E \wedge E \wedge \bar{E}^p \longrightarrow E \wedge \bar{E}^{p+1}$ has degree -1. As we have already remarked, we have a left inverse for $1 \wedge i \wedge 1$, given by $\mu \wedge 1: E \wedge E \wedge \bar{E}^p \longrightarrow E \wedge \bar{E}^p$. So the exact homotopy sequence of this cofibering is split short exact. By the inductive hypothesis and the Künneth theorem, the first non-zero homotopy group of $E \wedge E \wedge \bar{E}^p$ is

$$\pi_0(E \wedge E \wedge \bar{E}^p) = \pi_0(E) \otimes_Z \pi_0(E \wedge \bar{E}^p).$$

Therefore $\pi_r(E \wedge \bar{E}^{p+1}) = 0$ for $r < -1$, and $\pi_{-1}(E \wedge \bar{E}^{p+1})$ is isomorphic to the kernel of

$$\pi_0(E) \otimes_Z \pi_0(E \wedge \bar{E}^p) \longrightarrow \pi_0(E \wedge \bar{E}^p) .$$

But this map is an isomorphism by the remarks above, so its kernel is zero, and $\pi_r(E \wedge \bar{E}^{p+1}) = 0$ for $r < 0$. This completes the induction.

We have also assumed $\pi_r(Y) = 0$ for $r < n_0$. Since we may take $W_p = E \wedge \bar{E}^p \wedge Y$, we have $\pi_r(W_p) = 0$ for $r < n_0$.

Now I undertake to prove by induction over p that $\pi_r(Y_p) = 0$ for $r < n_0 - 1$. This is immediate, from the following exact sequence.

$$\cdots \longrightarrow \pi_{r+1}(W_p) \longrightarrow \pi_r(Y_{p+1}) \longrightarrow \pi_r(Y_p) \longrightarrow \cdots$$

So at this stage we have established a uniform bound $n_0 - 1$ such that $\pi_r(Y_p) = 0$ for $r < n_0 - 1$.

Next we need to construct a spectrum Y_∞, the E-homotopy inverse limit of the Y_p. The construction is easy. First we observe that we can assume without loss of generality that Y is E-complete, and therefore that all the Y_p are E-complete. This requires a word of justification; we have to see that when we replace Y by Y^E, we do not sacrifice the property that Y is connective. Recall that by the proof of (14.5), we can find a uniform bound ν and a cofinal set of E-equivalences $e: Y \longrightarrow Y'$ such that $\pi_r(Y') = 0$ for $r < \nu$. This shows that $[S, Y]_r^E = 0$ for $r < \nu$ and $\pi_r(Y^E) = 0$ for $r < \nu$.

Assume then that all the Y_p are E-complete. Then we can form the categorical product $\prod_{i=0}^{\infty} Y_i$ in C, and it is E-complete; for if $E_*(W) = 0$, and $f: W \longrightarrow \prod_{i=0}^{\infty} Y_i$ is a map, then all the components $p_i f: W \longrightarrow Y_i$ are zero, and so f is zero. It follows that $\prod_{i=0}^{\infty} Y_i$ is the categorical product not only in C, but also in the category of fractions F.

Now we construct a map $f: \prod_{i=0}^{\infty} Y_i \longrightarrow \prod_{i=0}^{\infty} Y_i$; the ith component of f is to be the difference of two maps, that is,

$$\left(\prod_{i=0}^{\infty} Y_i\right) \xrightarrow{p_i} Y_i$$

minus

$$\left(\prod_{i=0}^{\infty} Y_i\right) \xrightarrow{p_{i+1}} Y_{i+1} \longrightarrow Y_i \ .$$

We define Y_∞ so that we have the following cofibre sequence.

$$Y_\infty \longrightarrow \prod_{i=0}^{\infty} Y_i \xrightarrow{f} \prod_{i=0}^{\infty} Y_i \longrightarrow Y_\infty$$

It follows from 14.15 that Y_∞ is E-complete. Applying $[X, \ ]_*^E$, we see that for any X we have the following short exact sequence.

$$0 \longrightarrow \varprojlim_i{}^1 [X, Y_i]^E_{r+1} \longrightarrow [X, Y_\infty]^E_r \longrightarrow \varprojlim_i{}^0 [X, Y_i]^E_r \longrightarrow 0 .$$

Now I need to quote the following result, which I will prove later.

THEOREM 15.2. Let R be a subring of the rationals Q. Suppose Y_α, E are spectra such that

(i) $\pi_r(Y_\alpha) = 0$ for $r < n_1$, for some n_1 independent of α,

(ii) $\pi_r(Y_\alpha)$ is an R-module for all r, α,

(iii) $\pi_r(E) = 0$ for $r < n_2$, for some $n_2 \in Z$, and

(iv) $H_r(E)$ is a finitely-generated R-module for all r.

Then the canonical morphism

$$E \wedge (\prod_\alpha Y_\alpha) \longrightarrow \prod_\alpha (E \wedge Y_\alpha)$$

is an equivalence.

The canonical morphism is of course the one with components

$$E \wedge (\prod_\alpha Y_\alpha) \xrightarrow{1 \wedge p_\alpha} E \wedge Y_\alpha .$$

It can be shown by examples that the behaviour of $\wedge$ with respect to $\prod$ is in general very bad; one cannot hope for a much stronger theorem.

Now 14.17 shows that $\pi_r(Y_i)$ is an R-module, where R is as in 14.17. So 15.2 applies and shows that

$$E \wedge (\prod_{i=0}^{\infty} Y_i) \longrightarrow \prod_{i=0}^{\infty} (E \wedge Y_i)$$

is an equivalence. This shows that

$$E_*(\prod_{i=0}^{\infty} Y_i) = \pi_*(E \wedge (\prod_{i=0}^{\infty} Y_i)) \cong \pi_*(\prod_{i=0}^{\infty} (E \wedge Y_i)) \cong \prod_{i=0}^{\infty} E_*(Y_i)$$

under the obvious homomorphism. It follows that we have the following short exact sequence.

$$0 \longrightarrow \varprojlim_i{}^1 E_*(Y_i) \longrightarrow E_*(Y_\infty) \longrightarrow \varprojlim_i{}^0 E_*(Y_i) \longrightarrow 0$$

But by the construction the maps $E_*(Y_{i+1}) \longrightarrow E_*(Y_i)$ are zero. It follows immediately that $\varprojlim_i^0 E_*(Y_i) = 0$. It also follows that $\varprojlim_i^1 E_*(Y_i) = 0$ (see section 8, exercise (ii).) Therefore $E_*(Y_\infty) = 0$. It follows that $[X, Y_\infty]_*^E = 0$. Using the exact sequence above, we have

$$\varprojlim_i^0 [X, Y_i]_*^E = 0$$

$$\varprojlim_i^1 [X, Y_i]_*^E = 0 .$$

This proves 15.1 (iii). It remains to prove Theorem 15.2.

LEMMA 15.3. Suppose that R is a subring of the rationals, the G_α are R-modules and F is a finitely-generated R-module. Then

$$F \otimes_R (\prod_\alpha G_\alpha) \rightarrow \prod_\alpha (F \otimes_R G_\alpha)$$

and

$$\mathrm{Tor}_1^R (F, \prod_\alpha G_\alpha) \rightarrow \prod_\alpha \mathrm{Tor}_1^R (F, G_\alpha)$$

are isomorphisms.

<u>Proof.</u> R is a principal ideal ring. Take a resolution of F of the form

$$0 \rightarrow \sum_1^n R \xrightarrow{d} \sum_1^m R \rightarrow F \rightarrow 0 .$$

Form the following diagram.

$$\begin{array}{ccccccccc}
0 \rightarrow & \mathrm{Tor}_1^R(F, \prod_\alpha G_\alpha) & \rightarrow & (\sum_1^n R) \otimes \prod_\alpha G_\alpha & \xrightarrow{d \otimes 1} & (\sum_1^m R) \otimes \prod_\alpha G_\alpha & \rightarrow & F \otimes \prod_\alpha G_\alpha & \rightarrow 0 \\
& \downarrow & & \| & & \| & & \downarrow & \\
& & & \prod_1^n \prod_\alpha G_\alpha & & \prod_1^m \prod_\alpha G_\alpha & & & \\
& & & \| & & \| & & & \\
& & & \prod_\alpha \prod_1^n G_\alpha & & \prod_\alpha \prod_1^m G_\alpha & & & \\
& & & \| & & \| & & & \\
0 \rightarrow & \prod_\alpha (\mathrm{Tor}_1^R(F, G_\alpha)) & \rightarrow & \prod_\alpha ((\sum_1^n R) \otimes G_\alpha) & \xrightarrow[\prod_\alpha (d \otimes 1)]{} & \prod_\alpha ((\sum_1^m R) \otimes G_\alpha) & \rightarrow & \prod_\alpha (F \otimes G_\alpha) & \rightarrow 0
\end{array}$$

The result follows.

LEMMA 15.4. Suppose that R is a subring of the rationals, E is such that $H_r(E)$ is a finitely-generated R-module for all R, and the G_α are R-modules. Then

$$H_n(E; \prod_\alpha G_\alpha) \longrightarrow \prod_\alpha H_n(E;G_\alpha)$$

is an isomorphism.

Proof. First observe that since R is torsion-free, the ordinary universal coefficient theorem gives $H_r(E;R) \cong H_r(E) \otimes_Z R$; and since $R \otimes_Z R \longrightarrow R$ is isomorphism, and $H_r(E)$ is an R-module, the argument given in 15.1 (iii) (applied to R rather than $\pi_0(E)$) shows that $H_r(E) \otimes_Z R \longrightarrow H_r(E)$ is an isomorphism. So $H_r(E;R)$ is finitely-generated over R. Now consider the following diagram.

$$\begin{array}{ccccccccc}
0 \to & H_n(E;R) \otimes_R \prod_\alpha G_\alpha & \longrightarrow & H_n(E; \prod_\alpha G_\alpha) & \longrightarrow & \mathrm{Tor}_1^R(H_{n-1}(E;R), \prod_\alpha G_\alpha) & \to 0 \\
& \downarrow \cong & & \downarrow & & \downarrow \cong & \\
0 \to & \prod_\alpha H_n(E;R) \otimes_R G_\alpha & \longrightarrow & \prod_\alpha H_n(E;G_\alpha) & \longrightarrow & \prod_\alpha \mathrm{Tor}_1^R(H_{n-1}(E;R), G_\alpha) & \to 0
\end{array}$$

The two vertical arrows marked are isomorphisms by 15.3. The rows are exact by the ordinary universal coefficient theorem. The result follows by the short five lemma.

COROLLARY 15.5. (of Lemma 15.4). Theorem 15.2 is true in the special case in which the Y_α are all Eilenberg-MacLane spectra with homotopy groups in the same dimension q.

Proof. Let G_α be the R-module $\pi_q(Y_\alpha)$. Then $\prod_\alpha Y_\alpha$ is an Eilenberg-MacLane spectrum with homotopy group $\prod_\alpha G_\alpha$ in dimension q. We have the following commutative diagram.

$$\begin{array}{ccc}
\pi_r(E \wedge \prod_\alpha Y_\alpha) & \longrightarrow & \pi_r(\prod_\alpha E \wedge Y_\alpha) \\
\downarrow \cong & & \cong \downarrow \\
H_{r-q}(E; \prod_\alpha G_\alpha) & \longrightarrow & \prod_\alpha H_{r-q}(E; G_\alpha)
\end{array}$$

By 15.4 the lower horizontal arrow is an isomorphism. The result follows immediately from the theorem of J.H.C. Whitehead.

LEMMA 15.6. Suppose $A_\alpha \longrightarrow B_\alpha \longrightarrow C_\alpha \longrightarrow A_\alpha \longrightarrow B_\alpha$ is a cofibering for each α, where $C_\alpha \longrightarrow A_\alpha$ has degree -1. Then

$$\prod_\alpha A_\alpha \longrightarrow \prod_\alpha B_\alpha \longrightarrow \prod_\alpha C_\alpha \longrightarrow \prod_\alpha A_\alpha \longrightarrow \prod_\alpha B_\alpha$$

is a cofibering.

<u>Proof.</u> Construct a cofibering

$$\prod_\alpha A_\alpha \longrightarrow \prod_\alpha B_\alpha \longrightarrow D \longrightarrow \prod_\alpha A_\alpha \longrightarrow \prod_\alpha B_\alpha .$$

Then it admits the following map.

$$\begin{array}{ccccccccc}
\prod A_\alpha & \longrightarrow & \prod B_\alpha & \longrightarrow & D & \longrightarrow & \prod A_\alpha & \longrightarrow & \prod B_\alpha \\
p_\alpha \downarrow & & p_\alpha \downarrow & & \downarrow & & p_\alpha \downarrow & & p_\alpha \downarrow \\
A_\alpha & \longrightarrow & B_\alpha & \longrightarrow & C_\alpha & \longrightarrow & A_\alpha & \longrightarrow & B_\alpha
\end{array}$$

So we can construct the following diagram.

$$\begin{array}{ccccccccc}
\prod_\alpha A_\alpha & \longrightarrow & \prod_\alpha B_\alpha & \longrightarrow & D & \longrightarrow & \prod_\alpha A_\alpha & \longrightarrow & \prod_\alpha B_\alpha \\
1 \downarrow & & 1 \downarrow & & \downarrow & & 1 \downarrow & & 1 \downarrow \\
\prod_\alpha A_\alpha & \longrightarrow & \prod_\alpha B_\alpha & \longrightarrow & \prod_\alpha C_\alpha & \longrightarrow & \prod_\alpha A_\alpha & \longrightarrow & \prod_\alpha B_\alpha
\end{array}$$

Now the five lemma shows that the map $D \longrightarrow \prod_\alpha C_\alpha$ induces an isomorphism of homotopy, and the theorem of J.H.C. Whitehead shows that it is an equivalence. Since the upper line of the diagram is a cofibering, it follows that the lower line is a cofibering. This proves 15.6.

Proof of Theorem 15.2. We wish to show that

$$\pi_r(E \wedge \prod_\alpha Y_\alpha) \longrightarrow \pi_r(\prod_\alpha E \wedge Y_\alpha)$$

is an isomorphism, and we do this by induction over $r - n_1 - n_2$. The result is trivial if $r - n_1 - n_2 < 0$. Suppose as an inductive hypothesis that the result is true for smaller values of $r - n_1 - n_2$. We can construct a cofibering

$$K_\alpha \longrightarrow W_\alpha \longrightarrow Y_\alpha \longrightarrow K_\alpha \longrightarrow W_\alpha$$

in which $K_\alpha \longrightarrow W_\alpha$ has degree -1, $\pi_r(W_\alpha) = 0$ for $r < n_1 + 1$ and K_α is an Eilenberg-MacLane spectrum for the R-module $\pi_{n_1}(Y_\alpha)$ in dimension n_1. Using (15.6), we see that

$$E \wedge \prod_\alpha K_\alpha \longrightarrow E \wedge \prod_\alpha W_\alpha \longrightarrow E \wedge \prod_\alpha Y_\alpha \longrightarrow E \wedge \prod_\alpha K_\alpha \longrightarrow E \wedge \prod_\alpha W_\alpha$$

and

$$\prod_\alpha (E \wedge K_\alpha) \longrightarrow \prod_\alpha (E \wedge W_\alpha) \longrightarrow \prod_\alpha (E \wedge Y_\alpha) \longrightarrow \prod_\alpha (E \wedge K_\alpha) \longrightarrow \prod_\alpha (E \wedge W_\alpha)$$

are also cofiberings. Now consider the following diagram.

$$\begin{array}{ccc}
\pi_{r+1}(E \wedge \prod_\alpha K_\alpha) & \xrightarrow{1} & \pi_{r+1}(\prod_\alpha (E \wedge K_\alpha)) \\
\downarrow & & \downarrow \\
\pi_r(E \wedge \prod_\alpha W_\alpha) & \xrightarrow{2} & \pi_r(\prod_\alpha (E \wedge W_\alpha)) \\
\downarrow & & \downarrow \\
\pi_r(E \wedge \prod_\alpha Y_\alpha) & \xrightarrow{3} & \pi_r(\prod_\alpha (E \wedge Y_\alpha)) \\
\downarrow & & \downarrow \\
\pi_r(E \wedge \prod_\alpha K_\alpha) & \xrightarrow{4} & \pi_r(\prod_\alpha (E \wedge K_\alpha)) \\
\downarrow & & \downarrow \\
\pi_{r-1}(E \wedge \prod_\alpha W_\alpha) & \xrightarrow{5} & \pi_{r-1}(\prod_\alpha (E \wedge W_\alpha))
\end{array}$$

Maps 1 and 4 are isomorphisms by (15.5); maps 2 and 5 are isomorphisms by the inductive hypothesis. So map 3 is an isomorphism by the five lemma. This completes the induction and proves Theore 15.2.

16. APPLICATIONS TO $\pi_*(bu \wedge X)$; MODULES OVER $K[x, y]$

I would like to present some applications of the spectral sequence of §15, in which we can do the algebra without too much trouble. For this purpose I will consider the calculation of $\pi_*(bu \wedge X)$ for various spectra X. Of course, I am really interested in $\pi_*(bo \wedge X)$; however, it seems best if I do things for the most elementary case, which is the case bu, but undertake to use only methods which extend to the case bo. For a similar reason I will consider mostly the prime 2, but I will try to say only things which can also be said for the prime p.

If we apply the spectral sequence of §15 to compute $\pi_*(bu \wedge X)$, using say $E = HZ_2$, we obtain a spectral sequence of the following form.

$$\operatorname{Ext}^{s,t}_{A_*}(Z_2, (HZ_2)_*(bu \wedge X)) \underset{s}{\Longrightarrow} [S, bu \wedge X]^{HZ_2}_{t-s}.$$

However, in this case the Ext group simplifies very greatly. To explain how it simplifies, recall that in A_* we have a base consisting of the monomials $\xi_1^{r_1}\xi_2^{r_2}\ldots\xi_n^{r_n}$. The dual base in A is written $Sq^{r_1 r_2 \ldots r_n}$. This is consistent because $Sq^{r\,0\ldots 0}$ is Steenrod's Sq^r. In particular, Sq^{01} is the element of this dual base corresponding to the monomial ξ_2. We have $Sq^{01} = Sq^1 Sq^2 + Sq^2 Sq^1$. The elements Sq^1 and Sq^{01} generate an exterior subalgebra of A; we write B for this exterior subalgebra. It is a Hopf subalgebra. The algebra B is of course dual to a quotient B_* of A_*, namely the quotient of A_* by the ideal generated by $\xi_1^2, \xi_2^2, \xi_3, \ldots, \xi_n, \ldots$. Just as we can consider $(HZ_2)^*(X)$ as a module over B, we can consider $(HZ_2)_*(X)$ as a comodule over B_*.

For the case of an odd prime, the analogues of Sq^1 and Sq^{01} are the Milnor elements Q_0 and Q_1. These are the elements of the Milnor base for A corresponding to τ_0 and τ_1 in A_*. We have $Q_0 = \beta_p$, $Q_1 = P^1\beta_p - \beta_p P^1$. B is then the exterior subalgebra of A generated by Q_0 and Q_1; B_* is a quotient of A_* and is an exterior algebra generated by τ_0 and τ_1.

PROPOSITION 16.1. Assume X is connective. Then we have a spectral sequence

$$\operatorname{Ext}^{s,t}_{B_*}(Z_2, (HZ_2)_*(X)) \underset{s}{\Longrightarrow} [S, bu \wedge X]^{HZ_2}_{t-s} .$$

For the case of an odd prime we should take the precaution of splitting buQ_p into (p-1) similar summands and using only one of them on the right-hand side.

I will finish stating the results I need before I start to prove anything.

In order to use this spectral sequence to the best advantage we have to know something about the structure-theory of comodules over B_*. As long as our comodules are locally finite-dimensional we may as well dualise and consider the structure-theory of modules over B. Even if our comodules are not locally finite-dimensional, we can consider a B_*-comodule M as a B-module by the following construction: if

$$\psi m = \sum_i b_i' \otimes m_i'' , \quad b^* \in B,$$

set

$$b^* m = \sum_i \langle cb^*, b_i' \rangle m_i'' ,$$

where c is the canonical anti-automorphism of B.

The structure-theory works perfectly well for modules over the exterior algebra $K[x,y]$ on two generators x and y of distinct dimension. Here K is supposed to be a field; for some theorems one wants K to be a finite field, but not for anything in these lectures. We assume that the degrees of x and y are odd unless K has characteristic 2; in other words, we want $K[x, y]$ to be a Hopf algebra, with x and y primitive.

Some of the ideas of the structure-theory work for a finite-dimensional Hopf algebra A, more general than $K[x, y]$. Let M and N be left A-modules. We say they are stably isomorphic if there exists free modules F and G such that $M \oplus F \cong N \oplus G$. This is an equivalence relation. For $s > 0$ the groups

$$\mathrm{Ext}_A^{s,t}(M, K)$$

depend only on the stable isomorphism class of M; this is one reason why it is often sufficient to know only the stable isomorphism class of M.

We can form the sum and the tensor product of two modules. Here we give $M \otimes N$ the diagonal action, using the fact that A is a Hopf algebra. The sum and product pass to stable isomorphism classes. The product has a unit, namely the module 1 with K in degree 0.

We say that a stable class P is invertible if there is a stable class Q such that $PQ \simeq 1$.

We define Σ to be the module with K in degree 1. Σ is clearly invertible; its inverse is the module Σ^{-1} with K in degree -1.

We define I to be the augmentation ideal of A.

LEMMA 16.2. If A is a connected finite-dimensional Hopf algebra, then I is invertible.

We now return to the case $A = K[x,y]$. We observe that a module M has two very useful invariants:

$$H_*(M;x) = \mathrm{Ker}\ x/\mathrm{Im}\ x$$

$$H_*(M;y) = \mathrm{Ker}\ y/\mathrm{Im}\ y\ .$$

These are defined on stable isomorphism classes, and send sums to sums, products to products. The latter follows from the Künneth theorem.

THEOREM 16.3. Let M be a finite-dimensional module over $K[x,y]$ such that $H_*(M;x)$ and $H_*(M;y)$ both have dimension 1 over K. Then

(i) M is invertible,

(ii) the stable class of M is $\Sigma^a I^b$ for unique $a, b \in Z$.

Notice how one proves uniqueness. We have

$$H_*(\Sigma^a I^b;x) = \begin{cases} K & \text{in degree } a + b|x| = c, \text{ say} \\ 0 & \text{otherwise} \end{cases}$$

$$H_*(\Sigma^a I^b;y) = \begin{cases} K & \text{in degree } a + b|y| = d, \text{ say} \\ 0 & \text{otherwise} \end{cases} .$$

Since $|x| \neq |y|$, c and d determine a and b.

If we are to use Proposition 16.1 to compute $\pi_*(bu \wedge X)$, we need to know $(HZ_2)_*(X)$ as a comodule over B_*, or equivalently, $(HZ_2)^*(X)$ as a module over B. In particular, if we want to compute $\pi_*(bu \wedge bu \wedge \ldots \wedge bu)$ ($n+1$ factors), we need this information for $X = bu \wedge bu \wedge \ldots \wedge bu$ (n factors).

PROPOSITION 16.4. (i) The stable class of $(HZ_2)^*(bu)$, as a module over B, is

$$(1 + \Sigma^2)(1 + \Sigma^3 I)(1 + \Sigma^5 I^3) \ldots (1 + \Sigma^{2^r+1} I^{2^r-1}) \ldots$$

(ii) Let $(bu)^n = bu \wedge bu \wedge \ldots \wedge bu$ (n factors). Then the stable class of $(HZ_2)^*((bu)^n)$, as a module over B, is

$$(1 + \Sigma^2)^n (1 + \Sigma^3 I)^n (1 + \Sigma^5 I^3)^n \ldots$$

Of course part (ii) follows immediately from part (i).

For the next section, we need one last fact about bu. Recall that $\pi_2(bu) \cong Z$; let $t \in \pi_2(bu)$ be the generator. The homotopy ring $\pi_*(bu)$ is the polynomial ring $Z[t]$. We may identify $t \in \pi_2(bu)$ with its image in $H_2(bu)$ or $(HQ)_2(bu)$. The homology ring $(HQ)_*(bu)$ is the polynomial ring $Q[t]$. We define a numerical function $m(r)$ by

$$m(r) = \prod_p p^{\left[\frac{r}{p-1}\right]} .$$

Here p runs over prime numbers, and $[x]$ means the integral part of x. For example,

if	$r =$	1	2	3	4 ,
	$m(r) =$	2	12	24	720 .

PROPOSITION 16.5. The image of $H_*(bu)$ in $(HQ)_*(bu)$ is the Z-submodule generated by the elements

$$\frac{t^r}{m(r)}, \qquad r = 0, 1, 2, \ldots .$$

This completes the statement of results. Now I turn to the proofs.

Let A once more denote the mod 2 Steenrod algebra.

PROPOSITION 16.6. As an A-module, we have

$$(HZ_2)^*(bu) \cong A/A(Sq^1 + ASq^{01}) = A \otimes_B Z_2 .$$

For the case of an odd prime, we either write

$$(HZ_p)^*(bu) \cong \sum_1^{p-1} A/(AQ_0 + AQ_1) ,$$

or we split buQ_p into $(p-1)$ similar summands and take one of them.

For the case of bo, we have

$$(HZ_2)^*(bo) \cong A/(ASq^1 + ASq^2) .$$

<u>Proof of 16.6.</u> First we obtain information on the first k-invariant of bu, which lies in $H^3(H)$, which is Z_2 generated by $\delta_2 Sq^2$. The k-invariant must be 0 or $\delta_2 Sq^2$. We wish to find out which; and of course we do it by looking at the terms in the bu-spectrum. For each term in the bu-spectrum, the first k-invariant is given by the same stable operation. We choose to look at the third term of the bu-spectrum, which happens to be the first place where we can get the required information. The third term of the bu-spectrum is the space SU. Now $\delta_2 Sq^2 \neq 0$ in $H^6(Z, 3)$, but $H^6(SU) = 0$. We conclude that the first k-invariant of bu is $\delta_2 Sq^2$ rather than 0.

Now the Bott periodicity theorem gives us the following cofibering.

$$S^2 \wedge bu \xrightarrow{i} bu \xrightarrow{j} H$$

This leads to a long exact sequence

$$\longleftarrow (HZ_2)^n(bu) \xleftarrow{j^*} (HZ_2)^n(H) \xleftarrow{k^*} (HZ_2)^n(S^3 \; bu) \longleftarrow \quad \ldots$$

Let f^0 be the fundamental class in $(HZ_2)^0(H)$; then we have $(HZ_2)^*(H) \cong A/ASq^1$, under the map which takes a to af^0. The class j^*f^0 is the fundamental class in $(HZ_2)^0(bu)$; therefore we obtain a fundamental class f^3 in $(HZ_2)^3(S^3 \wedge bu)$. The information on the k-invariant says that

$$k^*f^3 = \beta_2 Sq^2 f^0 = Sq^{01} f^0$$

(since $Sq^2 Sq^1 f^0 = 0$). Thus $Sq^1(j^*f^0) = 0$ and $Sq^{01}(j^*f^0) = 0$. So certainly we get a homomorphism

$$A/(ASq^1 + ASq^{01}) \longrightarrow (HZ_2)^*(bu)$$

defined by

$$a \longmapsto a(j^* f^0) .$$

We recall that Sq^1 and Sq^{01} generate the exterior subalgebra $B \subset A$, and A is free as a right module over B. So we have the following short exact sequence.

$$0 \longleftarrow A/(ASq^1 + ASq^{01}) \longleftarrow A/(ASq^1) \longleftarrow A/(ASq^1 + ASq^{01}) \longleftarrow 0$$

Here the map on the right takes

$$\text{cls } x \longrightarrow \text{cls } xSq^{01} .$$

Indeed, we have the following diagram.

$$\begin{array}{ccccccc}
\cdots \xleftarrow{0} & [A/(ASq^1+ASq^{01})]_n & \longleftarrow & [A/(ASq^1)]_n & \longleftarrow & [A/(ASq^1+ASq^{01})]_{n-3} & \xleftarrow{0} \cdots \\
& \downarrow & & \downarrow & & \downarrow & \\
\cdots \longleftarrow & (HZ_2)^n(bu) & \longleftarrow & (HZ_2)^n(H) & \longleftarrow & (HZ_2)^n(S^3 \wedge bu) & \longleftarrow \cdots
\end{array}$$

Suppose as an inductive hypothesis that

$$[A/(ASq^1 + ASq^{01})]_r \longrightarrow (HZ_2)^r(bu)$$

is an isomorphism for $r < n$. Then for $(HZ_2)^r(S^3 \wedge bu)$ the same thing holds for $r < n+3$. Now the five lemma shows that

$$[A/(ASq^1 + ASq^{01})]_r \longrightarrow (HZ_2)^r(bu)$$

is an isomorphism for $r = n$. This completes the induction and proves 16.6.

Proof of 16.1. We have a spectral sequence

$$\text{Ext}^{s,t}_{A_*}(Z_2, (HZ_2)_*(bu \wedge X)) \underset{s}{\Longrightarrow} [S, bu \wedge X]^{HZ_2}_{t-s} .$$

Suppose to begin with that $(HZ_2)_*(bu \wedge X)$ is locally finite-dimensional over Z_2. Then Ext of comodules over A_* is the same as Ext of modules over A :

$$\mathrm{Ext}_{A_*}^{s,t}(Z_2, (HZ_2)_*(bu \wedge X)) \cong \mathrm{Ext}_A^{s,t}((HZ_2)^*(bu \wedge X), Z_2) .$$

The latter is the classical way of writing the E_2 term. Now of course the Künneth theorem gives us an isomorphism

$$(HZ_2)^*(bu \wedge X) \cong (HZ_2)^*(bu) \otimes_{Z_2} (HZ_2)^*(X) .$$

This is an isomorphism of A-modules, provided we make A act on the right-hand side by the diagonal action:

$$a(u \otimes v) = \sum_i (a_i' u) \otimes (a_i'' v)$$

where

$$\psi a = \sum_i a_i' \otimes a_i'' .$$

(The isomorphism is A-linear by the Cartan formula.) By 16.6 this gives

$$(HZ_2)^*(bu \wedge X) = (A \otimes_B Z_2) \otimes (HZ_2)^*(X)$$

where the right-hand side is again furnished with the diagonal action. On the other hand, if M is an S-module, then A acts on $A \otimes M$ by the left action

$$a'(a \otimes m) = a'a \otimes m$$

and on $(A \otimes_B Z_2) \otimes M$ by the diagonal action. We have an isomorphism

$$A \otimes_B M \longrightarrow (A \otimes_B Z_2) \otimes M$$

given by

$$a \otimes m \longrightarrow \sum_i a_i' \otimes a_i'' m.$$

So we find

$$(HZ_2)^*(bu \wedge X) \cong A \otimes_B (HZ_2)^*(X).$$

Now by a change-of-rings theorem we have

$$\begin{aligned}\mathrm{Ext}_A^{s,t}((HZ_2)^*(bu\wedge X), Z_2) &\\ &\cong \mathrm{Ext}_A^{s,t}(A\otimes_B (HZ_2)^*(X), Z_2)\\ &\cong \mathrm{Ext}_B^{s,t}((HZ_2)^*(X), Z_2).\end{aligned}$$

Finally, the assumption of local finite-dimensionality is unnecessary, provided we dualise the argument and work in homology the whole time. Using the corresponding lemmas for comodules and the "cotheorem" to the change-of-rings theorem, we find

$$\mathrm{Ext}_{A_*}^{s,t}(Z_2, (HZ_2)_*(bu\wedge X)) \cong \mathrm{Ext}_{B_*}^{s,t}(Z_2, (HZ_2)_*(X)) .$$

This proves 16.1.

The structure-theory for modules I defer for the moment, so the next thing is to prove 16.4, assuming the results of the structure-theory. I need one more result not yet stated.

LEMMA 16.7. (Adams and Margolis). Let M and N be modules over $K[x,y]$ which are <u>connective</u> (<u>bounded below</u>), i.e., there exists $n_0 \in Z$ such that $M_r = 0$ and $N_r = 0$ for $r < n_0$. Alternatively, let M and N be bounded above, i.e., M_r and N_r are zero for r greater than some n_0. Let $f: M \longrightarrow N$ be a map of modules such that

$$f_*: H_*(M;x) \longrightarrow H_*(N;x)$$

and

$$f_*: H_*(M;y) \longrightarrow H_*(N;y)$$

are isomorphisms. Then M and N are stably isomorphic.

Now we can continue to study bu. In 16.6 we said that by using the morphism $bu \xrightarrow{f^0 j} HZ_2$ we can regard $(HZ_2)^*(bu)$ as a quotient of $(HZ_2)^*(HZ_2) = A$. Dually, we can regard $(HZ_2)_*(bu)$ as a subobject of $(HZ_2)_*(HZ_2) = A_*$. In fact, for calculation it is usually convenient to apply the canonical anti-automorphism of A_*; in other words, instead of looking at the morphism

$$HZ_2 \wedge bu \xrightarrow{1 \wedge f^0 j} HZ_2 \wedge HZ_2,$$

and taking the induced map of homotopy, we look at

$$bu \wedge HZ_2 \xrightarrow{f^0 j \wedge 1} HZ_2 \wedge HZ_2$$

and take the induced map of homotopy.

LEMMA 16.8. $(f^0 j \wedge 1)_*$ identifies $\pi_*(bu \wedge HZ_2)$ with the subalgebra of A_* generated by

$$\xi_1^2, \xi_2^2, \xi_3, \xi_4, \ldots .$$

This is immediately equivalent to 16.6; $\text{Im}(f^0 j \wedge 1)_*$ is the annihilator of $Sq^1 A + Sq^{01} A$.

Similarly, one would identify $\pi_*(bo \wedge HZ_2)$ with the subalgebra of A_* generatedy by $\xi_1^4, \xi_2^2, \xi_3, \xi_4, \ldots$

In order to prove 16.4, on the structure of $(HZ_2)^*(bu)$ as a B-module, an obvious move is to compute the homology of $(HZ_2)^*(bu)$ for the boundaries Sq^1 and Sq^{01} (acting on the left). It is equivalent to compute the homology of $\pi_*(bu \wedge HZ_2)$ for the boundaries Sq^1 and Sq^{01} (acting on the right); these boundaries may be calculated as follows. Regard $\pi_*(bu \wedge HZ_2)$ as a subalgebra of A_*; let

$$\psi a = \sum_i a_i' \otimes a_i'' \ ;$$

then

$$aSq^1 = \sum_i a_i' \langle Sq^1, a_i'' \rangle$$

$$aSq^{01} = \sum a_i' \langle Sq^{01}, a_i'' \rangle .$$

These boundaries are derivations.

LEMMA 16.9 (i) The homology for Sq^1 is a polynomial algebra on one generator ξ_1^2.

(ii) The homology for Sq^{01} is an exterior algebra on generators $\xi_1^2, \xi_2^2, \xi_3^2, \ldots$.

Proof. (i) Decompose $\pi_*(bu \wedge HZ_2)$ as the tensor product of the following chain complexes.

$$(1) \quad 1, \xi_1^2, \xi_1^4, \xi_1^6, \xi_1^8, \ldots$$

$$(r) \quad 1, \xi_r^2 \leftarrow \xi_{r+1}, \quad \xi_r^4 \leftarrow \xi_r^2 \xi_{r+1}, \quad \xi_r^6 \leftarrow \xi_r^4 \xi_{r+1}, \quad \ldots .$$

$$(r \geq 2).$$

Each chain complex (r) has homology Z_2, generated by [1]. By the Künneth theorem, the homology of the tensor-product is the homology of (1). A similar proof works for (ii).

Proof of 16.4. We show that $\pi_*(bu \wedge HZ_2)$ contains a finite-dimensional submodule M_r such that $H(M_r; Sq^1)$ is Z_2, generated by $\xi_1^{2^r}$, and $H(M_r; Sq^{01})$ is Z_2, generated by ξ_r^2. It is sufficient to indicate the first few modules.

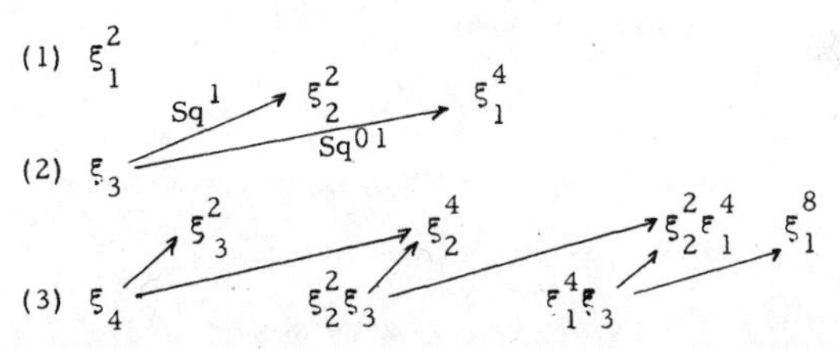

Since $\pi_*(bu \wedge HZ_2)$ is an algebra over B, we obtain a map

$$(1 + M_1)(1 + M_2) \ldots \longrightarrow \pi_*(bu \wedge HZ_2)$$

which induces an isomorphism of $H(\ ; Sq^1)$ and $H(\ ; Sq^{01})$, so that the two sides are stably isomorphic by 16.7. Dualising, we obtain the stable class of $(HZ_2)^*(bu)$ as

$$(1 + M_1^*)(1 + M_2^*) \ldots (1 + M_r^*) \ldots .$$

Here M_r^* satisfies the hypotheses of Theorem 16.3, which allows one to express it in the form $\Sigma^a I^b$. This proves 16.4.

All this work carries over to bo.

We now turn to the proof of 16.5. This is done essentially by the Bockstein spectral sequence, although I will not assume any knowledge of that. We recall that the Bockstein boundary

$$\beta_2 : (HZ_2)_n(bu) \longrightarrow (HZ_2)_{n-1}(bu)$$

is the boundary Sq^1 of 16.9.

<u>Proof of 16.5.</u> We separate the primes p. Let Q_p be the localisation of Z at p, that is, the subring of fractions a/b with b prime to p. We wish to prove that the image of $(HQ_p)_*(bu)$ in $(HQ)_*(bu)$ is the Q_p-subalgebra generated by t and t^{p-1}/p. Of course I give the proof for the case p = 2; the case of an odd prime is similar.

The spectrum bu has a (stable) cell-decomposition of the form

$$bu = S^0 \cup_\eta e^2 \cup \ldots$$

where η is the generator for the stable 1-stem, and the cells omitted have (stable) dimension ≥ 4. It follows that the Hurewicz homomorphism

$$Z \cong \pi_2(bu) \longrightarrow H_2(bu) \cong Z$$

is multiplication by 2; that is, $H_2(bu)$ is generated by $t/2 = T$, say. It follows immediately that the image of $H_*(bu) \longrightarrow (HQ)_*(bu)$ contains $T^r = (t/2)^r$. We wish to prove a result in the opposite direction.

The image of $H_{2r}(bu) \longrightarrow (HQ)_{2r}(bu)$ is a finitely-generated abelian group, and since it is non-zero, it is isomorphic to Z; let $h \in H_{2r}(bu)$ map to a generator. Let us write $\bar{h}$, $\bar{T}$ for the images of h, T in

$(HZ_2)_*(bu)$. Then we have

$$\beta_2 \bar{h} = 0 .$$

By 16.9, $(\mathrm{Ker}\, \beta_2 / \mathrm{Im}\, \beta_2)_{2r}$ is generated by ξ_1^{2r}. So we have

$$\bar{h} = \lambda \xi_1^{2r} + \beta_2 k$$

where $\lambda \in Z$ and $k \in (HZ_2)_{2r+1}(bu)$. That is,

$$\bar{h} = \lambda \bar{T}^r + \overline{\delta_2 k} ,$$

where $\delta_2 = (HZ_2)_{2r+1}(bu) \longrightarrow H_{2r}(bu)$ is the integral Bockstein. This gives

$$h = \lambda T^r + \delta_2 k + 2L$$

where $L \in H_{2r}(bu)$. For the images in $(HQ)_{2r}(bu)$ we have

$$h = \lambda(t/2)^r + 2\mu h$$

where $\mu \in Z$; that is,

$$h = \frac{\lambda}{1 - 2\mu}(t/2)^r$$

where $\frac{\lambda}{1-2\mu} \in Q_2$. This proves the result for the prime 2.

Now we turn to the structure-theory.

<u>Proof of 16.2.</u> Recall that A is a connected finite-dimensional Hopf algebra. So if M is an A-module, we can make its dual $M^* = \mathrm{Hom}_K^*(M, K)$ into an A-module. Also A^* is free on one generator. Recall also that I is the augmentation ideal of A, so that we have the following exact sequence.

$$0 \longrightarrow I \longrightarrow A \longrightarrow 1 \longrightarrow 0$$

Dualising, we have the following exact sequence.

$$0 \longrightarrow 1 \longrightarrow A^* \longrightarrow I^* \longrightarrow 0$$

Tensoring the first sequence with I^*, we have

$$0 \longrightarrow I \otimes I^* \longrightarrow A \otimes I^* \longrightarrow I^* \longrightarrow 0 .$$

Here A^* and $A \otimes I^*$ are free. By Schanuel's lemma, we have

$$(I \otimes I^*) + A^* \cong 1 + (A \otimes I^*) ,$$

so $I \otimes I^*$ is stably isomorphic to 1, and I is invertible. This proves 16.2.

To prove 16.3 (i) I need 16.7, the lemma of Adams and Margolis. First one proves a special case.

LEMMA 16.10. Let M be a module over $K[x, y]$ which is connective, i.e., bounded below; alternatively, let M be bounded above. Assume $H_*(M;x) = 0$, $H_*(M;y) = 0$. Then M is free.

<u>Proof of 16.10.</u> Since $H_*(M;x) = 0$, we have a short exact sequence

$$0 \longrightarrow M/xM \xrightarrow{i} M \xrightarrow{j} M/xM \longrightarrow 0$$

in which $i([m]) = xm$ and j is the quotient map. This leads to a long exact sequence of homology for the boundary y, namely

$$H_r(M;y) \longrightarrow H_r(M/xM;y) \longrightarrow H_{r+|y|-|x|}(M/xM;y) \longrightarrow H_{r+|y|}(M;y).$$

Since $H_*(M;y) = 0$, we have

$$H_r(M/xM;y) = H_{r+|y|-|x|}(M/xM;y) .$$

Since M is bounded either below or above, we have $H_r(M/xM;y) = 0$ either for $r < n_0$ or for $r > n_1$. Since $|y| - |x| \neq 0$, we can use the isomorphism

$$H_r(M/xM;y) \cong H_{r+|y|-|x|}(M/xM;y)$$

to prove by induction over r that

$$H_r(M/xM;y) = 0$$

for all r.

It is now immediate that M/xM is free over $K[y]$. That is, let b_α be elements in M whose images form a K-base in

$$\frac{M/xM}{y(M/xM)} \quad ;$$

then the images of b_α, yb_α form a K-base in M/xM. It follows that the elements $b_\alpha, yb_\alpha, xb_\alpha, xyb_\alpha$ form a K-base in M. This proves 16.10.

Proof of 16.7. Let $f: M \longrightarrow N$ be a map of modules, say bounded below, such that

$$f_*: H_*(M;x) \longrightarrow H_*(N;x)$$

and

$$f_*: H_*(M;y) \longrightarrow H_*(N;y)$$

are isomorphisms. By adding to M a free module F bounded below, we can extend f to $f' = (f, g): M \oplus F \longrightarrow N$ which is onto and also induces an isomorphism of $H_*(\;;x)$, $H_*(\;;y)$. Consider Ker f'; this is bounded below, and by the exact homology sequence we have $H_*(\text{Ker } f';x) = 0$, $H_*(\text{Ker } f';y) = 0$. So Ker f' is free by 16.10. But over $K[x, y]$ the free modules are injective, so we have

$$M \oplus F \cong N \oplus \text{Ker } f'$$

and M is stably isomorphic to N. This proves 16.7.

Proof of 16.3 (i). Let M be a finite-dimensional module over $K[x, y]$ such that $H_*(M;x)$ and $H_*(M;y)$ have dimension 1 over K. Then the same holds for M^*. Consider the evaluation map $M^* \otimes M \longrightarrow 1$. This is a map of modules over $K[x, y]$, and (using the Künneth theorem) it induces an isomorphism of $H_*(\;;x)$, $H_*(\;;y)$. By 16.7, $M^* \otimes M$ and 1 are stably isomorphic; so M is invertible. This proves 16.3 (i).

To prove 16.3 (ii) we need some more structure theory. First we put in evidence several examples of graded modules over $K[x, y]$. The first is called the lightning-flash. It has generators g_i in dimension

$(|y| - |x|)i$ $(i \in Z)$ and relations $yg_i = xg_{i+1}$.

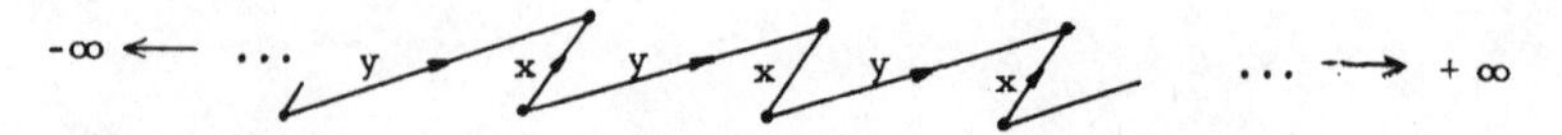

We can bring the lightning-flash to an end on the left either by taking the submodule generated by the g_i for $i \geq \nu$, or by taking a quotient module, factoring out the g_i for $i < \nu$.

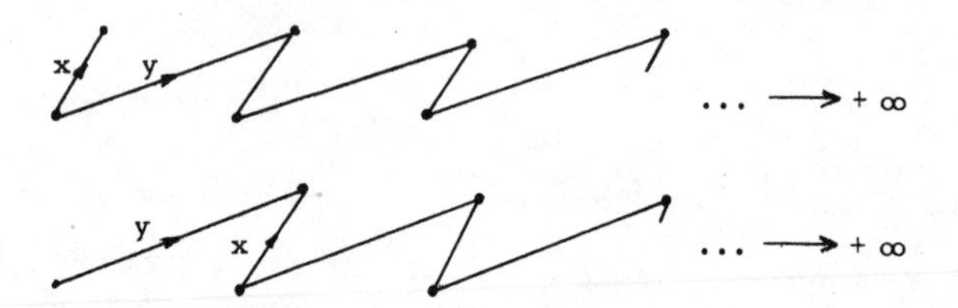

In the latter case $xg_\nu = yg_{\nu-1} = 0$. Similarly, we can bring the lightning-flash to an end on the right, either by taking the submodule generated by the g_i for $i \leq \nu$, or by taking a quotient module, factoring out the g_i for $i > \nu$.

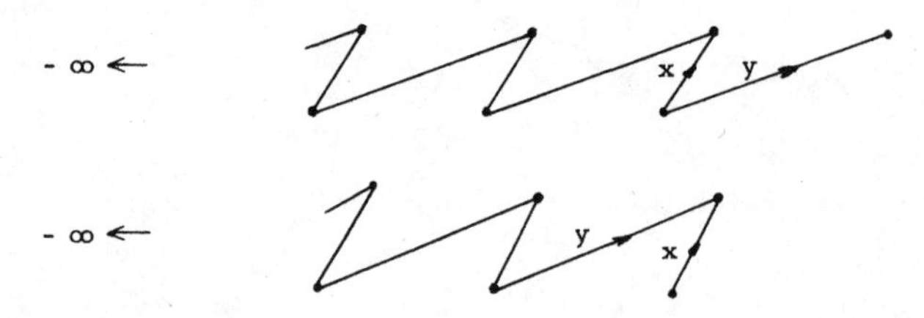

In the latter case $yg_\nu = xg_{\nu+1} = 0$.

If we want finite-dimensional modules, we can end the lightning-flash two ways on the left and two ways on the right, giving four sorts of module. Of course, for modules of one sort we can alter the length, e.g.,

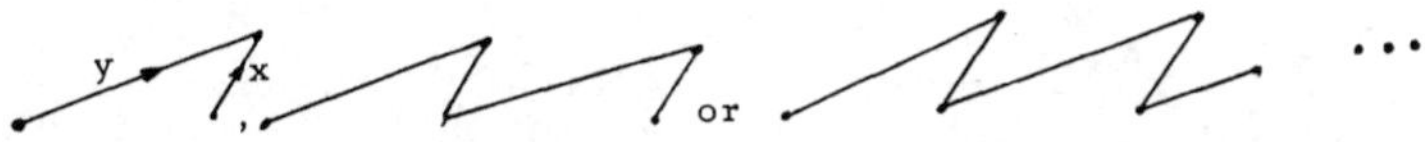

Also we might alter the grading, e.g., we might put the generator g_0 in degree 1 instead of degree 0.

We add to these four sorts of modules the free modules on one generator.

THEOREM 16.11. Let M be a module over $K[x, y]$ which is finite-dimensional over K. Then M is a (finite) direct sum of modules of these five types.

<u>First step.</u> Suppose $xyM \neq 0$. Then M is the direct sum of some module N and a free module on one generator.

<u>Proof.</u> Take $m_0 \in M_r$ such that $xym_0 \neq 0$. Then there is a linear functional $\theta: M_{r+|x|+|y|} \longrightarrow K$ such that $\theta(xym_0) = 1$. Let F be free on one generator f of degree r. Define maps of modules

$$F \xrightarrow{\alpha} M \xrightarrow{\beta} F$$

by $\alpha(f) = m_0$,

$$\beta(m) = \begin{cases} \theta(m)xyf & (m \in M_{r+|x|+|y|}) \\ \theta(xm)yf & (m \in M_{r+|y|}) \\ -\theta(ym)xf & (m \in M_{r+|x|}) \\ \theta(xym)f & (m \in M_r) \\ 0 & (\text{otherwise}) . \end{cases}$$

This shows $M \cong (\operatorname{Ker} \beta) \oplus F$.

Second step. $M \cong N \oplus F$, where F is free and N is annihilated by xy.

Proof. Choose a K-base for xyM. Let m_α be elements in M such that the elements xym_α are the chosen K-base in xyM. Either proceed as in the first step, or remark that this gives an injection $F \longrightarrow M$ and F is injective.

In what follows, then, we can assume that M is annihilated by xy, and we have to prove that M is a (finite) direct sum of modules of the four types.

By a base for a graded module, we mean a K-base of homogeneous elements.

We will say that a base $\{b_\alpha\}$ for M is good if it satisfies the following conditions.

(i) For each vector b_α in the base, xb_α is either zero or a vector in the base; and $xb_\alpha = xb_\beta \neq 0$ implies $\alpha = \beta$.

(ii) For each vector b_α in the base, yb_α is either zero or a vector in the base; and $yb_\alpha = yb_\beta \neq 0$ implies $\alpha = \beta$.

LEMMA 16.12. If $xyM = 0$ and M has a good base, then the conclusion of Theorem 16.11 follows.

Proof. Suppose M has a good base $\{b_\alpha\}$. Take the indices α as the vertices of a graph. It is a finite graph, since we are assuming M finite-dimensional over K. For each relation $xb_\alpha = b_\beta$ introduce one directed edge marked "x" and running from α to β. For each relation $yb_\alpha = b_\beta$ introduce one directed edge marked "y" and running from α to β. Divide the graph into connected components. It is clear that a vector cannot have edges arriving and departing, since

xx, xy, yx and yy act as zero on M. By the definition of a "good base", a vector cannot have more than two edges arriving (one x and one y), and of course it cannot have more than two edges departing (one x and one y). The connected components of the graph are therefore zigzags. (A zigzag cannot join up into a closed polygon, because we assume $\deg x \neq \deg y$.) Each connected component of the graph gives a submodule of M, which is of one of the 4 types described above; and M is their direct sum. This proves 16.12.

We define the indecomposable quotient $Q(M)$ of M by $Q(M) = M/(xM + yM)$. Over K we can if we wish choose a direct sum splitting

$$M = (Q(M) \oplus (xM + yM).$$

Both x and y map $xM + yM$ to 0, since we assume $xyM = 0$; they also map $Q(M)$ to $(xM + yM)$.

Let V be a finite-dimensional vector space over K, and let

$$0 \subset V_1 \subset V_2 \subset \cdots \subset V_n = V$$

be a filtration of V by a finite increasing sequence of vector subspaces. We say that a K-base $\{b_\alpha\}$ for V is <u>adapted</u> to the filtration if, for every i, those b_α which lie in V_i form a base for V_i.

LEMMA 16.13. Let M be a module over $K[x, y]$ such that (i) $xyM = 0$ and (ii) $Q(M)_r = 0$ for $r < a$ and for $r > c$. Then there are filtrations of yM_r for $c - \delta < r \leq c$ with the following property. For each r in the range $c - \delta < r \leq c$ let $\{b_{r\alpha}\}$ be a base of yM_r which is adapted to the filtration; then the set of elements $b_{r\alpha}$ can be extended to a good base of M.

Note. It is assumed that deg y > deg x, and δ has been written for deg y - deg x.

Note. In the range $c-\delta < r \leq c$ we have $xM_{r+\delta} = 0$, and therefore $yM_r = xM_{r+\delta} + yM_r$. So the vector space being filtered is the whole of the decomposable subspace of M in the dimension concerned.

COROLLARY 16.14. If M is as in 16.13, it has a good base.

Proof. Any filtered vector space

$$0 \subset V_1 \subset V_2 \subset \dots \subset V_n = V$$

has at least one adapted base; for one begins by choosing a base for V_1, extends it to a base for V_2, and so on by induction. So 16.13 provides a good base for M.

Proof of 16.13. The proof is essentially by induction over c - a; the result is true if $c < a$, for then $M = 0$.

Choose a direct sum splitting $M = Q(M) \oplus (xM + yM)$. Let N be the submodule of M generated by $Q(M)_r$ for $a \leq r < c$. The relations between N and M are as follows. We have $Q(N)_r = Q(M)_r$ except for $r = c$, in which case $Q(N)_c = 0$. Thus we have $xN_r = xM_r$ and $yN_r = yM_r$ except for $r = c$; that is,

$$(xN + yN)_r = (xM + yM)_r$$

except for $r = c + d$ and $r = c + e$, where $d = \deg x$, $e = \deg y$. In the first case we have

$$yN_{c-\delta} = xN_c + yN_{c-\delta} \subset xM_c + yM_{c-\delta},$$

and in the second case we have

$$0 = xN_{x+\delta} + yN_c \subset xM_{c+\delta} + yM_c = yM_c.$$

We assume, as our inductive hypothesis, that the lemma is true for N.

Then there are filtrations of yN_r for $c-\delta \leq r < c$ which have the property stated in the lemma. In particular, let the filtration of $yN_{c-\delta} = yM_{c-\delta}$ be

$$0 = V_0 \subset V_1 \subset V_2 \subset \dots \subset V_n = yM_{c-\delta} .$$

Adjoin to it the further subgroup

$$V_{n+1} = xM_c + yM_{c-\delta} .$$

We have a map $x\colon Q(M)_c \longrightarrow xM_c + yM_{c-\delta}$; so we can filter the vector space $Q(M)_c$ by the counterimages

$$0 \subset x^{-1}V_0 \subset x^{-1}V_1 \subset \dots \subset x^{-1}V_n \subset x^{-1}V_{n+1} = Q(M)_c .$$

We also have a map $y\colon Q(M)_c \longrightarrow yM_c$. We filter yM_c by taking the images

$$0 \subset yx^{-1}V_0 \subset yx^{-1}V_1 \subset \dots\dots \subset yx^{-1}V_n \subset yx^{-1}V_{n+1} = yM_c .$$

We now have filtrations on yM_r for $c-\delta < r \leq c$; those for $c-\delta < r < c$ arise from the inductive hypothesis, and that for $r = c$ has just been constructed. Suppose given bases $\{b_{r\alpha}\}$ in yM_r for $c-\delta < r \leq c$, adpated to the filtrations. We leave the bases as they are for $c-\delta < r < c$, and start work on the base $\{b_{c\alpha}\}$ for yM_c.

In $Q(M)_c$ we may choose elements b'_α such that $yb'_\alpha = b_{c\alpha}$ and $b'_\alpha \in x^{-1}V_m$ if and only if $b_{c\alpha} \in yx^{-1}V_m$. We may also choose elements b''_β in $Q(M)_c$ forming a base adapted to the following filtration.

$$0 \subset \operatorname{Ker} y \cap x^{-1}V_0 \subset \operatorname{Ker} y \cap x^{-1}V_1 \subset \dots \subset \operatorname{Ker} y \cap x^{-1}V_n \subset \operatorname{Ker} y \cap x^{-1}V_{n+1}$$

The elements b'_α and b''_β together form a base of $Q(M)_c$ adapted to the filtration

$$0 \subset x^{-1}V_0 \subset x^{-1}V_1 \subset \dots \subset x^{-1}V_n \subset x^{-1}V_{n+1} = Q(M)_c .$$

From among the elements b'_α and b''_β, let us for the moment omit those which lie in $x^{-1}V_0 = \text{Ker } x$ and those which do not lie in $x^{-1}V_n$. Then the remaining xb'_α and xb''_β form a base of $yN_{c-\delta}$ compatible with its filtration. By the inductive hypothesis, the bases in yN_r for $c-\delta \leq r < c$ form part of a good base for N. We now adjoin to this base for N the elements b'_α and b''_β in $Q(M)_c$, the elements $yb'_\alpha = b_{c\alpha}$ in yM_c, and the elements xb'_α, xb''_β for which b'_α, b''_β do not lie in $x^{-1}V_n$. We obtain a good base for M, containing the given elements $b_{r\alpha}$. This completes the induction, and proves 16.13.

This therefore completes the proof of 16.14. Theorem 16.11 follows from 16.14 and 16.12, so this completes the proof of 16.11.

Proof of 16.3 (ii). Let M be a finite-dimensional module over $K[x, y]$ such that $H_*(M;x)$ and $H_*(M;y)$ both have dimension 1 over K. Then by 16.11 it is a sum of modules of the types considered above. By inspecting $H_*(M;x)$ and $H_*(M;y)$, it can have only one summand which is not free, and this summand can only lie in two out of the four types. By the same argument applied to $\Sigma^a I^b$, each such summand is stably equivalent to some $\Sigma^a I^b$.

17. STRUCTURE OF $\pi_*(bu \wedge bu)$

Mahowald and others have been using methods which rely essentially on a calculation of $\pi_*(bo \wedge bo \wedge \ldots \wedge bo)$, where we take $(n+1)$ factors bo. I would like to give an introduction to this calculation; it seems best if I do things for the most elementary case, which is the case of bu, but undertake to use only methods which extend to the case bo. For similar reasons I will mostly consider the case of two factors $bu \wedge bu$; the case of $(n+1)$ factors is similar. Again, I will consider mostly the prime 2, but try to make only statements which can also be made for the prime p.

Some things can be said for a fairly general connective spectrum X. My standing hypotheses on X will be as follows. First, assume that for each r, $H_r(X)$ is a finitely generated group. This may be unnecessary for some purposes, but it is convenient. Secondly, for each prime p, consider $(HZ_p)^*(X)$ as a module over $B = Z_p[Q_0, Q_1]$, and assume that its stable class is $\bigoplus_i \Sigma^{a(i,p)} I^{b(i,p)}$, where $b(i,p) \geq 0$ and $a(i,p) + b(i,p) \equiv 0 \bmod 2$.

<u>Example.</u> Let $X = bu \wedge bu \wedge \ldots \wedge bu$ (n factors). We have checked the condition at the prime 2 by 16.4. We have not checked the condition at the prime $p > 2$, but I believe it holds. In any case, the results at the prime 2 follow from the assumptions at the prime 2.

Our assumptions on X have obvious consequences for the homology of X with integral coefficients.

LEMMA 17.1. (i) $H_*(X)$ is a direct sum of groups Z_2 and Z_p, and groups Z in even degree.

(ii) The same holds for $H_*(bu \wedge X)$.

<u>Proof.</u> (i) The argument is essentially by the Böckstein spectral sequence, but we do not need to assume any knowledge of that. By assumption, $H_r(X)$ is finitely-generated abelian group; so it is a direct sum of groups Z_{p^f} and Z. A group Z_{p^f} with $f \geq 2$ will introduce into $\mathrm{Ker}\,\beta_p/\mathrm{Im}\,\beta_p$ two groups Z_p in consecutive degrees, which is impossible; we have assumed $\mathrm{Ker}\,\beta_p/\mathrm{Im}\,\beta_p$ has one summand Z_p in each degree $a(i,p) + b(i,p)$, and that $a(i,p) + b(i,p)$ is always even. A

group Z in degree r will introduce into $\mathrm{Ker}\,\beta_p/\mathrm{Im}\,\beta_p$ a group Z_p in degree r, which is possible only if r is even.

(ii) The spectrum $bu \wedge X$ satisfies the assumptions made on X.

Of course we propose to obtain essential information on $\pi_*(bu \wedge X)$ from the spectral sequence 16.1. The two results which we obtain this way are as follows.

PROPOSITION 17.2. Assume that X is as above.

(i) The Hurewicz homomorphism

$$h: \pi_*(bu \wedge X) \longrightarrow H_*(bu \wedge x)$$

is a monomorphism.

(ii) The Hurewicz homomorphism

$$h: \pi_*(K \wedge X) \longrightarrow H_*(K \wedge X)$$

is a monomorphism.

(iii) The homomorphism

$$\pi_*(K \wedge X) \longrightarrow \pi_*(K \wedge X) \otimes Q$$

is a monomorphism.

Part (ii) follows immediately from part (i), by passing to direct limits.

Part (iii) follows from part (ii); we have $H_*(K) \cong \pi_*(K) \otimes Q$, and therefore $H_*(K \wedge X) \cong \pi_*(K \wedge X) \otimes Q$.

Given this proposition, one obviously tries to get a hold on $\pi_*(K \wedge X)$ by describing its image in $\pi_*(K \wedge X) \otimes Q$. It is also very reasonable to try to get a hold on $\pi_*(bu \wedge X)$ by describing its image in $\pi_*(bu \wedge X) \otimes Q$; the kernel of

$$\pi_*(bu \wedge X) \longrightarrow \pi_*(bu \wedge X) \otimes Q$$

may contain elements of order p, but no elements of order p^2; this follows of course from 17.1 and 17.2. The p-torsion subgroup of

$\pi_*(bu \wedge X)$ maps monomorphically to $(HZ_p)_*(X)$.

We shall also need another result. Consider the following diagram.

$$\begin{array}{ccc} \pi_*(bu \wedge X) & \longrightarrow & H_*(bu \wedge X) \\ \downarrow & & \downarrow \\ \pi_*(K \wedge X) & \longrightarrow & H_*(K \wedge X) = \pi_*(K \wedge X) \otimes Q \end{array}$$

THEOREM 17.3. Let X be as above. Suppose an element $h \in H_*(K \wedge X)$ lies both in the image of $H_*(bu \wedge X)$ and in the image of $\pi_*(K \wedge X)$. Then it lies in the image of $\pi_*(bu \wedge X)$.

The usefulness of this result will appear later.

I said it was reasonable to try to get a hold on $\pi_*(K \wedge X)$ by describing its image in $\pi_*(K \wedge X) \otimes Q$, and to try to get a hold on $\pi_*(bu \wedge X)$ by describing its image in $\pi_*(bu \wedge X) \otimes Q$. In the case $X = bu$ we see that $\pi_*(bu \wedge bu) \otimes Q$ is the polynomial algebra $Q[u, v]$, where $u \in \pi_2(bu)$ and $v \in \pi_2(bu)$ are the generators for the two factors. Similarly, we have

$$\pi_*(K \wedge bu) \otimes Q = Q[u, u^{-1}, v] .$$

We wish to describe the images of the maps

$$\pi_*(K \wedge bu) \longrightarrow \pi_*(K \wedge bu) \otimes Q = Q[u, u^{-1}, v]$$

$$\pi_*(bu \wedge bu) \longrightarrow \pi_*(bu \wedge bu) \otimes Q = Q[u, v] .$$

THEOREM 17.4. In order that a finite Laurent series $f(u, v) \in Q[u, u^{-1}, v]$ lie in the image of $\pi_*(K \wedge bu)$, it is necessary and sufficient that it satisfy the following condition.

Condition (1): for all $k \neq 0$, $\ell \neq 0$ in Z we have

$$f(kt, \ell t) \in Z[t, t^{-1}, k^{-1}, \ell^{-1}].$$

THEOREM 17.5. In order that a polynomial $f(u,v) \in Q[u,v]$ lie in the image of $\pi_*(bu \wedge bu)$, it is necessary and sufficient that it satisfy the following two conditions.

Condition (1): as in 17.4.

Condition (2); it lies in the subgroup additively generated by the monomials

$$\frac{u^i}{m(i)} \quad \frac{v^j}{m(j)} .$$

Here $m(r) = \prod_p p^{\left[\frac{r}{p-1}\right]}$, as in section 16. Of course, the subgroup specified is actually a subring.

It is very easy to prove that the conditions given in 17.4 and 17.5 are necessary, so I will do that now.

Proof that Condition (1) is necessary. Consider the following commutative diagram.

$$\begin{array}{ccc}
\pi_*(K \wedge bu) & \longrightarrow & \pi_*(K \wedge bu) \otimes Q = Q[u, u^{-1}, v] \\
\Big\downarrow \Psi^k \otimes \Psi^\ell & & \Big\downarrow \Psi^k \otimes \Psi^\ell \\
\pi_*(K \wedge bu) \otimes Z[k^{-1}, \ell^{-1}] & & \pi_*(K \wedge bu) \otimes Q \\
\Big\downarrow \mu & & \Big\downarrow \mu \\
\pi_*(K) \otimes Z[k^{-1}, \ell^{-1}] & \longrightarrow & \pi_*(K) \otimes Q \\
\| & & \| \\
Z[t, t^{-1}, k^{-1}, \ell^{-1}] & & Q[t, t^{-1}]
\end{array}$$

The right-hand vertical arrow carries $f(u,v)$ into $f(kt, \ell t)$. This proves that Condition (1) is necessary.

Proof that Condition 2 is necessary. Consider the following commutative diagram.

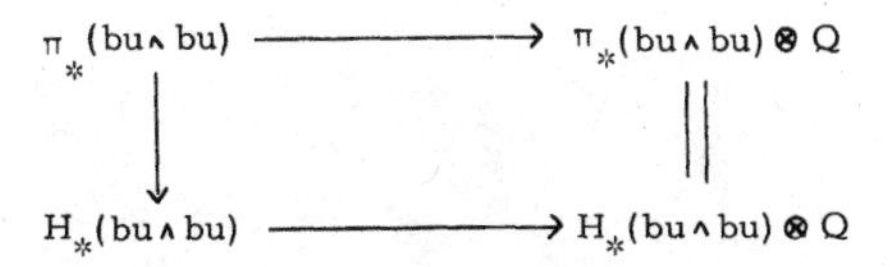

$$\begin{array}{ccc} \pi_*(bu \wedge bu) & \longrightarrow & \pi_*(bu \wedge bu) \otimes Q \\ \downarrow & & \| \\ H_*(bu \wedge bu) & \longrightarrow & H_*(bu \wedge bu) \otimes Q \end{array}$$

Here $H_*(bu \wedge bu)$ is described by the Künneth theorem, and the terms $\mathrm{Tor}_1^Z(H_i(bu), H_j(bu))$ map to zero in $H_*(bu \wedge bu) \otimes Q$, so the image of $H_*(bu \wedge bu)$ in $H_*(bu \wedge bu) \otimes Q$ is the same as the image of $H_*(bu) \otimes H_*(bu)$. By 16.5, this is the subgroup additively generated by the monomials

$$\frac{u^i}{m(i)} \quad \frac{v^j}{m(j)} .$$

This proves that Condition (2) is necessary.

<u>Proof of 17.5 from 17.3 and 17.4.</u> Suppose a polynomial $f(u, v)$ satisfies Conditions (1) and (2). Consider f as an element of $Q[u, u^{-1}, v] = \pi_*(K \wedge bu) \otimes Q = H_*(K \wedge bu)$. According to the proof we have just given, Condition (2) ensures that f lies in the image of $H_*(bu \wedge bu)$. By 17.4, Condition (1) ensures that f lies in the image of $\pi_*(K \wedge bu)$. Now 17.3 shows that it lies in the image of $\pi_*(bu \wedge bu)$. This proves 17.5.

<u>Remarks.</u> When we replace bu by bo, we replace $Q[u, v]$ by $Q[u^2, v^2]$ and $Q[u, u^{-1}, v]$ by $Q[u^2, u^{-2}, v^2]$; that is, we only use functions which are even in both variables. We also replace the ring $Z[t, t^{-1}, k^{-1}, \ell^{-1}]$ by $\pi_*(KO) \otimes Z[k^{-1}, \ell^{-1}]$; since we only need the components of degree congruent to 0 mod 4, this is essentially $Z[2t^2, t^4, t^{-4}, k^{-1}, \ell^{-1}]$. Condition (2) is unchanged.

In order to do calculations it is often desirable to know exactly what functions do satisfy the condition given. In such calculations it is usually convenient to separate the primes and consider the images of

$$\pi_*(K \wedge bu) \otimes Q_p \longrightarrow \pi_*(K \wedge bu) \otimes Q$$

$$\pi_*(bu \wedge bu) \otimes Q_p \longrightarrow \pi_*(bu \wedge bu) \otimes Q \, .$$

Of course I consider the prime 2. The analogue of Condition (1) reads as follows.

(1') For each pair of odd integers, k, ℓ, $f(kt, \ell t) \in Q_2[t, t^{-1}]$.

The analogue of Condition (2) reads as follows.

(2') $f(u,v) \in Q_2[u/2, v/2]$.

PROPOSITION 17.6. (i) The subring of finite Laurent series which satisfy (1') is free over $Q_2[u, u^{-1}]$ on generators

$$1, \frac{v-u}{3-1}, \frac{(v-u)(v-3u)}{(5-1)(5-3)}, \frac{(v-u)(v-3u)(v-5u)}{(7-1)(7-3)(7-5)} \cdots$$

(ii) The subring of polynomials which satisfy (1') and (2') is free over Q_2 on the following generators.

$$u^4, \frac{u^4(v-u)}{2}, \frac{u^4(v-u)(v-3u)}{2^3}, \frac{u^4(v-u)\ldots(v-5u)}{2^4}, \frac{u^4(v-u)\ldots(v-7u)}{2^7} \cdots$$

$$u^3, \frac{u^2(v-u)}{2}, \frac{u^3(v-u)(v-3u)}{2^3}, \frac{u^3(v-u)\ldots(v-5u)}{2^4}, \frac{u^3(v-u)\ldots(v-7u)}{2^7} \cdots$$

$$u^2, \frac{u^2(v-u)}{2}, \frac{u^2(v-u)(v-3u)}{2^3}, \frac{u^2(v-u)\ldots(v-5u)}{2^4}, \frac{u^2(v-u)\ldots(v-7u)}{2^6} \cdots$$

$$u, \frac{u(v-u)}{2}, \frac{u(v-u)(v-3u)}{2^3}, \frac{u(v-u)\ldots(v-5u)}{2^4}, \frac{u(v-u)\ldots(v-7u)}{2^5} \cdots$$

$$1, \frac{v-u}{2}, \frac{(v-u)(v-3u)}{2^2}, \frac{(v-u)(v-3u)(v-5u)}{2^3}, \frac{(v-u)(v-3u)(v-5u)(v-7u)}{2^4} \cdots$$

The principle in part (ii) is that one takes each product $(v-u)(v-3u)\ldots(v-(2n-1)u)$, multiplies it by u^i, and then divides it by the greatest power of 2 which will still leave it satisfying (1') and (2'). The greatest power of 2 which leaves it satisfying (1') is read off from (1), and is the 2-primary factor of $2^n(n!)$. The greatest power of 2 which leaves it satisfying (2') is 2^{n+i}.

Remark. For an odd prime p we replace the arithmetic progress progression 1, 3, 5, 7, of 17.6 by the sequence of positive integers prime to p. Alternatively, if one takes the precaution of splitting buQ_p into $(p-1)$ similar summands and taking one of them, one replaces $(v-u)(v-3u)(v-5u)\ldots$ by $(v^{p-1}-u^{p-1})(v^{p-1}-(p+1)u^{p-1})(v^{p-1}-(2p+1)u^{p-1})\ldots$ When one replaces bu by bo, one replaces $(v-u)(v-3u)(v-5u)\ldots$ by $(v^2-1^2u^2)(v^2-3^2u^2)(v^2-5^2u^2)\ldots$.

The proof of 17.6 is straight algebra, and will be given later.

We begin the proof of these results with a simple result on the homology of X, essentially comparable with 17.1.

LEMMA 17.7. Let X be as in 17.1-17.3, and let $\{c_i\}$ be any Z_2-base for the subquotient $\operatorname{Ker}\beta_2/\operatorname{Im}\beta_2$ of $(HZ_2)_{2r}(X)$ (e.g. arising from our assumed decomposition $(HZ_2)^*(X) \cong \oplus\, \Sigma^{a(i,2)}I^{b(i,2)}$.) Let $h_i \in H_{2r}(X)$ be any element whose image in $(HZ_2)_{2r}(X)$ is c_i. Then the elements h_i yield a Q_2-base for the image of $(HQ_2)_{2r}(X)$ in $(HQ)_{2r}(X)$.

Proof. Let k_j be a Z-base for $H_{2r}(X)$ mod torsion; then in $H_{2r}(X)$ m6d torsion we can write $h_i = \sum_j a_{ij}k_j$, where $a_{ij} \in Z$. When we pass to $(HZ_2)_{2r}(X)$, both the h_i and the k_j yield a Z_2-base for $\operatorname{Ker}\beta_2/\operatorname{Im}\beta_2$. So the h_i and k_j are equal in number, and $\det(a_{ij})$ is odd. The result follows.

Next I recall some results of homological algebra over $K[x,y]$. Consider the following short exact sequences.

$$0 \longrightarrow \Sigma^{|x|} \xrightarrow{x} \frac{K[x,y]}{yK[x,y]} \longrightarrow 1 \longrightarrow 0$$

$$0 \longrightarrow \Sigma^{|y|} \xrightarrow{y} \frac{K[x,y]}{xK[x,y]} \longrightarrow 1 \longrightarrow 0$$

They represent elements

$$\xi \in \operatorname{Ext}^{1,|x|}_{K[x,y]}(K,K) ,$$

$$\eta \in \operatorname{Ext}^{1,|y|}_{K[x,y]}(K,K) .$$

LEMMA 17.8. $\operatorname{Ext}^{**}_{K[x,y]}(K,K)$ is the polynomial algebra $K[\xi,\eta]$.

This is a completely standard calculation.

LEMMA 17.9. We have an epimorphism

$$\operatorname{Ext}^{s,t}_{K[x,y]}(I \otimes M, K) \longrightarrow \operatorname{Ext}^{s+1,t}_{K[x,y]}(M,K)$$

which is an isomorphism for $s > 0$.

This is trivial, since we have an exact sequence

$$0 \longrightarrow I \otimes M \longrightarrow A \otimes M \longrightarrow M \longrightarrow 0$$

with $A \otimes M$ free.

Now observe that as a matter of formal algebra, I can construct a free module over $K[\xi,\eta]$ on various generators, where I may assign bidegrees to the generators at will. In particular, given M as a locally-finite sum $M \cong \bigoplus_i \Sigma^{a(i)} I^{b(i)}$ with $b(i) \geq 0$, I take F to be a free module over K with generators a_i of bidegrees $s = -b(i)$, $t = a(i)$.

LEMMA 17.10. In degrees $s \geq 0$ we have an epimorphism

$$\operatorname{Ext}^{**}_{K[x,y]}(M, Z_2) \longrightarrow F$$

which is an isomorphism in degrees $s > 0$.

The case of one factor $\Sigma^a I^b$ follows immediately from 17.8 and 17.9; the factor Σ^a causes a trivial shift in the t-grading. Then one passes to sums.

Now I specialise to the case $p = 2$, $K[x, y] = B$, $a(i) = a(i, 2)$, $b(i) = b(i, 2)$. Then Lemma 17.10 computes for us the E_2-term of the spectral sequence 16.1, which converges to $\pi_*(bu \wedge X)$ at the prime 2.

LEMMA 17.11 (i) There is a homomorphism $E_r^{s,t} \longrightarrow E_r^{s+t, t+1}$ of the spectral sequence 16.1 which for $r = 2$ is multiplication by ξ and for $r = \infty$ is obtained by passing to quotients from multiplication by 2 in $\pi_*(bu \wedge X)$.

(ii) There is a homomorphism $E_r^{s,t} \longrightarrow E_r^{s+1, t+3}$ of the spectral sequence 16.1 which for $r = 2$ is multiplication by η and for $r = \infty$ is obtained by passing to quotients from multiplication by the generator $t \in \pi_2(bu)$ in $\pi_*(bu \wedge X)$.

For an odd prime we use t^{p-1} in part (ii). For bo we use the generator in $\pi_8(bo)$, and replace η by the generator in $\mathrm{Ext}_{A_1}^{4,12}(Z_2, Z_2)$.

Part (i) is absolutely standard. For part (ii), consider the morphism $S^2 \wedge bu \longrightarrow bu$ which corresponds to multiplication by the generator $t \in \pi_2(bu)$, consider its effect on the spectral sequence 15.1, and chase that effect through the change-of-rings theorem.

LEMMA 17.12. Let X be as in 17.1-17.3. Then the spectral sequence of 16.1 has all its differentials zero.

Proof. From 17.10 and our assumption that $a(i) + b(i) \equiv 0 \bmod 2$, it follows that $E_2^{s,t} = 0$ for $s > 0$ and $t-s \equiv 1 \bmod 2$; therefore the same holds for $E_r^{s,t}$. So it is sufficient to consider $d_r(e)$, where

$e \in E_r^{s,t}$ and $s = 0$, $t-s \equiv 1 \bmod 2$.

We suppose, as an inductive hypothesis, that $d_m = 0$ for $m < r$, so that

$$E_r^{s,t} \cong E_2^{s,t} \cong \mathrm{Ext}_B^{s,t}((HZ_2)_*(X), Z_2) .$$

Argument (i).

$$\xi d_r(e) = d_r(\xi e) = 0,$$

but multiplication by ξ is a monomorphism on Ext^s for $s > 0$, therefore on $E_r^{s,t}$, so $d_r(e) = 0$.

Argument (ii).

$$\eta d_r(e) = d_r(\eta e) = 0 ,$$

but multiplication by η is a monomorphism on Ext^s for $s > 0$, therefore on $E_r^{s,t}$, so $d_r(e) = 0$.

This completes the induction, and proves 17.12.

Remark. Argument (ii) becomes better than argument (i) when we replace bu by bo.

Proof of 17.2 (i). Let $\alpha \in \pi_*(bu \wedge X)$ be an element in the kernel of the Hurewicz homomorphism. Then certainly α maps to zero in $(HZ_p)_*(bu \wedge X)$, i.e., α has filtration at least 1 in the spectral sequence 16.1, and similarly for odd primes p. Also α maps to zero in $(HQ)_*(bu \wedge X) \cong \pi_*(bu \wedge X) \otimes Q$, so α is a torsion element. But by 17.10, 17.1 and 17.12 multiplication by 2 induces a monomorphism $E_\infty^{s,t} \longrightarrow E_\infty^{s+1,t+1}$ for $s > 0$, i.e., multiplication by 2 is a monomorphism on the subgroup of elements of filtration at least 1; and similarly for odd primes p. Therefore $\alpha = 0$. This proves 17.2 (i).

Remark. If we tried to compute $bu_*(X)$ by using the Atiyah-Hirzebruch spectral sequence

$$H_*(X;\pi_*(bu)) \Longrightarrow bu_*(X)$$

we would encounter non-trivial extensions; it would not be obvious how multiplication by 2 acts in $bu_*(X)$.

In order to prove 17.3, we pursue the proof of 17.2 a bit further. Let Y be a connective spectrum; then we may filter $\pi_*(Y)$ by the filtration subgroups F_s of 15.1 (with $E = HZ_2$). Also we may filter $H_*(Y)$ by the subgroups $F'_s = 2^s H_*(Y)$.

LEMMA 17.13 (i) The Hurewicz homomorphism

$$h: \pi_*(Y) \longrightarrow H_*(Y)$$

maps F_s into F'_s.

(ii) $h^{-1}F'_1 = F_1$.

Proof of (i). Let Y_s be as in §15, $\alpha \in \pi_*(Y_s)$. Suppose as an inductive hypothesis that in $Y_{s-\sigma}$ we have $h(\alpha) = 2^\sigma k_\sigma$ for some $k_\sigma \in \pi_*(Y_{s-\sigma})$. The map

$$Y_{s-\sigma} \longrightarrow Y_{s-\sigma-1}$$

induces the zero homomorphism $(HZ_2)_*(Y_{s-\sigma}) \longrightarrow (HZ_2)_*(Y_{s-\sigma-1})$, so k_σ maps to zero in $(HZ_2)_*(Y_{s-\sigma-1})$, and in $H_*(Y_{s-\sigma-1})$ we have $k_\sigma = 2k_{\sigma+1}$, $h(\alpha) = 2^{\sigma+1}k_{\sigma+1}$. This completes the induction and shows that in $H_*(Y) = H_*(Y_0)$ we have $h(\alpha) = 2^s k_s$.

Proof (ii). Suppose $h(\alpha) \in F'_1$. Then α maps to zero in $(HZ_2)_*(Y)$, so $\alpha \in F_1$. This proves 17.13.

LEMMA 17.14. Take $Y = bu \wedge X$, where X is as above. Then

(i) $E_\infty^{s*} = F_s/F_{s+1} \xrightarrow{h} F'_s/F'_{s+1}$ is a monomorphism for all s.

(ii) $F_s = h^{-1}F'_s$; in other words, the filtration in $\pi_*(bu \wedge X)$ is obtained exactly by pulling back the filtration in $H_*(bu \wedge X)$.

Proof. First we show that (ii) follows from (i). Suppose (i) true, and let $\alpha \in \pi_*(bu \wedge X)$, $h\alpha \in F'_s$. Suppose, as an inductive hypothesis, that $\alpha \in F_\sigma$ for some $\sigma < s$. Consider $F_\sigma / F_{\sigma+1} \xrightarrow{h} F'_\sigma / F'_{\sigma+1}$. We are assuming that this homomorphism is a monomorphism; it maps α to zero, so $\alpha \in F_{\sigma+1}$. This completes the induction, and shows that if $h\alpha \in F'_s$, then $\alpha \in F_s$. This proves part (ii).

We note that part (i) is true for $s = 0$, by 17.13 (ii). It is therefore sufficient to prove it for $s \geq 1$. It will now do no harm to replace F'_s by the image of $2^s(HQ_2)_*(bu \wedge X)$ in $(HQ)_*(bu \wedge X)$; for this does not alter F'_s/F'_{s+1} for $s \geq 1$, by 17.1.

We now divide the proof into three parts. First we exhibit a base for F_s/F_{s+1}; secondly, we exhibit a base for F'_s/F'_{s+1}; thirdly we show that with respect to these bases h is given by a non-singular triangular matrix.

The base for F_s/F_{s+1} is easy; if $s \geq 1$, then E_∞^{s*} has a Z_2-base consisting of the elements $\xi^{m_i}\eta^{n_i}g_i$ with $m_i + n_i = s + b(i)$, by 17.10 and 17.12. We turn to the base for F'_s/F'_{s+1}.

Take an element $\gamma_i \in \pi_*(bu \wedge X)$ representing $\xi^{b_i}g_i$. We can consider its image in $H_*(bu \wedge X)$; we see that there is an element $h_i \in H_*(X)$ such that the images of γ_i in $(HQ)_*(bu \wedge X)$ and $(HZ_2)_*(bu \wedge X)$ both have the form

$$h(\gamma_i) = 1 \otimes h_i \quad \text{mod lower terms}$$

where "lower terms" means terms

$$b_j \otimes x_j$$

with

$$b_j \in (HQ)_*(bu) \quad \text{or} \quad (HZ_2)_*(bu), \quad \deg b_j > 0,$$

$$x_j \in (HQ)_*(X) \quad \text{or} \quad (HZ_2)_*(X), \quad |x_j| < |h_i| .$$

Now by construction, the image of h_i in $(HZ_2)_*(X)$ is the ith basis element for $\mathrm{Ker}\,\beta_2/\mathrm{Im}\,\beta_2$. By 17.7, the elements h_i form a Q_2-base for the image of $(HQ_2)_*(X)$ in $(HQ)_*(X)$. Let $t/2$ be the generator for $H_2(bu)$, as above. Then F'_s/F'_{s+1} has a Z_2-base consisting of the elements

$$2^s(t/2)^{\nu}h_i \qquad (\nu \geq 0).$$

I claim that if $m_i + n_i = s + b(i)$, then the image of $\xi^{m_i}\eta^{n_i}g_i$ in F'_s/F'_{s+1} is

$$2^s(t/2)^{n_i}h_i \quad \text{mod lower terms.}$$

Here "lower terms" means terms $2^s(t/2)^{\nu}h_j$ with $\nu > n_i$, $\deg h_j < \deg h_i$. By construction, γ_i represents $\xi^{b(i)}g_i$, and its image in $(HQ)_*(bu\wedge X)$ is h_i mod lower terms of filtration ≥ 0. So $2^{m_i}t^{b_i}\gamma_i$ represents $\xi^{b(i)+m_i}\eta^{n_i}g_i$, and its image in $(HQ)_*(bu\wedge X)$ is $2^{m_i+n_i}(t/2)^{n_i}h_i$ mod lower terms of filtration $\geq n_i + m_i$. Now multiplication by ξ or 2 is a monomorphism on F_s/F_{s+1} and on F'_s/F'_{s+1}. So the image of $\xi^{m_i}\eta^{n_i}g_i$ is $2^s(t/2)^{n_i}h_i$ mod lower terms of filtration $\geq s$. This proves 17.14.

COROLLARY 17.15. (of the proof): Suppose $\alpha \in \pi_*(bu\wedge X)\otimes Q_2$ has filtration $\geq q$ and its image in $(HQ)_*(bu\wedge X)$ lies in $\sum_{i\geq q}(HQ)_{2i}(bu)\otimes(HQ)_*(X)$. Then the class of α in E_∞^{**} can be divided by η^q.

Proof. The result is empty for $q = 0$, so we may assume $q \geq 1$. Then the class of α in E_∞^{s*} is a linear combination of the basis elements

$$\xi^{m_i}\eta^{n_i}g_i .$$

I claim that every element appearing with a non-zero coefficient has $n_i \geq q$. For let the highest terms appearing be

$$\sum_i \lambda_i \xi^{m_i}\eta^{\nu} g_i$$

where not all the λ_i are zero; then in $(HQ)^*(bu \wedge X)$, α maps to

$$\sum_i \lambda_i 2^s (t/2)^{\nu} h_i$$

mod $2^{s+1}(HQ)_*(bu \wedge X)$ and lower terms, and hence $\nu \geq q$.

Since α has filtration $\geq q$, each term

$$\xi^{m_i}\eta^{n_i}g_i$$

which appears has $m_i + n_i \geq b(i) + q$, and there is an element of E_∞^{s-q*} mapping onto $\xi^{m_i}\eta^{n_i - q}g_i$. Therefore the class of α in E_∞^{s*} can be divided by η^q. This proves 17.15.

LEMMA 17.16. Let $\alpha \in \pi_*(bu \wedge X) \otimes Q_2$, and suppose

(i) α has filtration $\geq q$,

(ii) the image of α in $(HQ)_*(bu \wedge X)$ lies in

$$\sum_{i \geq q} (HQ)_{2i}(bu) \otimes (HQ)_*(X).$$

Then $\alpha = t^q \beta$ for some $\beta \in \pi_*(bu \wedge X) \otimes Q_2$.

Proof. Consider the subgroup of α which satisfy (ii), modulo the subgroup $t^q \pi_*(bu \wedge X) \otimes Q_2$. The quotient is evidently finite in each degree, for when we tensor with Q the result is zero. In particular, for each degree there is a filtration s such that all elements of

filtration $\geq$ s in $\pi_*(bu \wedge X) \otimes Q_2$ which satisfy (ii) lie in $t^q\pi_*(bu \wedge X) \otimes Q_2$.
Now we argue by downward induction over the filtration of α. Suppose the result is true for elements α' of filtration $> \sigma$, and α has filtration $\sigma \geq q$. Then by 17.15 the class of α in $E_\infty^{\sigma *}$ can be divided by η^q; that is, $\alpha = \alpha' + t^q\beta''$, where α' has filtration $\geq \sigma + 1$ and $\beta'' \in \pi_*(bu \wedge X) \otimes Q_2$. Here α' also satisfies (ii), so by the inductive hypothesis, $\alpha' = t^q\beta'$. Then $\alpha = t^q(\beta' + \beta'')$. This completes the induction and proves 17.16.

Proof of 17.3. Suppose an element $h \in H_*(K \wedge X)$ lies both in the image of $H_*(bu \wedge X)$ and in the image of $\pi_*(K \wedge X)$. Then it comes from an element

$$\alpha \in \pi_*(K(-2r, \ldots, \infty) \wedge X)$$

for some sufficiently large value of n. The image of α in $H_*(K \wedge X)$ lies in the image of $H_*(bu \wedge X)$. Now $H_*(K(-2n, \ldots, \infty) \wedge Z) \longrightarrow H_*(K \wedge X)$ is not a monomorphism, but the image of $H_*(K(-2n, \ldots, \infty) \wedge X) \longrightarrow H_*(K(-2n-2, \ldots, \infty) \wedge X)$ does map monomorphically to $H_*(K \wedge X)$. So by replacing 2n with 2n+2 if necessary, we may assume that the image of α in $H_*(K(-2n, \ldots, \infty) \wedge X)$ lies in the image of $H_*(bu \wedge X)$.

Now $K(-2n, \ldots, \infty) \simeq S^{-2n} \wedge bu$. By 17.14, the element $\alpha \in \pi_*(K(-2n, \ldots, \infty) \wedge X)$ has filtration $\geq$ n. Also its image in $(HQ)_*(K(-2n, \ldots, \infty) \wedge X)$ lies in the image of $HQ_*(bu \wedge X)$. Now 17.16 applies to show that $\alpha = t^n\beta$, that is, α lies in the image of $\pi_*(bu \wedge X) \otimes Q_2$. We proceed similarly for the odd primes. Therefore α lies in the image of $\pi_*(bu \wedge X)$. This proves 17.3.

To prove 17.4, we give means independent of the Adams spectral sequence for constructing elements in $\pi_*(K \wedge bu)$. Consider CP^∞. We have a canonical map from CP^∞ to BU, which we can consider as term 2 of the bu-spectrum. We get an element $x \in bu^2(CP^\infty)$. Then the Atiyah-Hirzebruch spectral sequence shows that $bu_*(CP^\infty)$ is free over $\pi_*(bu)$ on generators $\beta_i \in bu_{2i}(CP^\infty)$ such that

$$<x^i, \beta_j> = \delta_{ij} .$$

Consider again the canonical map from CP^∞ to BU, considered as term 2 of the bu-spectrum. Applying this to β_{i+1}, we obtain an element

$$b_i \in bu_{2i}(bu) .$$

For more detail see [2].

LEMMA 17.17 (Adams, Harris and Switzer). The image of b_n in $\pi_{2n}(bu \wedge bu) \otimes Q$ is

$$\frac{(v-u)(v-2u)\ldots(v-nu)}{(n+1)!}$$

The proof is essentially that of [2], Lemma 13.6, except for changes of detail.

Proof of 17.4. Separating components, we can assume that f is homogeneous, say of degree d. On multiplying $f(u,v)$ by a sufficiently high power of u, we can ensure that

$$g(u,v) = u^N f(u,v)$$

is a polynomial which has the following property:

$$g(k,1) \in Z \quad \text{for all } k \in Z.$$

The argument is essentially given in [2], p.102, but add one more power of u to take care of the case $k = 0$. Then it is elementary that $g(u,v)$ can be written as a Z-linear combination of the polynomials

$$\frac{u(u-v)(u-2v)\dots(u-nv)}{(n+1)!} v^{d+N-n-1} .$$

Take Lemma 17.5 and apply $c: bu \wedge bu \longrightarrow bu \wedge bu$; we see that

$$\frac{(u-v)(u-2v)\dots(u-nv)}{(n+1)!}$$

lies in the image of $\pi_*(bu \wedge bu)$. Clearly also u and $v^{d+N-n-1}$ lie in the image of $\pi_*(bu \wedge bu)$. Therefore $g(u, v)$ lies in the image of $\pi_*(bu \wedge bu)$. Dividing by u^N, we see that $f(u, v)$ lies in the image of $\pi_*(K \wedge bu)$. This completes the proof of 17.4, which therefore completes the proof of 17.5.

Proof of 17.6 (i). First I claim that the given polynomials do satisfy (1'). Consider the special case $\ell = 1$. Let f be the given product of degree n; then

$$f((2k+1)t, t) = t^n \frac{(2k)(2k-2)(2k-4)\dots(2k-2n+2)}{(2n)(2n-2)(2n-4)\dots 2}$$

$$= t^n \frac{k(k-1)(k-2)\dots(k-n+1)}{1 \cdot 2 \cdot 3 \cdot \dots \cdot n}$$

which lie in $Z[t]$. Now consider $f(kt, \ell t)$ with k and ℓ odd. The denominator of f contains only a finite number of powers of 2, say 2^m, so we may solve $\ell\lambda = 1 \bmod 2^m$; then

$\lambda^n(f(kt, \ell t) = f(k\lambda t, \ell\lambda t) = f(k\lambda t, t) \bmod Q_2[t]$, so this lies in $Q_2[t]$ by the special case $\lambda = 1$. Hence $f(kt, \ell t)$ lies in $Q_2[t]$ and f satisfies (1').

It is now clear that $Q_2[u, u^{-1}]$-linear combinations of the given polynomials also satisfy (1').

Conversely, let $f(u, v) \in Q[u, u^{-1}, v]$ satisfy (1'). We wish to write it as a $Q_2[u, u^{-1}]$-linear combination of the given polynomials. By separating homogeneous components, it is sufficient to consider the case in which $f(u, v)$ is homogeneous, say of degree n. Then we may write $f(u, v)$ as a Q-linear combination

$$f(u,v) = \lambda_0 u^n + \lambda_1 u^{n-1}\frac{v-u}{3-1} + \lambda_2 u^{n-2}\frac{(v-u)(v-3u)}{(5-1)(5-3)} \cdots .$$

Suppose as an inductive hypothesis that $\lambda_0, \lambda_1, \ldots, \lambda_{r-1}$ lie in Q_2. Then the sum of the remaining terms

$$g(u,v) = \lambda_r u^{n-r}\frac{(v-u)\ldots(v-(2r-1)u)}{((2r+1)-1)\ldots((2r+1)-(2r-1))} + \ldots$$

satisfies (1'). We may find λ_r by substituting $v = (2r+1)t$, $u = t$; we see that

$$g((2r+1)t, t) = \lambda_r t^r .$$

and $\lambda_r \in Q_2$. This completes the induction and proves 17.6 (i).

<u>Proof of 17.6 (ii).</u> We first observe that the given polynomials do satisfy (1') and (2'), and so do Q_2-linear combinations of them.

Conversely, let $f(u, v) \in Q[u, v]$ satisfy (1') and (2'); we wish to write it as a Q_2-linear combination of the given polynomials. By separating homogeneous components, it is sufficient to consider the case in which $f(u, v)$ is homogeneous, say of degree n. Then we may write $f(u, v)$ as a Q-linear combination

$$f(u,v) = \frac{\lambda_o}{2^{q_o}} u^n + \frac{\lambda_1}{2^{q_1}} u^{n-1}(v-u) + \frac{\lambda_2}{2^{q_2}} u^{n-2}(v-u)(v-3u) + \ldots ,$$

where $\lambda_r \in Q_2$. Here 2^{q_r} divides $r!2r$ by part (i); we wish to prove it also divides 2^n. Suppose, as an inductive hypothesis, that this is true for $r' > r$. Then the sum of the remaining terms

$$g(u,v) = \frac{\lambda_o}{2^{q_o}} u^n + \ldots + \frac{\lambda_r}{2^{q_r}} u^{n-r}(v-u)\ldots(v-(2r-1)u)$$

also satisfies (1') and (2'). But now $\dfrac{\lambda_r}{2^{q_r}}$ is the coefficient of $u^{n-r}v^r$, so $q_r \leq n$. This completes the induction, and proves 17.6 (ii).

REFERENCES

[1] J.F. Adams, Lectures on generalised cohomology, in Lecture Notes in Mathematics, vol. 99 (1969), Springer-Verlag Berlin-Heidelberg-New York.

[2] ____________, Quillen's work on formal groups and complex cobordism, Mathematics Lecture Notes, University of Chicago, Chicago, 1970.

[3] M.F. Atiyah and F. Hirzebruch, Riemann-Roch theorems for differentiable manifolds, Bull. AMS 65(1959), 276-281.

[4] E.H. Brown, Cohomology theories, Ann. of Math. 75 (1962), 467-484; Abstract homotopy theory, Trans. Amer. Math. Soc. 119 (1965), 79-85.

[5] P.E. Conner and E.E. Floyd, The Relation of Cobordism to K-theories, Lecture Notes in Mathematics, vol. 28 (1966), Springer-Verlag Berlin-Heidelberg-New York.

[6] S. Eilenberg and N.E. Steenrod, Foundations of Algebraic Topology, Princeton University Press, Princeton 1952.

[7] E.L. Lima, Duality and Potnikov invariants, Thesis, University of Chicago, Chicago, 1958.

[8] S. Mac Lane, Natural associativity and commutativity, Rice University Studies 49, No. 4, 1963.

[9] W.S. Massey, Exact couples in algebraic topology, Ann. of Math. 56 (1952), 363-396; 57 (1953), 248-286.

[10] J.W. Milnor, The Steenrod algebra and its dual, Ann. of Math. 67 (1958), 150-171.

[11] R. Mosher and M. Tangora, Cohomology Operations and Applications in Homotopy Theory, Harper and Row, New York, 1968.

[12] S.P. Novikov, Rings of operations and spectral sequences, Doklady Akademii Nauk SSSR 172 (1967), 33-36.

[13] __________, Izvestija Akademii Nauk SSSR, Serija Matematiceskaja 31 (1967), 885-951.

[14] E.H. Spanier, Algebraic Topology, McGraw-Hill, New York, 1966.

[15] N. E. Steenrod and D.B.A. Epstein, Cohomology Operations, Princeton University Press, Princeton, 1962.

[16] R. Stong, Notes on Cobordism Theory, Math. Notes, Princeton University Press, Princeton, 1968.

[17] G.W. Whitehead, Generalized homology theories, Trans. Amer. Math. Soc. 102 (1962), 227-283.

[18] __________, Homotopy groups of joins and unions, Trans. Amer. Math. Soc. 83 (1956), 55-69.